ECONOMY–ENERGY–ENVIRONMENT SIMULATION

BEYOND THE KYOTO PROTOCOL

Economy & Environment

VOLUME 20

The titles published in this series are listed at the end of this volume.

Economy–Energy–Environment Simulation
Beyond the Kyoto Protocol

Edited by

Kimio Uno
Faculty of Policy Management
Keio University at Shonan Fujisawa Campus, Japan

SPRINGER-SCIENCE+BUSINESS MEDIA, B.V.

Library of Congress Cataloging-in-Publication Data is available.

ISBN 978-90-481-5945-1 ISBN 978-0-306-47549-8 (eBook)
DOI 10.1007/978-0-306-47549-8

Printed on acid-free paper

Originally published by Kluwer Academic Publishers in 2002
Softcover reprint of the hardcover 1st edition 2002

Contents

Preface

In the policy arena, as well as in the academic world, a new challenge is having to deal with the global community. We are increasingly aware that the world is linked through economy–energy–environment interactions. We are increasingly aware, at the same time, that the emergence of the global community does not imply an integrated harmonious world; rather, it is a community where countries/regions of different interests and values face each other directly. Global governance has to be achieved through actions of national governments under different motives and constraints. We need to have an analytical tool that is capable of producing a global picture, yet with detailed country resolution.

If the world is a better place now compared to 100 years ago in terms of per-capita income, this is due to the industrialization that continued throughout the 20th century. We entered the 21st century knowing that the human aspiration that translates into ever-increasing production may not be tenable in the long run. Sustainability of the global community is at stake. In contrast to incremental decision making through the market mechanism that should lead to some optimal state under some assumptions such as perfect knowledge, smooth movement of resources, no externalities, and so forth, we need to have an analytical tool to provide us with details of the future state of the world. Whether our mode of production, our demand for energy, and our impact on the environment can be sustained over a long period of time cannot be answered based on mere intuition. There are many actors involved. The inter-linkages are complex. Technology is changing.

In a previous publication we pointed to several research agendas (Uno and Bartelmus, eds. 1998):

> Physical-monetary linkages and indicators; International linkages; Intertemporal comparisons; Technological change; Quality of life and social dimensions; and shift from accounting to modelling.

The current work is a response to this self-imposed agenda. The framework presented here is a database and modelling-simulation system that is capable of projecting the coming decade. It provides detailed country resolution, identifying about 60 countries/regions of the global community. It has a detailed industrial classification, distinguishing 36 branches as a general rule. The focus is placed on the interaction running through economy, energy, and global environment. Physical-monetary linkage is inevitable because production is in general measured in terms of value added and other monetary indicators, including energy and other resource inputs into various activities, whereas energy service has to be specified in physical measurements such as kilocalories or tons of oil equivalent. The impact on global environment has to be in physical units such as volume of CO_2 emissions. International linkages are a must because the economic system is not confined geographically in one country; the country's activities are linked to others through the international flow of goods and services, of money, and of technology. If one country or group of countries imposes energy tax, for instance, the energy-intensive industrial sector may shift to other economies. Sustainability is a long-term issue, rather than being a problem to be tackled with policy measures with a time horizon of one to several years. We must be looking 10 years hence, 50 years, or even farther into the future. It is imperative to obtain the exact time path of economy–energy–environmental development in order to know what the future holds without policy measures or technological change, and what needs to be done now to improve the situation in the future. Technology is a key variable in the long run, determining the mode of production, energy shift and conversion efficiency, and emissions of global warming gases. International technology transfer is one way of softening the sharp conflict between economic development and environmental impact. The empirical evidence pertaining to our performance in economy, energy, and environment is recorded in statistics. More often than not, statistics are collected by national statistical offices following an accounting framework. International organizations sometimes provide collection of data from national sources, but they rarely collect raw data themselves. The statistical accounts compiled annually provide the empirical basis for analysis, but meaningful analysis would have to be based on modelling and policy simulation. The social dimension is not dealt with explicitly in the model framework presented here, but one chapter is devoted to the discussion of the impact of CO_2 tax on employment. It is hoped that the variables included in the current framework will serve as the basis for deriving various social indicators.

Part I is an introduction providing an overview on the sustainability issues in the economy–energy–environment spheres. Part II introduces the tools of analysis employed in this work, pointing out some of the characteristics of our approach, and describing the modelling framework including energy demand, conversion, and fuel substitution; macroeconomic performance and inter-industry relations; and international trade. Object-oriented design of the statistical database and modelling tool is discussed. Part III covers the economic and energy prospects of the global community, and the cases of G7 industrialized countries, developing economies in South-East Asia, China, and Russia are analysed. Finally, the impact of the imposition of CO_2 tax on production and employment is examined. Part IV is a methodology section. Major works by other research groups are examined in terms of the time span, regional and branch coverage, energy sources, assumptions on technology and energy intensity improvements, and the inherent logic in modelling/scenario writing. An attempt is made to compare their numerical results. The Appendix demonstrates the possible integration of a medium-term model such as the one we employ in the current study into a very long-term view covering the period from around 1950 to the year 2100.

This research was supported by Grant-in-Aid for Scientific Research from the Japan Society for the Promotion of Science of the Ministry of Education, Science and Technology of the Japanese Government (Category Code (A)(2), Project Number 10044036). The project was conducted during 1998-2000 with Kimio Uno of Keio University (Institution Number 32612) as the project leader. The publication was supported by Grant-in-Aid for Publication of Scientific Research Results from the Japan Society for the Promotion of Science (Project Number 135331). The subsidy was made available during 2001 with Kimio Uno as the responsible editor.

Last but not least, as the project leader and the editor, I would like to express my appreciation of the project members who have contributed their valuable time to this work. I should particularly mention the effort that went into constructing the infrastructure for the project that included development of simulation programs, a database system, graphic user interface, statistical data, meta data, and model prototypes. I cannot do justice to the effort that constitutes the backbone of the project, but is inevitably less visible in a publication such as this one that places focus on the contents. Construction of infrastructure is indeed crucial in carrying out a project of this scope and, more importantly, making future development possible. The group has also inherited a simulation engine and other important intellectual assets from our predecessors, for which we extend our particular appreciation.

Kimio Uno

Part I

Introduction

1
Economy–energy–environment: COMPASS approach

KIMIO UNO

1. From single-country policy to global governance

In the policy analysis arena there is a keen awareness of the need for global coverage with detailed country resolution. On the one hand we are faced with policy challenges at the global level; on the other hand government decisions at the national level comprise the only way in which resources can be committed to, in order to carry out the policy measures.

Introduced here is a COmprehensive Model for Policy ASSessment (COMPASS) which is an econometric simulation model designed to capture the interaction among 3E spheres encompassing economy, energy, and environment. COMPASS is designed for simulation experiments up to 2010-2020. The model achieves international linkage for 60 countries/regions, covering 99.5% of the world GDP. The country models are based on time-series of a 36-sector input-output framework, the SNA, and financial accounts. They are linked to each other through time-series of 25-commodity trade matrices which explicitly describe annual trade flows among world regions, achieving a global coverage. The model also generates energy balances and CO_2 emissions for individual countries. Thus, the COMPASS provides multi-sector, multi-country information on production, trade, balance of payments, saving and investment flows, energy demand and supply, and environmental impact. It is capable of generating consistent output and price series, together with interest rates, exchange rates, and energy prices. An empirically based econometric method is employed throughout, enabling users to trace the dynamic time path of policy measures, such as increased energy price through the introduction of tradable permits. The model framework we employ may be termed an input-output+econometric model (IOE) following West (1995). The modelling of this magnitude became possible by adopting an object-oriented database approach, which comprises a consistent naming convention, accounting rules, and logical structure. The graphic user interface facilitates access to the database, operation of the model, and policy simulation experiments.

The Kyoto Protocol of the UNFCCC, adopted in 1997 by COP-3, set the sage for implementing an international format aimed at curtailing CO_2 emissions. The Protocol stipulates a reduction target for Annex I countries to be achieved by around 2010. The problem is that, given the recent trends in energy use, it seems extremely difficult to contain energy consumption to the target level.

K. Uno (ed.), Economy–Energy–Environment Simulation, 3–30.

When signatories to the Framework Convention on Climate Change met in Kyoto in December 1997 at the third Conference of Parties to the Convention (COP-3), the initial position was zero reduction for the United States, EU 15%, and Japan 5% globally but 2.5% for herself. Agreement was reached that the United States, EU, and Japan should reduce the emission of global warming gases by 7%, 8%, and 6%, respectively, by the year 2010. The reduction targets take into account the portion achieved through technology transfer to developing countries. Reduction in developing economies was postponed to future discussion, although a proposal was made requesting their voluntary reduction targets. The COP-4, convened in Buenos Aires in November 1998, ended by adopting an action plan spelling out the future negotiation schedule, without going into substantial discussion on the tradable permits of global warming gases. There still remains the equity issue concerning the initial allocation of the permits, particularly for developing economies and for economies in transition. Technology transfers to developing countries, compensation for oil-producing countries for possible economic losses, and the mechanism for joint implementation are also being disputed. It was agreed that a concrete stipulation pertaining to emission trading be decided in the year 2000. When in fact environment ministers of the Group of Eight met in Japan in preparation for the G8 Summit in Okinawa in July 2000, and as a prelude to the COP-6 meeting in the Hague in November 2000, they could not agree to include a specific time frame in the joint communiqué. While all European Union countries, Japan, and Russia agreed, the two North American countries were reluctant to include the 2002 deadline (Chasek, 2001; Grubb *et al.*, 1999; Mitchell, 1996; Kawashima, 1995; Spector *et al.*, 1994). Thus, we may say that there is a real danger that the COP could degenerate once again into a political process.

One way out of this would be an information system which allows for various policy experiments revealing what would happen if nothing is done and/or if international coordination is realized in order to mitigate emission of GHGs. Our purpose, therefore, is a construction of a simulation model with a dynamic framework. Individual countries have to be explicitly treated in the model as they are the agents that implement policy measures. After all, policy measures will have to be implemented by national governments, sometimes in the face of public resentment or opposition from industrial circles. International repercussion would have to be sought in order to attain sustainability as the global community as a whole, but country resolution of simulation models is indispensable in the political arena.

Compared to the analytical need for which environmental accounting is expected to provide crucial information, current attention is almost solely being devoted to a single-country context. Global linkage would have to be sought in view of the global nature of the most pressing environmental issues such as climate change, resource exhaustion, and the impact of agricultural and forestry activities triggered by international trade (Uno, 2000).

2. What the models should achieve

There are already sufficient models which are operational in estimating the cost of emission reduction. Repetto and Austin (1997) provide a list of focal points of the available models as follows:

(1) The extent to which substitution among energy sources, energy technologies, products, and production methods is possible.
(2) The extent to which market and policy distortions create opportunities for low-cost (or no-cost) improvements in energy efficiency.
(3) The likely rate of technological innovation and the responsiveness of such change to price signals.
(4) The availability and likely future cost of non-fossil, backstop energy sources.
(5) The potential for international 'joint implementation' of emissions reductions.
(6) The possibility that carbon tax revenues would be recycled through the reduction of economically burdensome tax rates.

In addition, there are possible damages (or benefits) which can be avoided by the introduction of carbon tax. They are:

(7) Economic damage from climate change.
(8) The reduction of other air pollution damage associated with the burning of fossil fuels.

Of these, the COMPASS is capable of treating (1), (3), (5), (6) and (8) in whole or in part. Substitution between energy and other factors of production, particularly capital, has been discussed extensively in the literature. Substitution among energy sources can also be analysed in a model depicting energy balances encompassing various sources of primary energy. The rate of technological innovation, or item (3) above, is described in our framework as the changes in input coefficients over time. This does not automatically mean that technological change can be endogenously explained in the model framework. Rather, scenarios may be developed based on technology inventories, which then can be interpreted as alternative input coefficients. Such technology inventories in environmental spheres are available from IPCC (Intergovernmental Panel for Climate Change), IEA (International Energy Agency), UNEP (United Nations Environmental Program), and IIASA (International Institute for Applied Systems Analysis), in addition to various country sources. The assessment of joint implementation, which is item (5) above, is particularly suitable for the model framework at hand. In the arsenal of economic theory, technology is best described in an input-output framework in terms of input coefficients. An alternative technology described in input coefficients, whether they are the ones actually achieved in one country or the ones obtained by hypothetical scenarios, can be shifted to different countries in order to observe economy-wide repercussions. As for (6) above, one can incorporate a channel linking carbon tax revenues to, for example, reduction in social security contributions (Meyer and Ewerhart, 1998). It is also possible to deal with (8) in terms of changes in the imputed damages attributable to, for example, SO_2 (Uno, 1995).

The 1992 'Earth Summit' in Rio de Janeiro called for an integrated approach in achieving long-term sustainable development, and formulated Agenda 21 as its action plan. The UN's System of Integrated Environmental and Economic Accounting, dubbed as SEEA, EU's National Accounting Matrix including Environmental Accounts, or NAMEA, and related statistical systems have been devised during preparations for the Rio Conference. Experience in empirical application and theoretical issues are described in detail in a volume which is an outcome of the Special Conference of the International Association for Research in Income and Wealth (IARIW) held in 1996 (Uno and Bartelmus, 1998). In order to facilitate exchange of views and further development, a group of experts formed a forum called the London Group, named after its first meeting in London (London Group, 1994, 1995, 1996, 1997). The fifth meeting took place in Fontevraud, France, in 1998, followed by the sixth meeting in Canberra, Australia. The thrust of the discussion is directed towards refinement of the SEEA and empirical application of the NAMEA. In other words, development of an environmental accounting so far, however, remains a single-country, static framework.

The SEEA has established itself as the internationally accepted conceptual framework for linking the environmental concerns to the System of National Accounts (Inter-Secretariat Working Group on National Accounts, 1993). Its advantage clearly lies in the fact that it is designed as a satellite to be used in conjunction with the SNA. Likewise, the framework being developed in the European Union focuses on flows and stocks of environment-related goods and services based on the SNA definition (EUROSTAT, 1997; European Commission, 1997). Recent discussion on the integrated environmental and economic accounting in the United States by an expert panel concluded that such effort is solidly based in mainstream economic analysis and that developing a set of comprehensive non-market economic accounts is a high priority for the nation (Nordhaus and Kokkenberg, 1999). The environmental indicators approach, such as represented by listing of indicators (OECD, various years a; United Nations Development Programme, 1997) and decomposition analysis based on indicators (Unander and Schipper, 1998), can be conceived of dealing with summary expression of full accounts. One can pick up numerous other approaches to environmental accounting, but it is now clear that they are composed of more or less similar data sets, if not exactly identical in their formats.

No-one denies the usefulness and urgency of such an effort. Only through such attempts can the conceptual consistency of the framework be guaranteed, and inter-temporal and international comparability of the empirical measurement be obtained.

At the same time, one can look at environmental accounting from its applicability for policy analysis. The development of SNA followed a path where the effort to guarantee conceptual consistency and the need to respond to policy analysis stimulated each other. Environmental concerns cover a wide range of issues, as, for example, reflected in the sustainability issue contained in Agenda 21 (United Nations, 1993a) or the discussion surrounding global climate change (IPCC, 1996a, b).

In addition to domestic policy measures such as carbon tax, which in turn stimulate technological progress or lifestyle change, the Protocol introduced flexibility mechanisms such as Joint Implementation among Annex I parties (Article 6), Clean Development Mechanism among both Annex I and Non-Annex I parties (Article 12), and Emissions Trading among Annex B parties (Article 17). These instruments implemented at a global level (or at least among more industrialized countries/regions) are expected to result in most cost-effective means of realizing the reduction target (Grubb *et al.*, 1999; United Nations, 1997).

The point here is the fact that both the issues at hand and the policy measures go beyond national boundaries. The policy needs clearly point to a simulation model designed to gauge the implication of policy scenarios in a global as well as single-country context, and a statistical framework that is capable of providing globally linked empirical data.

3. Empirical data

The feasibility of a 3E model encompassing economic, energy, and environmental spheres is largely dependent on the availability and organization of the empirical data. The COMPASS is intended to be the basis for policy dialogue. It is therefore deemed important to adopt a transparent framework both theoretically and empirically. The model is based on open data published mostly by international organizations. Limiting the scope of environmental accounting to the global warming issue, the framework should cover economy, energy, and environmental spheres such as the following:

Economy: Final demand, production, income. Input structure, output structure, final demand, value added (including direct taxes and subsidies), output prices. Technological change (=input coefficients), technology transfer, lifestyle shift. Trade balance, trade shares, exporting country, importing country for each branch. Saving-investment balance, flow of funds. Prices, exchange rates, interest rates.

Energy: Total final consumption. Energy conversion. Total primary energy demand. Alternative primary energy sources, fuel substitution. Conversion efficiency.

Environment: CO_2 emissions, emission trade, CO_2 taxes. CO_2 sink.

Those data sets can be set in a logical order as described in Figure 1-1. The COmprehensive Model for Policy ASSessment (COMPASS) is an econometric simulation model designed to capture the interaction among 3E spheres encompassing economy, energy, and environment.

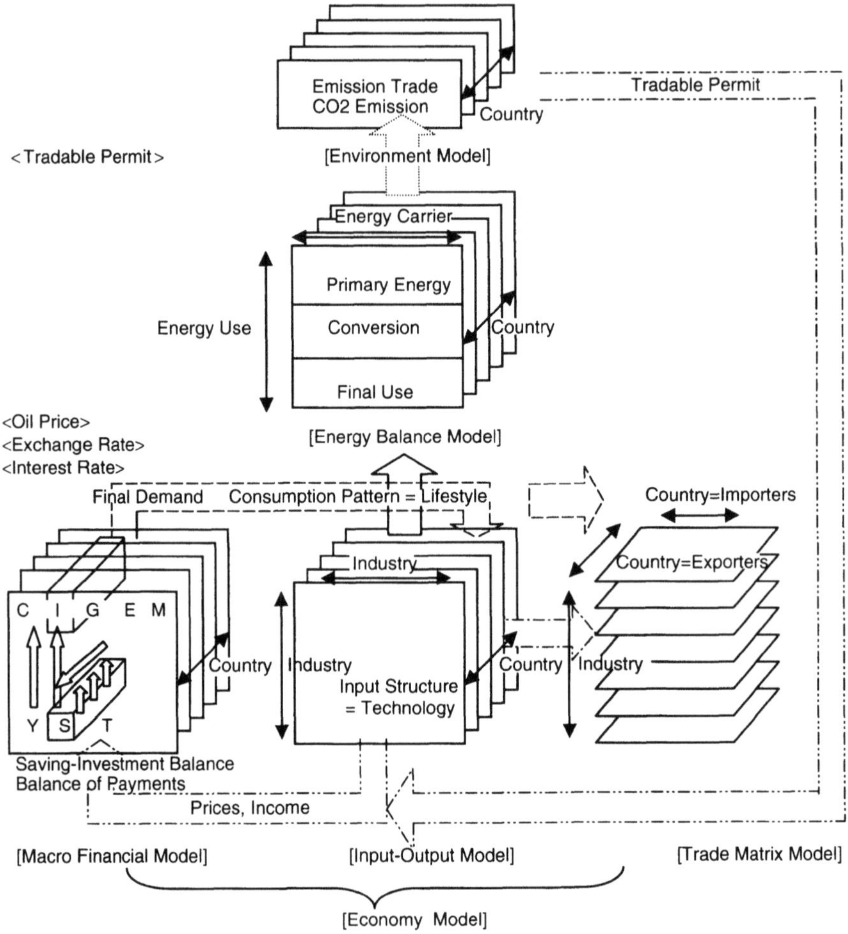

Figure 1-1. Structure of the EEE (economy–energy–environment) COMPASS (COmprehensive Model for Policy ASSessment). The figures represented in the diagram show the relationship among the data sets and refer to one point in time. The model has a dynamic framework.

The regional structure of the model is provided in Table 1-1. This table lists the regions (the term region is preferred over country here due to disputes surrounding the statehood in some cases), their code names in the COMPASS framework, and the model types describing the pertinent regions. The model covers explicit 54 regions and seven additional regions representing 'rest of Asia', etc. The coverage is intended to shed light on all OECD countries, all APEC countries (Peru and Vietnam have recently joined the APEC and are not covered by the current version of the COMPASS), all major energy producers, and most of the bigger developing countries. The model, therefore, is virtually closed globally regarding production of goods and services, foreign trade, financial flows, and energy demand and supply.

The country models are based on time-series of a 36-branch (25 for non-OECD Asian regions) input-output framework, the SNA, financial accounts, and energy balance tables. The term 'branches' is used when referring to disaggregated production activities and the term 'sectors' is reserved to refer to economic agents such as households, enterprises, and government. Branches are linked internationally through time-series of 25-commodity trade matrices that explicitly describe annual trade flows among world regions.

One can envisage COMPASS as a group of regional models linked internationally. Alternatively, one may view it as composed of layers of branch models that are combined to form an integrated global whole, making it possible to carry out branch- (sector)-specific analysis. In order to go beyond the limits of comparative static approach where accounts from mere two data points are compared, and to describe the dynamic time path of final demand, intermediate demand, and international repercussion through trade, relative prices would have to be determined endogenously by the system.

3.1. Time-series input-output tables

Input-output tables are indispensable in the construction of multi-sector models in that they provide consistent structural data ranging from final demand, intermediate inputs (whereas input coefficients describing production technology), value added and income generation, and prices (potentially including CO_2 tax or price hikes reflecting emission trade). One should mention two important developments in this field. One is the publication of input-output tables for the OECD member countries covering the period around 1970 to 1990 (OECD, 1995). Another is a series of bilateral and internationally linked tables for Asian countries by a Japanese source (Institute of Developing Economies 1982, 1992, 1998). The input-output and other pertinent data for China and Russia were provided by country experts. The tables for Asian countries are regularly recompiled into a standard format by the Institute of Developing Economies (1982, 1992, 1998). Among Non-Annex I countries, China is included in the IDE source. Russia and India compile input-output tables regularly, and their conversion to an internationally comparable scheme is not a difficult task. Reflecting the original structure of the tables, the input-output tables for OECD countries have 36 industrial branches, whereas those for Asian countries have 25 in the current version of the COMPASS. Details are depicted in Table 1-2.

A comprehensive framework which describes the structure of an economic system is provided by input-output tables which comprise a part of the SNA. The SEEA also recognizes this as the basis for further extension, although most widely used versions of the SEEA do not employ an input-output framework. Thus, an input-output framework seems to be the natural starting point if environmental accounting should also incorporate inter-industry relations, the relation between final demand (household) and intermediate inputs (industry), international repercussion (imports and exports), and changes in technology and

Table 1-1. COMPASS regional coverage

Serial No.	Region	Region Code	Trade Matrix Code	Region Status				Total Primary Energy Supply (TPES), 1995 (unit: million tonnes of oil equivalent)				
				OECD	APEC	Annex I	OPEC	Coal	Crude Oil	Pet Prod.	Gas	Total
1	Belgium	BE	# = 1	OECD		Annex I		9.10	27.05	−6.07	10.61	52.40
2	Denmark	DK	# = 2	OECD		Annex I		6.45	10.45	−1.14	3.12	20.29
3	France	FR	# = 3	OECD		Annex I		16.06	82.99	3.57	29.57	241.39
4	Germany	DE	# = 4	OECD		Annex I		91.03	106.08	29.63	66.4	338.98
5	Greece	GR	# = 5	OECD		Annex I		8.39	17.33	−3.16	0.04	23.64
6	Ireland	IE	# = 6	OECD		Annex I		3.13	2.32	3.35	2.33	11.35
7	Italy	IT	# = 7	OECD		Annex I		12.33	89.16	5.34	44.84	161.54
8	Netherlands	NL	# = 8	OECD		Annex I		9.19	64.12	−36.74	34.08	73.29
9	Portugal	PT	# = 9	OECD		Annex I		3.60	13.71	−0.02	0	19.25
10	Spain	ES	# = 10	OECD		Annex I		19.51	57.13	−1.47	7.72	103.29
11	UK	GB	# = 11	OECD		Annex I		48.64	100.50	−15.88	65.05	224.49
12	Austria	AT	# = 12	OECD		Annex I		3.29	9.36	1.64	6.31	26.29
13	Finland	FI	# = 13	OECD		Annex I		6.05	9.19	−0.73	2.84	28.85
14	Norway	NO	# = 14	OECD		Annex I		1.02	13.87	−5.86	3.46	23.53
15	Sweden	SE	# = 15	OECD		Annex I		2.79	18.29	−2.29	0.68	50.92
16	Switzerland	CH	# = 16	OECD		Annex I		0.19	4.83	7.63	2.19	25.19
22	Former USSR/RU	USSR	# = 17					193.39	255.70	−34.01	453.03	954.54
	Russia				APEC	Annex I		108.96	191.55	−42.05	306.45	621.48
23	Japan	JP	# = 18	OECD	APEC	Annex I		82.59	230.41	39.16	52.02	497.03
24	Iran	IR	# = 19				OPEC	0.95	56.91	−5.34	28.60	82.47
25	Iraq	IQ	# = 20				OPEC	0.00	24.62	−2.85	3.24	25.09
26	Kuwait	KW	# = 21				OPEC	0.00	47.31	−35.77	4.13	15.67
27	Oman	OM	# = 22				OPEC	0.00	0.00	0.00	0.00	0.00
28	Qatar	QA	# = 23				OPEC	0.00	3.92	−3.09	7.04	7.87
29	Saudi Arabia	SA	# = 24				OPEC	0.00	95.61	−11.90	32.94	83.66
30	Untd Arab Em	AE	# = 25				OPEC	0.00	16.74	−8.17	18.86	24.46
31	Brunei	BN	# = 26		APEC			0	0	0	0	0
32	Hong Kong, China	HK	# = 27		APEC			5.60	0.00	7.49	0.00	13.67

33	India	IN	# = 28					141.98	65.43	16.68	15.59	436.70
34	Indonesia	ID	# = 29		APEC		OPEC	6.36	44.91	−3.09	30.16	124.92
35	Korea Rp	KR	# = 30	OECD	APEC			26.85	89.64	4.46	8.32	148.02
36	Malaysia	MY	# = 31		APEC			1.61	20.36	−1.07	11.10	34.87
37	Philippines	PH	# = 32		APEC			1.16	16.44	1.68	0.01	35.37
38	Singapore	SG	# = 33		APEC			0.02	52.47	−32.50	1.45	21.44
39	Thailand	TH	# = 34		APEC			6.90	27.15	7.46	8.70	71.07
40	Taiwan	TW	# = 35		APEC			17.16	29.42	4.05	3.76	64.38
41	China	CN	# = 36		APEC			658.14	147.95	10.49	16.71	1058.64
42	Australia	AU	# = 37	OECD	APEC	Annex I		37.54	35.15	−0.79	16.74	94.52
43	New Zealand	NZ		OECD	APEC	Annex I		1.19	5.12	0.59	3.83	15.44
44	Papua New Guinea	PG			APEC							
45	Canada	CA	# = 38	OECD	APEC	Annex I		25.36	83.83	−5.66	67.09	232.38
46	USA	US	# = 39	OECD	APEC	Annex I		475.34	824.36	−19.94	508.67	2088.47
47	Brazil	BR	# = 40					11.73	67.01	7.34	4.26	156.36
48	Chile	CL			APEC			2.40	8.71	1.15	1.38	18.68
49	Mexico	MX	# = 41	OECD	APEC			5.43	84.84	−0.11	26.31	133.76
50	Venezuela	VE	# = 42				OPEC	0.21	54.51	−36.88	25.08	47.87
51	Algeria	DZ	# = 43				OPEC	0.52	25.65	−18.06	15.70	24.30
52	Libya	LY	# = 44				OPEC		18.49	−6.75	4.05	15.91
53	Angola	AO	# = 45									
54	Nigeria	NG	# = 46				OPEC	0.09	14.59	−1.08	4.40	83.18

Note: Serial numbers 17-21 are Turkmenistan, Uzbekistan, Kazakstan, Azerbeijan, and other former USSR republics. Serial numbers 55-61 (trade matrix code 47-53) are Rest of Europe, Asia, Oceania, North America, Latin America, Africa, and nes.

Table 1-2. Industrial branch classification of COMPASS

COMPASS IO Models		COMPASS APEC IO Models		Statistical Codes			
No.	Description	No.	Description	Short Code	Standard Code	ISIC Definition	World Trade
1	Agriculture, forestry and fishery			AFF	AG_FO_FI	1	AFF
		1	Paddy	PAD	PADDY		
		2	Other agriculture	OAG	OTH_AGRI		
		3	Livestock	LIV	LIVESTOC		
		4	Forestry	FOR	FORESTRY		
		5	Fishery	FIS	FISHERY		
2	Mining and quarrying			MAQ	MIN_QUAR	2	MAQ
		6	Crude petroleum, natural and liquid gas	CRP	CR_NA_LI		
		7	Other mining and quarrying	OMQ	OTH_MINI		
3	Food, beverages and tobacco	8	Food, beverage and tobacco	FBT	FO_BE_TO	31	FBT
4	Textile, apparel and leather	9	Textile, apparel and leather	TAL	TE_AP_LE	32	TAL
5	Wood and furniture products	10	Wood products and furniture	WAF	WOO_FURN	33	WAF
6	Paper, paper products and printing	11	Paper, paper products and printing	PAP	PA_PP_PR	34	PAP
		12	Chemical products	CHE	CHE_PROD		
7	Industrial chemicals			CHI	IND_CHEM	351 + 352 − 3522	CHI
8	Drugs and medicine			CHD	DRU_CHEM	3522	CHD
9	Petroleum and coal products	13	Petroleum and petro products	PCP	PET_COAL	353 + 354	PCP
10	Rubber and plastic products	14	Rubber products	RPP	RUB_PLAS	355 + 356	RPP
11	Non metallic mineral products	15	Non metallic mineral products	NMP	NOM_MINP	36	NMP
		16	Metal products and primary metals	PMP	MEP_PRIM		
12	Iron and steel			IAS	IRO_STEE	371	IAS
13	Non ferrous metals			NFM	NON_FERR	372	NFM
14	Metal products			MEP	MET_PROD	381	MEP
		17	Machinery	MAC	MACHINER		
15	Non electrical machinery			NEM	NOE_MACH	382 − 3825	NEM
16	Office and computing machinery			OAC	OFF_COMP	3825	OAC
17	Electrical apparatus nec			ELM	ELE_MACH	383 − 3832	ELM
18	Radio, TV and communication eq.			RTC	RA_TV_CO	3832	RTC
		18	Transportation equipment	TET	TRA_EQUI		

19	Shipbuilding and repairing			TES	SHI_REPA	3841	TES
20	Other transportation means			TEO	OTH_TRAM	3842 + 3844 3849	TEO
21	Motor vehicles			TEM	MOT_VEHI	3843	TEM
22	Aircraft			TEA	AIRCRAFT	3845	TEA
		19	Other manufacturing products	AOM	ALO_MANU		
23	Professional goods			PRG	PRO_GOO	385	PRG
24	Other manufacturing			OMA	OTH_MAN	39	OMA
35	Other producers			OTP	OTH_PROD		OTP
25	Electricity, gas and water	20	Electricity, gas, water supply	EGW	EL_GA_WA	4	EGW
26	Construction	21	Construction	CON	CONSTRUC	5	CON
		22	Trade and transportation	TRT	WRT_TRAN		
27	Wholesale and retail trade			WRT	WHO_RETA	61 + 62	WRT
28	Restaurants and hotels			RAH	RES_HOTE	63	REH
29	Transport and storage			TRA	TRA_STOR	71	TRS
		23	Other services	OSE	OTH_SERV		
30	Communication			COM	COMMUNIC	72	COM
31	Finance and insurance			FAI	FIN_INSU	81 + 82	FII
32	Real estate and business services			RAB	REA_BUSI	83	RBS
33	Social and personal services			SPS	CO_SO_PE	9	CPS
34	Government services	24	Government services	GOV	GOV_SERV		GOS
36	Statistical discrepancy	25	Others and discrepancy	SDY	STA_DESC		DSC

lifestyle. Various policy instruments such as emission trading, environmental tax, and the disbursement of revenues can be incorporated (Pearce, 1991; Sand, 1992; Sands, 1995; Sorrell and Skea, 1999). These can be linked to various related statistics such as emission factors (e.g. SO_2, NO_X, waste, water) and employment. It also provides a convenient linkage with various physical indicators.

3.2. *System of national accounts*

The economic flow between generation of income and final demand is not explicitly described in the input-output tables. This gap is filled by the SNA data. This highly standardized data set is available for most countries in the world and forms the basis for country models. Since input-output tables are available only with some time intervals, SNA data are also employed in estimating the economic structure in the interim years.

The national accounts data are published by the United Nations (United Nations, various years a). Of the original SNA tables, the model incorporates Tables 1.1 and 1.2 (final demand in current prices and constant prices, respectively), 1.3 (cost components of GDP), 1.4 (general government current receipts and disbursements), 1.5 (corporate and quasi-corporate enterprises current income and outlays), 1.6 (households and non-profit institutions current income and outlays), 3.13 (general government capital accumulation account), 3.23 (corporate and quasi-corporate enterprise capital accumulation account), and 1.7 (external transactions).

Needless to say, the SNA reveals a saving-investment balance for individual sectors. The external transactions account describes a country's linkage *vis-à-vis* the rest of the world in terms of merchandise trade, service trade, factor income, and current transfers.

3.3. *International financial statistics*

Financial variables such as exchange rates and interest rates are obtained from the IMF's *International Financial Statistics*. The items relating to international transactions in this source provide further details on direct investment and portfolio investment, which have gained importance as the global linkage through financial flows becomes increasingly powerful. These and other financial variables included in the model framework help determine the financial solvency of a nation. Foreign direct investment is also a vehicle for technology transfer. The COMPASS distinguishes four institutional sectors, namely: the government sector (*SEG*), the corporate sector (*SEC*), the household sector (*SHE*), and the foreign sector (*SEF*). The corporate sector *SEC* is split into financial and non-financial. The financial corporate sector (*SECB*) consists mainly of: the monetary authority (*MOA*), the deposit money banks (*DMB*), and the other banking institutions and non-bank financial institutions (*SOB*). (See

Chapter 2 for naming convention and Chapter 3 for macro-finance model specification.) They are covered by this statistical source, providing a clue to construct a flow-of-funds account for individual countries. Thus the COMPASS has as its statistical basis several accounting frameworks, a fact which provides a solid basis for consistency among various variables. The cases in question include input-output tables, the SNA, flow of funds, trade matrices, and energy balances.

3.4. Time-series trade matrices

International linkage has to do with commodity trade and financial flows. Trade data are compiled by the United Nations at the detailed commodity level involving several thousand items. International linkage can best be represented by trade matrices identifying exporters and importers for individual commodities. This is not an easy task because reports on bilateral trade from exporters and importers do not necessarily match, due to omissions and/or timing of recording international transactions which necessarily involve transportation. Adjustment to compile consistent matrices is carried out by at least two organizations; one is Statistics Canada and the other is the Institute of Developing Economies in Japan. Financial flows must be examined through SNA (saving and investment) and balance of payment data from the International Monetary Fund.

The country models are linked by time-series of international trade statistics compiled by the UN (United Nations, various years b). Trade matrices have been compiled from the most detailed trade data. The correspondence between production branch classification and trade classification is shown in Table 1-1 above. There are 24 distinct sectors as represented by the COMPASS world trade code plus one dummy sector representing statistical discrepancy.

3.5. Energy balances

Energy data are compiled by the International Energy Agency for both OECD and non-OECD countries (IEA various years, a, b). The energy data are in physical terms (tons of oil equivalent, or toe). IEA also publishes energy price data (IEA various years, c).

The time-series energy balance tables, which is the energy database distinguishing energy sources and energy uses, distinguish the following sectors.

(1) Total primary energy supply (TPES): indigenous production, import, export, stock change.
(2) Energy conversion: electricity plants, etc.
(3) Total final consumption (TFC):
  (3.1) Industry sector: iron and steel; chemical and petrochemical; non-ferrous metals; non-metallic minerals; transport equipment; machinery; mining

and quarrying; food and tobacco; paper, pulp and print; wood and wood products; construction; textile and leather; non-specified.
(3.2) Transport sector: road, rail, etc.
(3.3) Other sector: agriculture, commerce and public service, residential, non-specified.

The energy data are in physical terms (tons of oil equivalent, or toe).

Finally, the accounts should not end as a snapshot at a point of time. From the policy analysis point of view, the focal point is changes over time and, more importantly, the 'engine' which is causing such changes. The 'engine' in this case may be changes in relative prices, technological progress made possible by R&D and diffusion of new vintage of capital, or policy measures such as taxes and subsidies. Efficiency improvement, fuel shift, lifestyle change, are some examples. However, these factors are typically not an integral part of environmental accounts.

We next examine the possibility of developing an environmental accounting framework that is suitable for policy analysis with a focus on CO_2 emission. This calls for linkage between energy data in value terms and in physical terms. Input-output tables provide a starting point with the most general framework capable of integrating or establishing linkages with statistical framework in specific fields. This occupies a central core in the SNA. Input-output analysis is conducted to provide structural information of an economy in itself or in conjunction with macroeconomic analysis providing information on the level and composition of final demand. The framework is also suitable to deal with technological change (via changes in the input coefficients) and lifestyle (via composition of private consumption such as commodities, energy, and services of various kinds). This can be used for price analysis due to the fact that the framework incorporates indirect taxes and subsidies, making it possible to incorporate energy tax, etc. Input-output analysis can also have a dynamic dimension.

Table 1-3 compares the branch classification of the energy data with the one coming from the input-output tables. The activity classification scheme of the energy balances is shown at the left side of the table. The standardized input-output framework with 36 branch classification, which has been produced for selected OECD countries, is shown at the right corner (shaded). It is then rearranged in order to show the correspondence between the classification scheme for the two data sets. The result is shown in the middle of the table. We note the following points:

(1) One can distinguish three branches in the input-output framework which are directly related to energy. Indigenous production of primary energy carriers corresponds to mining activities related to coal, crude oil, and petroleum in 'mining and quarrying' in the input-output framework. Two sectors in the input-output framework, namely 'electricity, gas and water' and 'petroleum and coal products', correspond to energy conversion activities described in the energy balances.

(2) There is a good match in industrial use of energy, although some of the input-output branches are lumped together in the energy balances.
(3) 'Residential' use of energy in energy balances corresponds to energy components embedded in 'private domestic consumption'.

We note some of the points that are deemed essential in understanding the rearranged input-output table in the context of energy analysis:

(1) Energy conversion activity is represented clearly as the relation among 'energy production mining', 'petroleum and coal products' and 'electricity, gas, and water'. The last of these is understood to include 'heat'. The water supply would have to be removed from this branch in order to focus solely on energy-related activities.
(2) The input structures of energy-related sectors are shown. The input coefficients which are derived from the value table represent the production technology, and provide a clue for accommodating bottom-up technology information in the entire economic structure.
(3) Details of final consumption of energy, attributable to industry (manufacturing, agriculture, etc.), transportation, and private consumption (household) then follow. It is important to note that technology and lifestyle are represented as the input coefficients in various industry branches and converter (i.e., a vector showing the composition of supplying branches) of private consumption, respectively. Energy consumption attributable to the household sector (consumer durables, heating and cooling, and automobiles) can be examined in an economy-wide context. It is often pointed out that changes in technology and lifestyle are the keys in an attempt towards sustainability.
(4) Energy supply is represented by changes in stock, exports, imports, and indigenous production. It is noted that, in terms of the input-output framework, 'changes in stock' assumes a positive sign when the inventory is being increased, but, in terms of energy supply' the same inventory increase should be regarded as negative supply. The same thing applies to 'exports' and 'imports'. In the input-output framework 'exports' is a component of final demand and assumes a positive sign, and 'imports' is the portion of a country's demand being met by foreign supplies and assumes a negative sign. In contrast, in energy balances, 'exports' represent the portion not being made available to domestic use while 'imports' comprise the supply for domestic use.
(5) Policy variables embedded in the input-output framework include 'indirect taxes and subsidies' which is a part of the value-added sphere. This is where taxes on energy or CO_2 emission can be added. A fiscal scheme in which revenue from such taxes can be dispersed as subsidies for employment can also be represented. Depending on the country's trade policy, 'imports' may also include import duties. Trade liberalization entails removal of such measures.

Table 1-3. Branch classification of energy balance tables and input-output tables

IEANo.	Energy Balance Branches	IO No.	Rearranged IO Branches	IO No.	Original IO Classification: Industrial Branches
1	Indigenous production		Output in mining and quarrying		**Final Demand**
2	Imports	IMP	Imports of goods and services	HHC	Private domestic consumption
3	Exports	EXP	Exports of goods and services	GOV	Government consumption
4	Intl. marine bunkers			GFCF	Total gross fixed capital formation
5	Stock changes	energy STK	energy only, Changes in stocks	STK	Changes in stocks
6	**Total Primary Energy Supply (TPES)**			EXP	Exports of goods and services
	Energy Conversion			DFD	Total final demand
7	Transfers			IMP	Imports of goods and services
8	Statistical differences			GO	Gross output
9	Electricity plants	25	Electricity, gas and water		**Intermediate Demand**
10	CHP plants		(excl. water)	1	Agriculture, forestry and fishery
11	Heat plants			**2**	**Mining and quarrying**
12	Gas works			3	Food, beverages and tobacco
13	Petroleum refineries	9	Petroleum and coal products	4	Textiles, apparel and leather
14	Coal transformation			5	Wood products and furniture
15	Liquefaction			6	Paper, paper products and printing
16	Other transformation			7	Industrial chemicals
17	Own use			8	Drugs and medicines
18	Distribution losses			**9**	**Petroleum and coal products**
19	**Total Final Consumption (TFC)**			10	Rubber and plastic products
20	**Industrial sector**			11	Non-metalic mineral products
21	Iron and steel	12	Iron and steel	12	Iron and steel
22	Chemical and petrochemical	7	Industrial chemicals	13	Non-ferrous metals
23	of which petrochemical feed	8	Drugs and medicines	14	Metal products
		10	Rubber and plastic products	15	Non-electrical machinery
24	Non-ferrous metals	13	Non-ferrous metals	16	Office and computing machinery
25	Non-metallic minerals	11	Non-mettalic mineral products	17	Electric apparatus, nec
26	Transport equipment	19,	20, 21, 22	18	Radio, TV and comm. equipment
27	Machinery	14,	15,16,17,18, 23	19	Shipbuilding and repairing
28	Mining and quarrying		Energy use in Mining and quarrying	20	Other transport

29	Food and tobacco	3	Food, beverages and tobacco	21	Motor vehicles
30	Paper, pulp, and print	6	Paper, paper products and printing	22	Aircraft
31	Wood and wood products	5	Wood products and furniture	23	Professional goods
32	Construction	26	Construction	24	Other manufacturing
33	Textile and leather	4	Textiles, apparel and leather	**25**	**Electricity, gas and water**
34	Non-specified industry	24	Other manufacturing	26	Construction
35	**Transport sector**	29	Transport and storage	27	Wholesale and retail trade
36	37, 38, 39, 40, 41			28	Restaurants and hotels
42	Non-specified services			29	Transport and storage
43	**Other sectors**			30	Communication
44	Agriculture	1	Agriculture, forestry and fishery	31	Finance and insurance
45	Commerce and public serv.	25,	27, 28, 30, 31, 32, 33, 34	32	Real estate and business services
46	Residential		part of HHC part of private domestic consumption	33	Community, social, personal services
47	Non-specified	{ 35	Other producers	34	Producers of government services
		36	Statistical discrepancy	35	Other producers
48	**Non-energy use**			36	Statistical discrepancy

Note: Italics characters represent terms from the energy balance tables.

Energy balance tables consist basically of three parts: total primary energy supply (TPES), energy conversion (basically into 'electricity' and 'petroleum and coal products', and total final consumption (TFC). In addition, there is a small portion of energy carriers going into non-energy use. Some of the analytical points relating to energy balances are as follows:

(1) Fuel substitution can be represented as the time-series change in the individual shares in primary energy supply. We have witnessed declines in coal, increase in petroleum, and emergence of gas. Depending on the country's energy policy, a considerable portion of energy supply is borne by the nuclear sector. Introduction of non-polluting renewable sources of energy is a focal point in recent years, including combustion of urban wastes for power and heat generation. The shift can be explained by the relative prices of various energy carriers and R&D effort directed to this end.
(2) Energy conversion efficiency can be observed as the ratio between inputs of primary energy and output secondary energy (i.e. electricity). The efficiency has improved in some countries owing to R&D activities and capital formation embodying new technology.
(3) Energy intensity which is represented by energy inputs per unit production in individual economic branches or for the whole economy is also represented in the balance table. This parameter is typically explained by the relative price of energy *vis-à-vis* the price of output in respective branches.

Energy balance tables are intended to describe the energy supply being made available for domestic use, whereas input-output tables identify demand regardless of whether it originates domestically or across national boundaries. Reflecting the difference in point of view, export of energy assumes a negative sign in energy balance tables whereas it represents a positive final demand item in an input-output framework, and *vice-versa* for energy imports. The energy transformation from primary to secondary appears with a negative sign and electricity output has a positive sign in the IEA *Energy Balances*. In the input-output framework, transformation is represented with a positive sign, and electricity output with a negative sign in order to be able to say that the sum of transformation and final consumption equals primary energy supply.

4. The philosophy of the COMPASS approach

The purpose of the project is to construct a robust test-bed for simulation experiments concerning global policy issues. The COMPASS approach is based on a set of guiding principles which include the following:

Policy relevance: COMPASS is a policy model. Efforts have been made to identify major players which determined the scope of the model in terms of regions, economic branches, and relevant variables in economy-energy with its implication on energy prices, energy conservation, trade patterns, international relocation of energy-intensive activities, employment profiles, and economic performance in general. This may also lead to re-examination of the tax system as a whole (Oates, 1995).

Global closure: COMPASS is an international model. It distinguishes as many countries/regions as possible, within the constraints of data availability and other unavoidable factors. As a policy model rather than a mere academic exercise it is necessary to provide information to decision makers in individual countries/regions. However, more often than not we are faced with policy issues with global implications. Trade liberalization, international capital movements, and global environmental concerns are a few examples. In a single-country model we are obliged to assume economic trends in trading partners, exchange rate, world prices for various commodities, and international financial flows. With inclusion of economies representing more than 99% in terms of world GDP, trade flows, and energy production and consumption, the model is closed globally with respect to production of goods and services (as represented in trade balance), financial resources (as represented in saving-investment balance), and energy (production and final use). The COMPASS generates consistent international commodity prices and domestic prices at the disaggregated level. It also generates financial flows both within a country and internationally, making it clear whether or not an investment fund is forthcoming from either domestic or international sources. It has a very detailed energy model, using physical quantities as well as values, tracing the domestic flow from final use through conversion to supply of primary energy, which then goes into international trade of coal, oil, and gas, determining energy prices.

Bottom-up combined with top-down: The macroeconomic performance results from activities at the branch level. COMPASS is based on input-output tables and corresponding trade matrices precisely for this reason. Also, behavioural equations are intended to describe the actual decision rules at the branch level. The model traces the response of economic agents to price signals (price of goods and services at branch level, exchange rate, interest rate, and energy prices), but perfect knowledge and market clearing are not assumed *a priori*. On the contrary, the model reveals the disequilibrium which may emerge in the economic system and how it is resolved over time through adjustment on the part of individual economic agents induced by changing relative prices.

Full integration: COMPASS is intended to be a test-bed for various policy experiments. A model focusing on one particular issue is only able to provide partial views. The scope of COMPASS in contrast extends over economy–energy–environment interactions. The economic part consists of a macro-financial model, an input-output model, and a world trade model. It covers the flow of goods and services both domestically and internationally. It deals with saving-investment balance at local as well as global levels. With full integration of various aspects, any single issue can be analysed from a system-wide perspective.

Accounting consistency: Most models are built around the variables of interest. This has an advantage of making the model small and simple. COMPASS, in contrast, tries to generate various accounting tables which provide a system-wide perspective and are widely used by policy-related institutions. The cases in question include SNA with sectoral balances, balance of payments, flow of funds, input-output tables, energy balances, and trade matrices against which individual variables can be examined from the system-wide point of view.

Transparency: In order to serve as a test-bed for various policy experiments, COMPASS is based on generally accepted theory employing internationally open data. Different theoretical orientation or sensitivity analyses regarding a particular policy variable can be accommodated by changing the relevant portion of the model framework.

Operationality: In some modelling environments, data and parameters are parts of a very large programme, making it difficult to make the necessary adjustment when a model specification is altered. Easy operation is also an indispensable attribute required of a model which is used as a platform for various simulation experiences. The model framework may be very large and complex, but the operation has to be simple when one considers the number of simulation runs to test, for example, sensitivity of the world economy to various trade arrangements. COMPASS employs identical logical sequences in model development and simulation runs regardless of whether one is dealing with macro and financial part, input-output model, energy and environment block, or trade linkages.

Coping with time path: In policy analysis it is imperative to be able to tell the course of development over time if certain measures are taken. A framework which only describes what will happen when equilibrium is attained does not provide the required answer. There are models which project into the year 2100, for instance, but they have to rely on heavy assumptions on the parameters. COMPASS has a dynamic structure which provides information along the time horizon. It is based on econometrically estimated parameters from 1970 to around 1995, and it is capable of projecting annually into the future until 2010 (at the maximum 2020).

Coping with non-linearity: Econometric and other quantitative models are based on the assumption of linearity and COMPASS is no exception. In reality, however, there are non-linear relations, but one would be misguided if one concluded that econometric models cannot properly describe non-linearity and hence cannot adequately describe the reality. There are various non-linear relations which can be linearly approximated. Johnston (1963) lists semilog transformation (corresponding to a constant rate of growth), double-log transformation (corresponding to the assumption of constant elasticity), reciprocal transformation (corresponding to an assumption of an asymptotic level), logarithmic, reciprocal transformation (to accommodate threshold levels, etc.). In addition, proper disaggregation, such as attempted in COMPASS, reduces the cases of seemingly non-linear relations at the aggregate level. For example, energy consumption may show a kink at the macroeconomic level whereas at a disaggregated level individual branches behave the same as before but the weights are changing (e.g. manufacturing shrinking and services becoming dominant).

Coping with structural change: Models are often accused of their inability to deal with structural change. This is precisely why COMPASS is constructed at disaggregated levels. It also explicitly treats economic sectors such as household, industry, government, and the rest of the world in three distinct spheres: (1) goods and services, (2) financial flows, and (3) energy. Thus, it is possible to examine the structural change of the economy (input-output relations), trade structure (trade matrices explicitly stating exporters and importers for an individual commodity), balance of payments (trade and capital flows), energy (primary energy supply through energy conversion to final use) and policy tools (taxes and subsidies), among others.

Coping with lack of historical data: There is no way of overcoming the lack of past data. COMPASS bypasses the problem by borrowing and estimating. Borrowing the historical experience elsewhere has its limits, of course, but may shed light on the future course of events in the countries/regions undergoing rapid industrialization, for example. Estimating the past data based on available column sums or row sums of a matrix is acceptable and even desirable because there can exist only one set of variables which satisfies the accounting balance.

Model as a black box: As models become larger and more complex they are often treated as a black box, and only the main results are published. Model results are only as good as the assumptions that are fed to the model, but the influence of such exogenous information to the solution is often not discussed. Looking at the main results is of course interesting, but one has to be able to obtain information on how the rest of the economy looks. COMPASS graphic user interface (GUI) allows model-builders and model-users to examine macro-financial models, input-output models, energy models, and trade models in terms of their structure, assumptions, and econometrically obtained parameters.

Impact on employment: Trade liberalization and environmental preservation may affect employment levels and change occupational profiles. Policy makers in many countries are worried about the impact on employment just as they are concerned about the trade balance and the levels of GDP. The model presented here does not provide precise information on employment, although this point can be remedied. See Chapter 11.

5. Comparisons with other models

The COMPASS focuses on medium-term analysis, not short-term (cyclical) or long-term (general equilibrium) analyses, but traces how the paths would evolve in different regions and in different industrial branches responding to the introduction of policy measures (such as tradable permit or carbon tax), technological change (such as new input structure and better conversion efficiency), and lifestyle shift (such as service orientation or energy conservation effort).

The COMPASS framework, because of its branch disaggregation and regional breakdown, and because of its simultaneous equation system, allows bottom-up and top-down interaction. The inclusion of branch prices (36 branches), final demand prices (consumption, investment, government spending, exports, and imports), exchange rate, interest rate, and various energy prices as endogenous variables, and the adjustment mechanism by economic agents based on relative prices are described explicitly.

The model treats world petroleum prices as exogenous variables, but equilibrium world petroleum prices are calculated within the model in order to monitor the market forces behind the assumed price levels. Technology factors such as production technology (described by input coefficients), energy substitution (among different energy carriers), and changes in energy intensity are fully explained endogenously. The model can project input-output tables and energy balances in the future, providing structural information of the world community.

COMPASS has feedback channels running through economy–energy–environment spheres. Its framework is disaggregated almost to the maximum extent that is feasible from the data point of view. COMPASS incorporates, and is capable of endogenously generating, a relevant accounting scheme, including national income accounting, flow of funds (being tested) and saving-investment balance, input-output tables, trade matrices, and energy balance tables. It can also generate prices for individual industrial branches, final demand, and energy. Exchange rates and interest rates are endogenously determined. Technological factors are represented as changes in input coefficients and changes in energy intensity for an industry over time because they represent new production methods requiring different material use. COMPASS also provides information at detailed country levels, an advantage over various other models.

An object-oriented approach is maintained throughout the project, ranging from naming convention, statistical database, model-building software, and simulation engine. Consistent use of meta data makes the construction and maintenance of large-scale models mainly systematic (Soltes, 1999). Because of this approach, the inclusion of other regions in the model framework can be systematically accomplished.

Table 1-4 presents a comparison of COMPASS with other models with a focus on the interaction involving economy, energy, and environment; for details see Chapter 12.

6. Policy analysis

Some of the characteristics of COMPASS are summarized below, after cross-checking with other models. The model can be used as a test-bed for various policy scenarios.

6.1. Lifestyle

Lifestyle can be regarded as a reflection of changing personal values and income levels. Although an econometric model is ill-suited for dealing with people's values, it can at least deal with changing consumption patterns which presumably reflect changing lifestyle in addition to income effects. COMPASS accommodates this variable by allowing the model-builders to apply different branch make-up of private consumption vector in the final demand quadrant of the input-output tables (Parikh *et al.*, 1991; Rayner and Malone, 1998; Redclift and Benton, 1994; Schipper *et al.*, 1996; Wier *et al.*, 2001. See also Van Asselt *et al.*, 1995). Needless to say, one can specify a consumption function for individual branches if appropriate data (such as panel data providing consumption expenditures by household income bracket) are available if the purpose of the analysis is to measure the consumption pattern under changing income and price levels.

One way to mitigate the collision between the human aspiration and the global environment is to seek an alternative lifestyle which is more amenable to sustainability. The empirical foundation in this regard will be final consumption expenditures by function for government and private sectors provided by Tables 2.3 and 2.4 of the SNA. A different expenditure pattern signifies a different environmental impact, e.g. in terms of energy consumption. Particularly important is the household expenditures on durable goods, which form a linkage point between consumption and technology, such as represented by energy efficiency. Consumer durable goods as the stock determines a large portion of energy consumption that appears in the account as expenditure on fuel and power.

6.2. Alternative technology

Despite the fact that technological factors have become the central focus of the policy debate, the treatment of technology in economic theory is rather blurred. COMPASS is based on a time-series of input-output tables, and the input-output framework describes the input structure (which is the production technology *per se*) for individual branches of the economy. Technological change can be described by the changes in input coefficients over time. In the case of energy, the model includes extra variables such as substitution among primary energy sources and conversion efficiency.

In the COMPASS framework, technology is represented at four major points:

(1) input coefficients in the input-output models,
(2) conversion efficiency of primary energy to secondary energy in the energy models,
(3) the selection among different energy carriers also in the energy models,
(4) changes in energy intensity over time.

The input coefficients describe the composition of material and service inputs into respective branches of the economy. In other words they describe the

Table 1-4. Methodological comparison of economy–energy–environment projections

Model Names	Reference	Modeling Logic	Projection Period	Economy-Energy Linkage	Prices	Regions	Energy Sources	Industrial Branches	Commodity Trade Linkage	Emission Trading
3E COMPASS	**Uno *et al.* (2002)**	**IO + Econometric**	**2010 (annual)**	**Endogenous**	**Endogenous**	**61**	**6**	**36**	**Yes (53*53 matrices** for 26 commodities)	**Yes**
Adam Rose-Brand Steven	Rose & Stevens (1999)	General equilibrium, optimization	25 years	Economy exogenous	No	14	1	1	No	Yes
AIM	National Institute for Environmental Studies	Top-down world, bottom-up technology	2100	Economy exogenous	No	19	No	Varied	No	No
APEC Energy Outlook	APERC (1996)	Scenarios	2010	Economy exogenous	Exogenous	6	6	3	No	No
ASF	IPCC (2000)	Demand-supply adjustment by price	2100	Economy exogenous	Yes	9	Numerous	1	No	No
Decomposition	Unander & Schipper (1999)	Decomposition of indicators + scenarios	2010	Economy exogenous	No	7	Various	Various	No	No
DEMETER	van der Zwaan *et al.* (2000)	Optimization	2100	Partly endogenous (leaning-by-doing)	Yes (taxes, subsidies)	No	Carbon energy	No	Yes	Yes
ECN Study	Sijm *et al.* (2000)	Cost curves and demand curves	2010	Economy exogenous	No	30	Not explicit	Not explicit	No	Yes
EDGE Model	Jensen *et al.* (2000)	General equilibrium	2030	Endogenous	Yes (emission)	8	5	7	Yes	Yes
G-cubed	Bagnoli, McKibbin, & Wilcoxen (1996)	General equilibrium, econometric	2020	Endogenous	Yes	8	4	12	Yes (8*8 matrices for 12 commodities)	No
GemWTraP	Bernard & Vielle (1999)	General equilibrium, optimization	2020	Economy exogenous	Yes	7	5	12	No	Yes

GREEN	OECD (1994b)	IO + Econometric	2010 2050	Endogenous	Endogenous	12	5	11	Yes, bilateral flows	No
GTAP	Hertel & Tsigas (1997)	General equilibrium, I-O	No time series	Endogenous	Yes	24	5	37	Yes (31 commodities)	No
IEA Energy Model	Vouyoukas (1993)	Econometric	2005 (annual)	Economy exogenous	Exogenous	10	14	3	No	No
IFs: International Futures	Hughes (1999)	System dynamics	2100	Economy exogenous	No	20	6	5	Yes (pooled approach)	No
IIASA and WEC	Nakicenovic *et al.* (1998)	Scenarios	2100	Economy exogenous	No	11	5	No	No (assumption)	No
IMAGE 2.0	Alcamo (1994)	Elasticity	2100	Economy exogenous (= IPCC)	No	13	6	5	No	No
IPCC Special Report	IPCC (2000)	Scenarios	2100	Economy exogenous	No	No	6	No	No	No
MARIA	IPCC (2000)	Non-linear optimization	2100	Economy exogenous	Yes	8	9	3	No	No
MARKAL Models	Loulou & Kanudia (2000)	Partial equilibrium	Varies	International permits	Yes	Stand alone	Varies	Varies	Some models with regional linkage	Yes
MARKAL MATTER	Gielen & Kram (2000)	Linear programming	2060	Price dependent demand	Yes	No	25	No	No	No
MERGE	Manne, Mendelsohn, and Richels (1995)	General equilibrium, optimization	2200	Economy exogenous (= IPCC)	No	5	2	3	No	No
MESSAGE	IPCC (2000)	Dynamic linear programming	2100	Economy exogenous	No	10	Numerous	1	No	No
MIDAS	Capros *et al.* (1996)	Bottom-up technology	10 years	Economy exogenous (= HERMES)	Yes	4 independent	5	11	No	No
MiniCAM	IPCC (2000)	Partial equilibrium	2100	Economy exogenous	Yes	11	7	3	No	No

Table 1-4. continued

Model Names	Reference	Modeling Logic	Projection Period	Economy-Energy Linkage	Prices	Regions	Energy Sources	Industrial Branches	Commodity Trade Linkage	Emission Trading
MS-MRT	Bernstein *et al.* (1999b)	General equilibrium, optimization	2030	Economy exogenous	No	10	4	6	Yes	Yes
New Earth	Nishio, Fujii & Yamaji (2000)	Non-linear optimization	2020	Economy exogenous	No	10	9	No	No	Yes
The PRIMES	European Commission (1995)	Optimization type, Partial equilibrium	2030	Endogenous	Yes	12	4	12	No	No
Rains-Asia	Resource Management Association (1996)	Scenarios	2020	Economy exogenous	No	23	17	6	Not explicit	No
RICE	Nordhaus & Yang (1996)	General equilibrium, optimization	2200	Economy exogenous	No	10	1	1	No	No
WEPS	Energy Inf. Admin., USDE (1997a)	Scenarios	2020	Economy exogenous, elasticity assumptions	No	16	6	1	No	No
World Energy Outlook	International Energy Agency (1998b)	Scenarios	2020	Economy exogenous	Yes	10	8	4	No	No
World Model	Duchin & Lange (1994)	IO + Scenarios, bottom-up technology	2020	Economy exogenous	Relative price of oil	16	3	44	Yes (39 commodities)	No
WorldScan	Bollen *et al.* (1999)	General equilibrium, optimization	2100	Endogenous	Yes	12	4	11	Yes (two-way trade)	Yes

Note: For detailed discussion of each projection, see Chapter 12. Quantitative results are summarized in Table 12-25.

prevailing production processes. One can therefore experiment econometrically on how the changes in relative prices of one input (say, energy) affect the corresponding coefficients. Alternatively, based on information from a technology inventory which provides perspectives for innovation, one can try to insert input coefficients representing future technology scenarios. Internationally, experiments on technology transfers can be conducted by transplanting the actual input structure in industrialized countries to developing economies (Rosenberg, 1997; Sakurai *et al.*, 1996; Wallace, 1995).

In addition, it will be possible to include consumer durables in the model framework, and the effect of improved energy efficiency of automobiles, air conditioners, refrigerators, etc. can be examined.

6.3. Tradable permits

The COP is a very complex political process and it is not realistic to replicate the installment of various rules and fine-tuning of numerical targets to be implemented in a precise time span. The purpose of the simulation experiments regarding tradable permits must be as follows.

(1) Calculate the base trend without introduction of new policy measures such as tradable permits and carbon tax. The COMPASS, with its empirically estimated parameters incorporated in a multi-sector, multi-country framework and global closure regarding production and trade on the real side and saving and investment on the financial side, is a powerful tool for this purpose.

(2) Experiment with tradable permits based on the CO_2 emission targets which are given in the Kyoto Protocol. With some exceptions regarding developing economies and the economies in transition, every country will be initially granted that amount of permit that is equal to the actual emission. Subsequently the permit will be reduced following the path leading to the reduction target. On the individual regional level, government will be in a position to impose an energy tax (if the country is a buyer) or allocate the surplus (if the country is a seller). It is likely that most of the industrialized countries will find themselves buying the permits. A rising tax rate reduces energy demand and, thus, CO_2 emissions. On the global level, new energy demand and corresponding price of the permit is calculated. If the emission of the country is higher, it has to purchase permits on the international market; if the emissions are lower, the country can sell to the market. The price of the permits is set at a level where market clears by equilibrating demand and supply. Simulation will be conducted regarding the direct effect of changing energy prices. The effect may vary among regions, giving room for conflicts of interest.

(3) Examine the secondary effect of the introduction of tradable permits (and other policy measures) on industrial structure, trade patterns, government finance, employment, etc.

It should be remembered that simulation experiments are not forecasts. They are the basis for our awareness of the required policy measures and the needs for new technological development.

Part II

The framework of analysis

2
Object-oriented database and modelling system

DIRK VANWYNSBERGHE and FRANK HOHMANN

1. Introduction

This chapter sets up the time-series conventions for the database and the modelling engines used in COMPASS in the framework of international cooperation. The project provides the users with a set of tools that allow users, to understand the whole process and to be able to achieve the model in a well-defined number of steps within the limits of predefined time horizon and budget. The project offers a set of predefined prototype working models that are dubbed as Quickdyme versions for IO, Quickmac for SNA and Financial macro modelling, and Worldtrad for world trade linkage at commodity level. They allow users to complete the data set-up, test the consistency of the basic data, build more and more realistic models in transparent and standardized steps and to present and analyse the simulation results.

Why use a database concept? Data can be presented in two concepts: a *data set* concept (a file concept), and a *data base* concept (either a relational database system or an object-oriented database system). Although the *difference between set and base* is fundamental, the state of the art of the technology for IO modelling does not allow using a *unified* concept for the entire project scope. Why? As we shall see, *from the client's point of view, the raw data source side and the productivity side*, one needs a database approach. However, *from the model-builder and model engine's viewpoint*, only data sets are used for both input and output.

The COMPASS model links economy, energy, and environment (3E) in a consistent way. The data used in the framework come from a number of selected sources. Such a global system uses, for representative comparison and efficiency reasons as much as possible, widely recognized and standardized sources of data. Thus, data come in the first place from international sources (such as UN, IMF, World Bank, OECD, and IEA, among others), but also from a number of precious national sources (e.g. the Institute of Developing Economies in Japan). All these sources use database systems and transfer their data by classical database interface output formats. None of these is directly readable by the available model engines. Furthermore, one of the important pillars of the success of such a project in general, and of this one in particular, is the unified and standardized way in which the data are supplied, used and delivered. Such a way can only be achieved in a database concept.

K. Uno (ed.), Economy–Energy–Environment Simulation, 33–53.

Why are data sets still there? On the other hand, the individual model-builders and model engine constructors consider data as numbers in a file read by the model in one format or another. As model updates and simulation experiments always involve data, an unbridgeable gap tends to be innocently built in from the start by the model building. This gap has never been recognized by the model-builders who basically have no experience with database-organized information.

A data set file contains mainly numerical data and series names in a specific format for both the numerical data and for the name of the series. Special characters may be introduced to indicate the end of a series, e.g. a semicolon. Typically such a series will contain numerical values for each of the time periods but nothing else on the non-numerical level. Sometimes the file name may give some hint as to the type of data and the data source. Only the maker of the file will know what the data are, what the units are and so on. Another user cannot find any such information except in forms not linked to the data points. In a nutshell, everything is in the head of the maker and nothing is really shared with the potential user.

The solution is theoretically simple: use a database concept (RDMS or its improved version) for the whole scope of the project, meaning from data preparation, and model-building step to the output analysis (decision) step. Unfortunately, today almost all (if not all) modelling engines need data sets as input, and all of them use data sets (files) as output. This conceptual difference, together with the limited/unreliable performance of data-set-oriented software applications, has almost killed the classical model market and replaced it by robust but naïve and over-simplified database explorations combined with spread-sheet extrapolations. In day-to-day business obtaining the results is the dominant concern rather than how you are structured logically internally. Models are great tools but to the decision-makers they look like infantile instruments.

2. Does IO go the dinosaur's way?

Today's interactive computer applications basically consist of objects on a screen that are responding to standardized kind of actions from the user. These standards do not depend on the designer of the application or on the type of application, but are only reflecting the state-of-the-art technology in graphical user interfaces (GUI). With the introduction of the remote terminal all computer applications started to exploit the interaction between subjects (e.g. user, designer), an input device, and the application software. Nowadays the input device consists of the objects on the screen that can be activated by keystrokes, mouse clicks and very soon by voice. In the earlier days these objects on the screen were very primitive and rather cryptic, and the possible action of the user was limited, and to a large extent depended on the execution process of the application program. The program managed the users rather than the other way round. The application had a pre-programmed procedural execution and in the best case a structural program did achieve the predefined task. The users could not do anything that had not been pre-programmed, nor could they vary the run stream by any creative intervention. This solution puts too many constraints on the application both from the designer's viewpoint and from the user's viewpoint. The situation has changed drastically in recent years and is still in a turbulent revolution.

The input-output analysis has an inherent problem when looked at from this perspective. Input-output models were no exception in the past, but nowadays lag far behind the above-described fundamental changes. Building and using an input-output model is still a lifetime task for the few individuals who have the time, money and patience to struggle through that process. Maintaining and updating the models is such a labour-intensive and exclusive task that most of them are in constant construction and almost never reach the operational phase. The lucky few, once there, are so difficult to operate that in most cases only the designer or model-builder can use them. Why is this so?

We see the problem coming from three different angles:

(1) The IO technology is complicated because it runs through from IO table, static model, dynamic model, to econometric model. Standards in the naming are not available or not integrated with macroeconomics, social accounting, energy and environment.
(2) The data needed both in the numerical sense and in the conceptual sense are very demanding and difficult to handle. IO started with numbers and has never moved out from this short-sighted view. Most of the meta data are not at all recorded or referenced and only exist in the brain, and in some cryptic notes made by the model-builder.
(3) The available computer software is not designed as an engine and does not pay much attention to a user-friendly data preparation, model design, model operation, or an update of both the data and the model. It is usually built together with one particular model and will probably never reach the operational phase of its lifecycle. Maintenance and update of the software are typically not part of the project design.

The result is that only a few experts are able to handle IO models; transfer of their know-how is difficult; with a result that fewer and fewer junior people can be found. The misunderstanding of the IO model has also brought some mistrust – not least from colleague economists – and the usage of this powerful technology is far below its potential.

Is there another way of doing things to remedy these drawbacks? The object-oriented future is the way to go. We believe there is indeed a way that is based on an object-oriented approach in the design, the database, the programming and the use of standardized IO models. Clearly the origin of the problem cannot be with individuals, but with the conceptual approach to models and their extremely poor implementation as objects. One can define an object as the realization of every working concept. It consists of many smaller objects in itself. The user does not have to know how an object performs its tasks, but just how to send messages so that the object executes by itself the appropriate method that gives the expected result for the action of the user. The action of the user is limited to keystrokes and clicks. The user action will create events. The events will execute methods that will manipulate data (known as properties of the object on which the action is directed).

It should be possible to bring IO models to the state-of-the-art user interfaces for data preparation, for the model building, for model using and for the analysis phase of all data input and output. The major steps solved in the project are at three levels: the software level: *Compass GUI;* the IO model level: *Compass IO;* the policy and management conclusion level: *Compass PM.* What is of most importance to see is how each of these fields integrates into the whole COMPASS project, and how each step is necessary to achieve the following one. The policy and management considerations cannot be studied without such an integrated and productive approach.

3. The database (DB) and the database management system (DBMS)

Data can be presented in two concepts: a *data set* concept (system-wise a file concept), and a *data base* concept (system-wise either a relational database system or an object-oriented database system). Although the difference between set and base is fundamental, the state-of-the-art technology for IO modelling does not allow us to use a *unified* concept for the entire project scope. Why?

A *relational database* stores information in the form of *tables* that contain *related information for the same topic.* A table is not stored in a file as such, but all tables and their relations are stored in a special file: the database file. One can say that *the type of information* is stored as *the column headers* also called fields, and that each *individual data point* will be a *record or row* in that table. Each record or row will contain the relevant value for each of the column headers that fit together for that individual data point. It should be clear that not only numerical data but also non-numerical data are stored in the same table. The order of the rows or records in a table does not matter and a retrieval of the data will be in a random order. However, *sort* and *search values* (alphanumerical ones with placeholders) can be put on each column of the table to sort and to look for a specific type of information. Also, such values can be used as *filters to retrieve* only the information that satisfies these criteria. The database management system, and not the application program, will keep track of all changes and the update; insertion and deletion of data will be done by that system. The database is independent of the operating system (platform) used and can be accessed by standard software other than the application. A copy of the database will not work on another computer if not properly installed. The way the data are stored is unusable for human beings, but extremely well adapted for computer treatment through the DBMS. Therefore, the non-DBMS user needs an interface to retrieve, input, update, delete explore, analyse, export and use the data in all kind of applications. Compass GUI is such software specially built for international modelling and linkage.

3.1. Client expectations: database-oriented

The model users, distinct from model-builders, are exclusively used to database-delivered information (not in a file but in an information system and today even in a decision system). They do not want, or do not have the time or the resources, to become involved with any of the data set considerations nourished so privately by the model-builders.

3.2. Data sources: database dominance

The COMPASS model links economy, energy, and environment in a consistent way. The data used come from a number of *selected sources*. Such a system uses for comparison and efficiency reasons as much as possible widely recognized and standardized *international* data. All these sources use database systems and transfer their data by classical database interface output formats. None of these is readable by the available IO model engines.

3.3. Productivity interest: database oriented

Furthermore, one of the important pillars of the success of such a project in general, and of this one in particular, is the unified and standardized way in which the data *are supplied, used and delivered.* Such a method can only be in a database concept.

3.4. IO model world: data set dominance

On the other hand, the individual model-builders and model-engine constructors consider data as numbers in a file read by the model in one format or another. As model update, simulation and output always involve data, an unbridgeable gap was unwittingly built in from the start by the model-builders. This gap was never recognized by the model-builders who basically have no experience with database-organized information (except negative ones). A data set file contains numerical data for a given series as single time-series or as matrix. The series will be presented in a special format for the numerical data and for the name of the series. Special characters may be introduced to indicate the end of a series. Typically such a series will contain numerical values for each of the time periods and nothing else on the non-numerical, information level. Only the maker of the file will know the data at the meta level, i.e., what the data are, what the units are, what the sources are, and so on. Another user cannot find any information on that except in a form not linked to the data points. Some comments may be added to the file for information, but they do not make part of the data used by the engine.

3.5. Communication interest: data set preference

From a communication point of view, data files are very efficient for transfer both on hard carriers and on networks. The export of database results is mainly done by files; however, the structure of the information usually reflects a database concept, and not a time-series concept, requiring some interfaces.

3.6. The theoretical solution

The solution is theoretically simple: use a database concept (RDMS or a better one) for the whole scope of the project, meaning from the model-building step to the output analysis (decision) step. Unfortunately, today almost all (if not all) modelling engines need data sets as input and all of them use data sets (files) as output. This conceptual difference, together with the limited capabilities and/or unreliable performance of data set-oriented software applications, have almost killed the classical model market and replaced it by robust but naïve and over-simplified database explorations combined with spread-sheet extrapolations.

3.7. The COMPASS compromise: a central database with automated data set interface for different engines

The data approach is the first step to show maturity for many reasons. COMPASS started with a new approach on the data side: first taking into account the user's understanding and preferences as the major objective, and second taking the technical limitations only as the constraints to be worked around.

The basic raw data are harvested in the basic raw data system (BRDS). This contains the full set of all available international data of the selected data sources and for the selected data domains and the equivalent national data for some countries with insufficient international coverage (for whatever reason), e.g. China and the Russian Federation, among others. For efficiency reasons the basic raw data from the BRDS are automatically transferred to the model raw data system (MRDS) that contains only the data needed for the model system. Here data correction and data manipulation is already done. Up to that level all data are stored in database management systems and use standard database engines. However, for the modelling, one has to create the specific data input sets needed by the model engines that is in a particular case. For this reason the MRDS creates in an automated way the data sets for the different modelling engines used. They are referred to as modelling engine data files (MEDF) and are stored in specific formats and in predefined directories designated by the model system and the model engines. The models themselves will then read this engine data set and create the model system data files (MSDF) for each specific model engine.

4. Standardization in IO: naming conventions

The naming conventions proposed here are for the starting versions of the models, and more elaborate models will have these variables as totals of more detailed ones developed by the experienced user. Only the classical variables should be standardized fully. Names of the series refer in the first place to a classical input-output table and should be easy to remember and have as much relation as possible to the English definition. We should do away with the pure mathematical conventions once used like *x* or *q* for gross output – there is no *x* or *q* in the definition, *y* for value added – there is also no *y* in value added, and so on. Exceptions should only exist for generally excepted concepts such as gdp or GDP. Names of other variables (non-input-output series) used in the database and model could be chosen more freely as long as they use the standard rules and do not conflict with the input-output names. However, as a number of models for different countries are developed within this software, using the same names for all countries does give the result a greater portability.

The naming principles are for the *modelling software developers* and for the *model developers* starting to use the versions of the database and model simulation engine known under the name Quickdyme. The modelling software will internally use these names. If the user also uses the same names for his data input, then no C++ code will have to be changed in the Quickdyme versions of the model. Experienced users can proceed as they did before. These names will be the ones used in the central database for all countries and in the different versions of Quickdyme. So understanding number 1 says that *if the user accepts* the naming conventions, then no C++ code must be changed for the all Quickdyme versions because externally used names are fully consistent with internally used names.

Time-series can be *single series* like an exchange rate, *vectors* (n × 1) like the exports for each of the branches in the model, or *matrices* (n × n) like the intermediate flows from each branch of origin to each branch of destination. The naming rules should be so that the user (and the system) sees immediately if the variable is a single series, a vector or a matrix. The vector series of dimension 1 × 1 are vectors for the software but single series for the user. They will have special names to be recognized immediately; we minimize their use to an absolute minimum.

Names for *vectors* and *matrices* have a *three-digit* root for the meaning, e.g. intermediate demand, household consumption, investment, etc.; *prefixes* for the different types within the root, such as shares, coefficients, prices, bridges, etc.; and *suffixes* for detail in the root, e.g. rural household consumption, urban household consumption, or for groups such as heavy industry, high-technology branches. Such names are: *prefix_root_suffix*.

The *single series* will use much the same conventions for vectors and matrices, but should have a name (like the root) *that is more than 5 characters long* so that one can immediately see that this is not a vector or a matrix.

Three-digit roots can never end in an integer, as branch numbers will be added automatically by the software. So suffixes for branches such as *1, 2, 3, n* are not needed, and if integers are used in the root for other reasons, then they should end in an '_' being part of the root. So forget integers in the root.

Names should immediately show if the variables are in real terms or in nominal terms (in more general terms, whether the variables in question are not affected by the price system or are affected – employment is like a vector in real terms), and whether they have unit dimensions or not (e.g. coefficients or shares, or growth rates, or percentage changes). A special rule will be worked out for series that are mixed, e.g. valued added in current prices divided by output in constant price.

The simpler the system is, the better it can be applied, and the shorter the names – if they have meaning for the user – the better. No full logical system has been found so far, so a final list will have to be agreed on. This list will be kept constantly up to date and new names – including exceptions – will be added. All new ideas are welcome.

Some series involve a mixture of current and constant prices such as value added in current prices per unit of output in constant prices. So we need a way to formalize this type of data. We will use the prefix (see below for prefixes) to express the first series in the root and the second series can be either silent (not referenced in the root, and then it should always be in the same form as the first one), or the second series might be referenced in the root, and then it can speak for itself. An example:

> *C_voo*: unit value added defined as value added in current prices per unit of output in constant prices. Here the '*C_*' is in CAPS to indicate that the first digit of the root (it must be small because this is a vector) is in current prices, the second digit is *o* to indicate the operation 'over' or element-by-element divide and small because it goes together with the second series, in constant prices, output giving the '*o*' as third digit of the root.
> *C-vOO*: would stand for value added in current prices 'over' output in current prices. We do not use this series, but it is given to clarify the rules.

The same rules are valid for the '*s_*' and '*S_*' referring to the first digit of the root being in constant or in current prices. The second series, if present in the root, can declare itself by the CAPS or small attribute. The same principle can be used for '*b_*' and '*B_*' for the first digit of the root.

As a general rule one could say that, in this type of mixed series made by operations, if the second series is not mentioned it is always in the same form as the first one. So if the first one is constant the 'silent' second one is also constant and *vice-versa*. If this is not the case the root should mention the two series involved. If more than two series are involved, as in the definition of domestic demand = output + imports – exports, we will omit the operators, e.g.:

> *omx*: domestic demand in constant prices defined as output (*o*) + imports (*m*) – exports (*x*);
> *opm*: domestic demand in constant prices defined as output (*o*) plus (*p*) imports (*m*);
> *oPM*: would stand for the same series but in current prices.

The concept behind this special series is that the user knows the definitions rather well so that the operators can be omitted. The basic included series should as much as possible be mentioned.

5. Naming format

The proposed format is '*prefix_root_suffix*', prefixes and suffixes can be omitted but if present should always be linked to the root by '_'.

5.1. *The root*

The root is three digits long for *matrices* and *vectors*, at least six digits for *single series*. Digits 4 and 5 are left open for future extensions of the root. The first digit in the three-digit root is in CAPS for a matrix (e.g. '*A*') and small (e.g. '*a*') for a vector. All other remaining digits of the root after the first one are in CAPS for *current prices* and small for *constant prices*. We have more constant price series in the input-output model than current price series.

The meaning of the first digit of the root is as follows:

a: intermediate
f: final
c: private or household consumption
g: public or governmental
m: imports
i: investment
n: iNventory changes
x: eXports
h: otHer final demand
o: output
v: value added
w: wages
l: other Labour income and alike (social security contributions)
u: operating sUrplus
s: subsidies
t: indirect Taxes
d: depreciation
r: other or Residual of value added
k: capital stock
e: employment
p: productivity
y: income
z: discrepancy
d: Domestic *demand* distinguished from domestic d as *origin* of goods.

So final demand has *f*(*c,g,i,n,x,h,z*); *m* stands for imports, value added has *v*(*w,l,u,t,s,d,r*), *o* stands for output and *k* for capital stock while *e* indicates employment and *p* productivity. The final demand minus imports, much used in input-output analysis, has to be formed by the *f* as first digit; as explained later the second cannot be *m* (imported) nor *d* (domestic), so we propose *fim* standing for final demand minus imports.

The meaning of the second digit in the root indicates either the origin of the goods and services, e.g. :

m: for iMported
d: for Domestic
t: for Total = imported + domestic.

The operation done on the series involved:

p: plus (+),
i: mInus (−)
o: over meaning an element-by-element division
x: meaning an element-by-element multiplication.

So to give the vector in constant prices of total resources = output + imports we can use the name *opm*, another free character to help the user to remember the name easily.

The meaning of the last digit in the root is:

t or ***T***: if the vector or matrix represents an *aggregate or a total*. In the simple model only the total series can be included, but at a later phase detail may be added by the builder, e.g. for household consumption, governmental consumption, investment, etc. The series ending with *t* in the root will always continue to exist and the detail will be added to the root by '*_p*' where *p* stands for partition in that root (see later: suffixes) So we have *fdt* for the final demand domestic total vector in constant prices, *fmt* for the final demand imported total in constant prices and *ftt* for the sum of *fmt* and *fdt*.

tot or ***TOT***: in the case of the *calculated total* of a vector – of which the root may or may not already be ending in at (like *fdt, fmt, ftt* defined above-, then the word 'tot' is added to the root to stress the fact that this single series is a vector (1 × 1) in vam software, e.g. *fdttot, fmttot, ftttot*.

5.2. *The prefixes*

The use of prefixes should be seen as types that are made on the same root, e.g.:

*p_* for prices and deflator, p90_out would stand for 1990-based output price indices, and p89_xtt for 1989-based export prices;
*c_* for coefficients, so *c_Att* would be the technical coefficient matrix, *c_Amt* the import coefficient matrix and *c_Adt* the domestic coefficient matrix;
*s_* for shares, so *s_ctt* would be the budget shares for household consumption based on the constant prices;
*d_* for percentage changes; and
*g_* for growth rates.

5.3. The suffixes

In general suffixes should be seen as subcategories of the root, e.g. partitions, or other details of a group:

- *_p*: for a partition of the root meaning that the sum over all '*_p*' for the same root will give the total for this root (always goes together with a series ending in *t* in the third digit of the root);
- *_d*: for detail of the root that is not a partition; and
- *_g*: for groups in the root with overlapping possibilities.

Other usage will be added later. The basic understandings for naming conventions can be compared to the grammar rules in a language. There will, however, as in every language, be some exceptions to the rules. We tried to minimize these.

6. GUI

This section presents some of the results already achieved, and highlights the development going on in the Compass GUI part of the project. There are a number of very senior IO model-builders unable to approach the performance of a junior person with no economic training, no database understanding but only standard user interface techniques. In less than one hour the junior person outperformed all of them, reading in IO tables for 10 different countries for four years each, domestic and imported flows, and outputting all those data to the application software in machine-readable form. The reason was that the possible actions on each object were clear from the graphical environment settings. In particular the display objects, the action objects and the clicking of the objects informed the user (but not the so-called specialist) of all possibilities and options.

The user can perform mouse actions (all clicking, dragging and dropping) and keyboard actions (alphanumerical input, short keys, and hot key and function key input) all conforming to the Windows standard. He is presented standard Windows user interfaces such as *edit lines, drop down and combo boxes, check boxes, tabs and so on.* In particular the use of the shift key and the Ctrl key is normalized on Windows. This together with a standard menu and toolbar approach allows any new user with Windows experience to navigate through the application with any prior information and most of the training to be done by Windows.

Figure 2-1 represents the conceptual design of the COMPASS database. Meta data are the key to an integrated relational database based on an object-oriented principle. The need for meta data stems from the following factors:

(1) Various organizations compile statistical data sets, and very often definitions, measuring units, etc., are not an integral part of the numerical information.

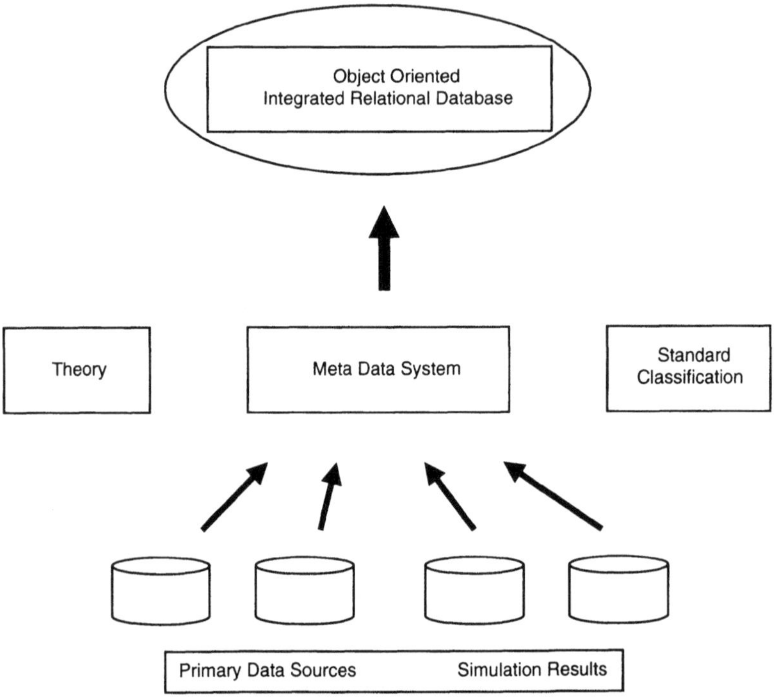

Figure 2-1. Conceptual scheme of COMPASS database.

(2) Data are published with various reporting intervals, such as monthly, quarterly, annual, five-year intervals, or varying intervals.
(3) Data take various forms such as matrix, vector, or scalar and even within the same kinds of data they come in various formats.
(4) They are made available in various information media, including printed form, floppy disks, CD-ROMs, and on-line service.
(5) Data refer to various measuring units such as value (yen, dollar, ECU, etc.; current prices vs. constant prices, etc.; thousand or million or billion, etc.), physical unit (ton, toe or ton of oil equivalent, cubic metres, litres, barrels, kcal, kW, kWh, joule, etc.), index (with various base years), ratio (income per capita, output per hour, etc.), among others.

Meta data are needed to fill statistical discrepancies among data sets, to establish linkage among them, and to enable data conversion. It is often necessary to bridge conceptual differences among data sets (examples are sample survey *vs.* aggregate value such as household data *vs.* national accounts data; establishment base *vs.* activity base *vs.* commodity base, such as production based on business survey, input-output tables, and international trade).

The COMPASS database is a relational database in the true sense of the word: each data point is stored in the database management system and is accessible in any output format. Thus, for example, one can look at the input-output data in a matrix form for a particular point in time or in a time-series of a particular cell or a vector. This provides enormous flexibility for the data users. This is in contrast to many so-called 'databases' that are actually sequential files. This results in a fixed-output format and poses a severe limit to the cross-disciplinary use of the data.

The user access to the database is facilitated by GUI. Figure 2-2 shows how the users may choose from among data spheres (input-output, macro-finance, trade, and energy; ENE is chosen in the example) and regions. By clicking on Show Result, the users have easy access to the data. It is noted that, in the case of COMPASS, data generated by simulation are also stored in the database and can be accessed in exactly the same manner as the historical series. Needless to say, the data thus obtained can be exported to spreadsheets. Figure 2-3 shows model-building steps. As above, users choose from among the spheres and regions (country/region lists are provided; Japan is chosen in the example). The model-building process in broken down into easy-to-understand steps (in the case of a macro-financial model, they consist of the following:

(1) prepare model data: according to model specification, standard query language (SQL) is automatically issued and data are retrieved from the database;
(2) check model data: data are checked for accounting/definitional consistency;
(3) model final demand: final demand equations are run;
(4) model income: likewise, income equations are run;
(5) model balance of payments: balance of payments equations are run that link macroeconomic variables to financial ones;
(6) model monetary market: financial equations are run.

These steps vary according to the spheres to look at because model structures are different, but principles are the same. This is followed by 'Make Stand-alone Model', 'Make COMPASS Linkage Run', and 'Show Results'. In the process, essential files in each step are shown so that users are able to look at individual programs. They are the files actually used in the model-building and simulation runs so it is possible to alter the programs, including the years to look at, functional forms, logical flow, etc. Thus, although COMPASS is a fairly large system with complex logical flow and data manipulation, it is also a transparent system in which each step is open to users.

Figures 2-4 to 2-7 are the results of the data query in respective fields. Figure 2-8 shows that, behind the scenes, SQL is issued by the system to gain access to the data. User query through the GUI is translated into SQL automatically, and data are retrieved from the database that are stored on hard disks.

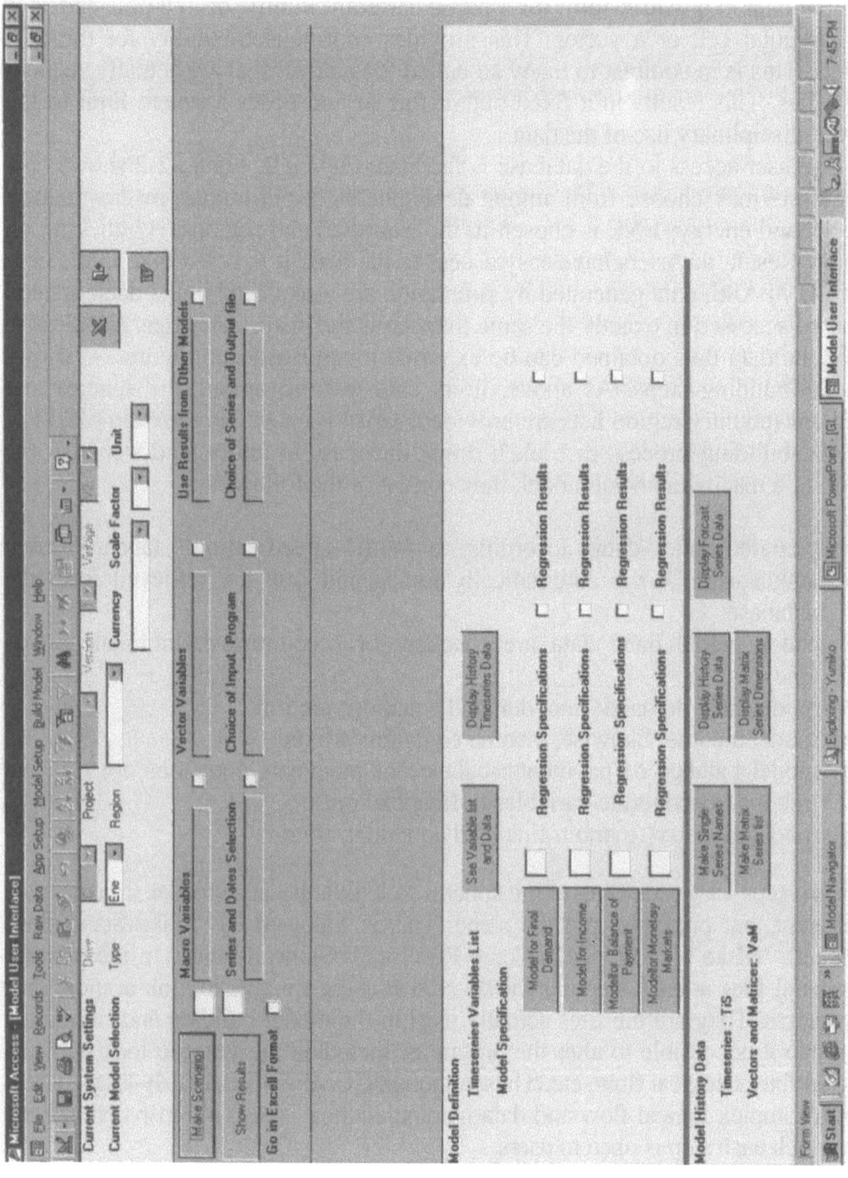

Figure 2-2. GUI for database access.

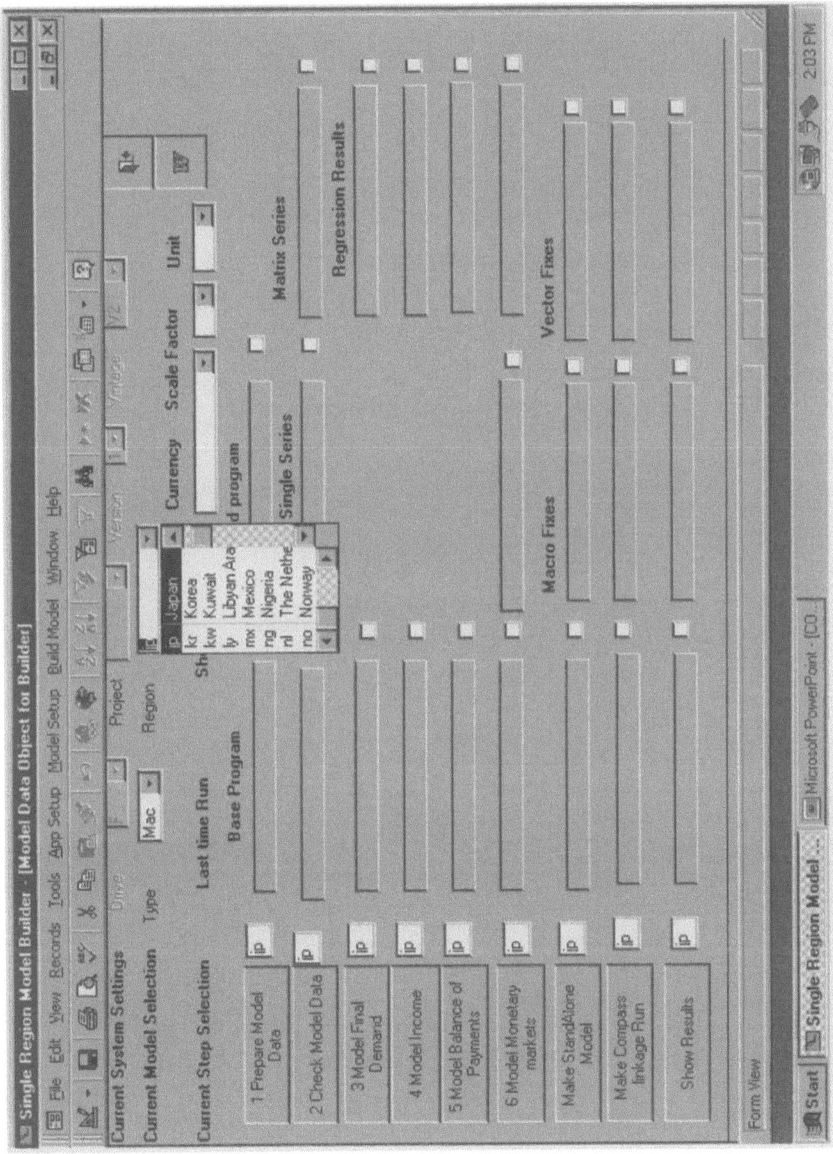

Figure 2-3. GUI for model building.

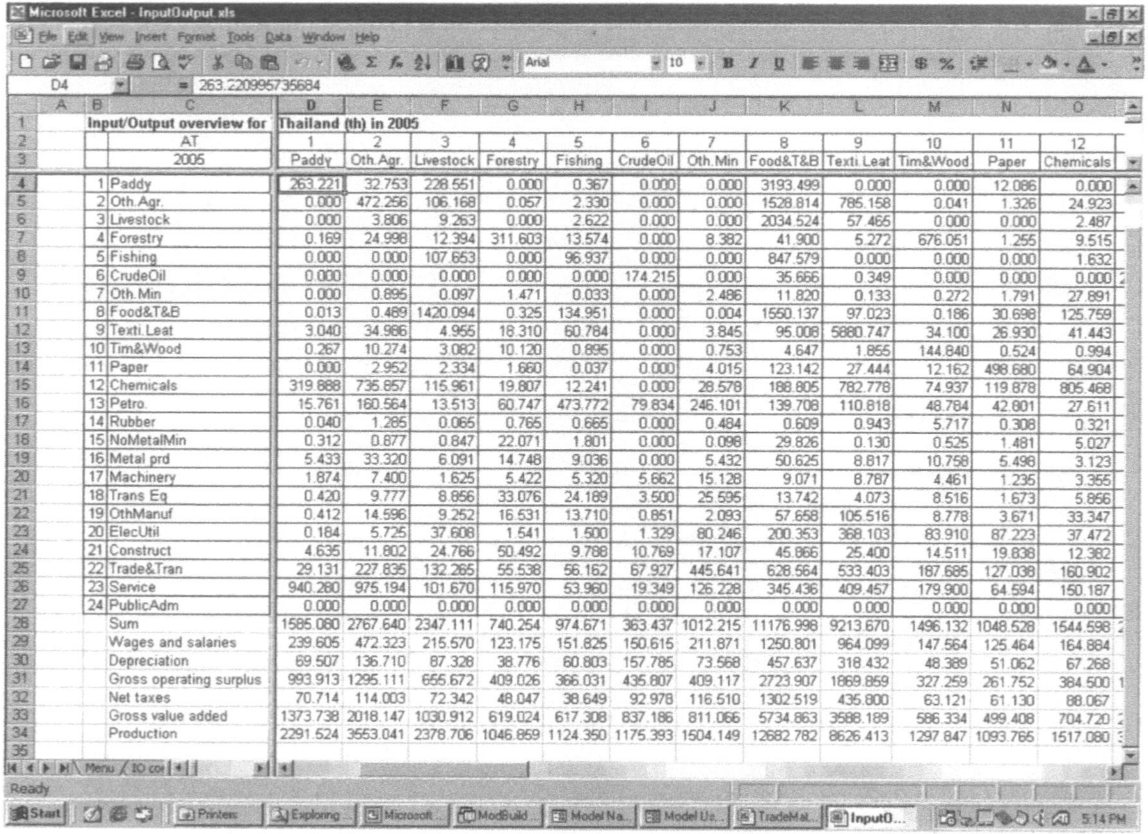

Input/Output overview for Thailand (th) in 2005

AT	1	2	3	4	5	6	7	8	9	10	11	12
2005	Paddy	Oth.Agr.	Livestock	Forestry	Fishing	CrudeOil	Oth.Min	Food&T&B	Texti.Leat	Tim&Wood	Paper	Chemicals
1 Paddy	263.221	32.753	228.551	0.000	0.367	0.000	0.000	3193.499	0.000	0.000	12.086	0.000
2 Oth.Agr.	0.000	472.256	106.168	0.057	2.330	0.000	0.000	1528.814	785.158	0.041	1.326	24.923
3 Livestock	0.000	3.806	9.263	0.000	2.622	0.000	0.000	2034.524	57.465	0.000	0.000	2.487
4 Forestry	0.169	24.998	12.394	311.603	13.574	0.000	8.382	41.900	5.272	676.051	1.255	9.515
5 Fishing	0.000	0.000	107.653	0.000	96.937	0.000	0.000	847.579	0.000	0.000	0.000	1.632
6 CrudeOil	0.000	0.000	0.000	0.000	0.000	174.215	0.000	35.666	0.349	0.000	0.000	0.000
7 Oth.Min	0.000	0.895	0.097	1.471	0.033	0.000	2.486	11.820	0.133	0.272	1.791	27.891
8 Food&T&B	0.013	0.489	1420.094	0.325	134.951	0.000	0.004	1550.137	97.023	0.186	30.698	125.759
9 Texti.Leat	3.040	34.986	4.955	18.310	60.784	0.000	3.845	95.008	5880.747	34.100	26.930	41.443
10 Tim&Wood	0.267	10.274	3.082	10.120	0.895	0.000	0.753	4.647	1.855	144.840	0.524	0.994
11 Paper	0.000	2.952	2.334	1.660	0.037	0.000	4.015	123.142	27.444	12.162	498.680	64.904
12 Chemicals	319.888	735.857	115.961	19.807	12.241	0.000	28.578	188.805	782.778	74.937	119.878	805.468
13 Petro.	15.761	160.564	13.513	60.747	473.772	79.834	246.101	139.708	110.818	48.784	42.801	27.611
14 Rubber	0.040	1.285	0.065	0.765	0.665	0.000	0.484	0.609	0.943	5.717	0.308	0.321
15 NoMetalMin	0.312	0.877	0.847	22.071	1.801	0.000	0.098	29.826	0.130	0.525	1.481	5.027
16 Metal prd	5.433	33.320	6.091	14.748	9.036	0.000	5.432	50.625	8.817	10.758	5.498	3.123
17 Machinery	1.874	7.400	1.625	5.422	5.320	5.662	15.128	9.071	8.787	4.461	1.235	3.355
18 Trans Eq	0.420	9.777	8.856	33.076	24.189	3.500	25.595	13.742	4.073	8.516	1.673	5.856
19 OthManuf	0.412	14.596	9.252	16.531	13.710	0.851	2.093	57.658	105.516	8.778	3.671	33.347
20 ElecUtil	0.184	5.725	37.608	1.541	1.500	1.329	80.246	200.353	368.103	83.910	87.223	37.472
21 Construct	4.635	11.802	24.766	50.492	9.788	10.769	17.107	45.866	25.400	14.511	19.838	12.382
22 Trade&Tran	29.131	227.835	132.265	55.538	56.162	67.927	445.641	628.564	533.403	187.685	127.038	160.902
23 Service	940.280	975.194	101.670	115.970	53.960	19.349	126.228	345.436	409.457	179.900	64.594	150.187
24 PublicAdm	0.000	0.000	0.000	0.000	0.000	0.000	0.000	0.000	0.000	0.000	0.000	0.000
Sum	1585.080	2767.640	2347.111	740.254	974.671	363.437	1012.215	11176.998	9213.670	1496.132	1048.528	1544.598
Wages and salaries	239.605	472.323	215.570	123.175	151.825	150.615	211.871	1250.801	964.099	147.564	125.464	164.884
Depreciation	69.507	136.710	87.328	38.776	60.803	157.785	73.568	457.637	318.432	48.389	51.062	67.268
Gross operating surplus	993.913	1295.111	655.672	409.026	366.031	435.807	409.117	2723.907	1869.859	327.259	261.752	384.500
Net taxes	70.714	114.003	72.342	48.047	38.649	92.978	116.510	1302.519	435.800	63.121	61.130	88.067
Gross value added	1373.738	2018.147	1030.912	619.024	617.308	837.186	811.066	5734.863	3588.189	586.334	499.408	704.720
Production	2291.524	3553.041	2378.706	1046.859	1124.350	1175.393	1504.149	12682.782	8626.413	1297.847	1093.765	1517.080

Figure 2-4. Sample output of the COMPASS database: input-output table simulation for Thailand in 2005.

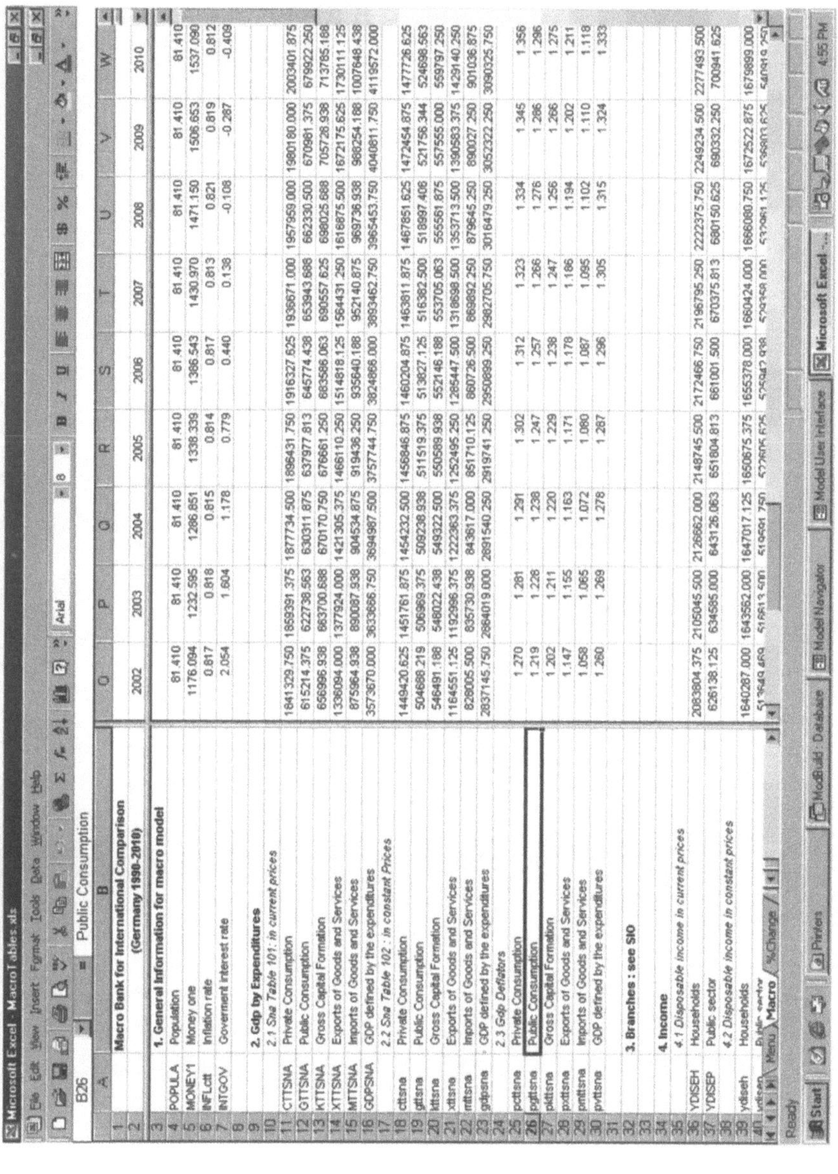

	A	B	O	P	Q	R	S	T	U	V	W
1		**Macro Bank for International Comparison**									
2		**(Germany 1990-2010)**	2002	2003	2004	2005	2006	2007	2008	2009	2010
3		**1. General Information for macro model**									
4	POPULA	Population	81.410	81.410	81.410	81.410	81.410	81.410	81.410	81.410	81.410
5	MONEY1	Money one	1176.094	1232.595	1286.851	1338.339	1386.543	1430.970	1471.150	1506.653	1537.090
6	INFLctt	Inflation rate	0.817	0.818	0.815	0.814	0.817	0.813	0.821	0.819	0.812
7	INTGOV	Goverment interest rate	2.054	1.604	1.178	0.779	0.440	0.138	-0.108	-0.287	-0.409
8											
9		**2. Gdp by Expenditures**									
10		*2.1 Sna Table 101: in current prices*									
11	CTTSNA	Private Consumption	1841329.750	1859391.375	1877734.500	1896431.750	1916327.625	1936671.000	1957959.000	1980180.000	2003401.875
12	GTTSNA	Public Consumption	615214.375	622738.563	630311.875	637977.813	645774.438	653943.688	662330.500	670981.375	679922.250
13	KTTSNA	Gross Capital Formation	656996.938	663700.688	670170.750	676661.250	683586.063	690557.625	698025.688	705728.938	713785.188
14	XTTSNA	Exports of Goods and Services	1336094.000	1377924.000	1421305.375	1466110.250	1514818.125	1564431.250	1616875.500	1672175.625	1730111.125
15	MTTSNA	Imports of Goods and Services	875964.938	890087.938	904534.875	919436.250	935640.188	952140.875	969736.938	988254.188	1007648.438
16	GDPSNA	GDP defined by the expenditures	3573670.000	3633666.750	3694987.500	3757744.750	3824866.000	3893462.750	3965453.750	4040811.750	4119572.000
17		*2.2 Sna Table 102 : in constant Prices*									
18	cttsna	Private Consumption	1449420.625	1451761.875	1454232.500	1456846.875	1460204.875	1463811.875	1467851.625	1472454.875	1477726.625
19	gttsna	Public Consumption	504688.219	506969.375	509238.938	511519.375	513827.125	516382.500	518997.406	521756.344	524698.563
20	kttsna	Gross Capital Formation	546491.188	548022.438	549322.500	550589.938	552146.188	553705.063	555561.875	557555.000	559797.250
21	xttsna	Exports of Goods and Services	1164551.125	1192996.375	1222363.375	1252495.250	1285447.500	1318698.500	1353713.500	1390583.375	1429140.250
22	mttsna	Imports of Goods and Services	828005.500	835730.938	843617.000	851710.125	860726.500	869892.250	879645.250	890027.250	901036.875
23	gdpsna	GDP defined by the expenditures	2837145.750	2864019.000	2891540.250	2919741.250	2950899.250	2982705.750	3016479.250	3052322.250	3090325.750
24		*2.3 Gdp Deflators*									
25	pcttsna	Private Consumption	1.270	1.281	1.291	1.302	1.312	1.323	1.334	1.345	1.356
26	pgttsna	Public Consumption	1.219	1.228	1.238	1.247	1.257	1.266	1.276	1.286	1.296
27	pkttsna	Gross Capital Formation	1.202	1.211	1.220	1.229	1.238	1.247	1.256	1.266	1.275
28	pxttsna	Exports of Goods and Services	1.147	1.155	1.163	1.171	1.178	1.186	1.194	1.202	1.211
29	pmttsna	Imports of Goods and Services	1.058	1.065	1.072	1.080	1.087	1.095	1.102	1.110	1.118
30	pvttsna	GDP defined by the expenditures	1.260	1.269	1.278	1.287	1.296	1.305	1.315	1.324	1.333
31											
32		**3. Branches : see SIO**									
33											
34		**4. Income**									
35		*4.1 Disposable income in current prices*									
36	YDISEH	Households	2083804.375	2105045.500	2126662.000	2148745.500	2172466.750	2196795.250	2222375.750	2249234.500	2277493.500
37	YDISEP	Public sector	626138.125	634585.000	643126.063	651804.813	661001.500	670375.813	680150.625	690332.250	700941.625
38		*4.2 Disposable income in constant prices*									
39	ydiseh	Households	1640287.000	1643562.000	1647017.125	1650675.375	1655378.000	1660424.000	1666080.750	1672522.875	1679899.000
40	ydisep	Public sector	[illegible]	[illegible]	[illegible]	[illegible]	[illegible]	[illegible]	[illegible]	[illegible]	[illegible]

Figure 2-5. Sample output of the COMPASS database: macroeconomics simulation for Germany to 2010.

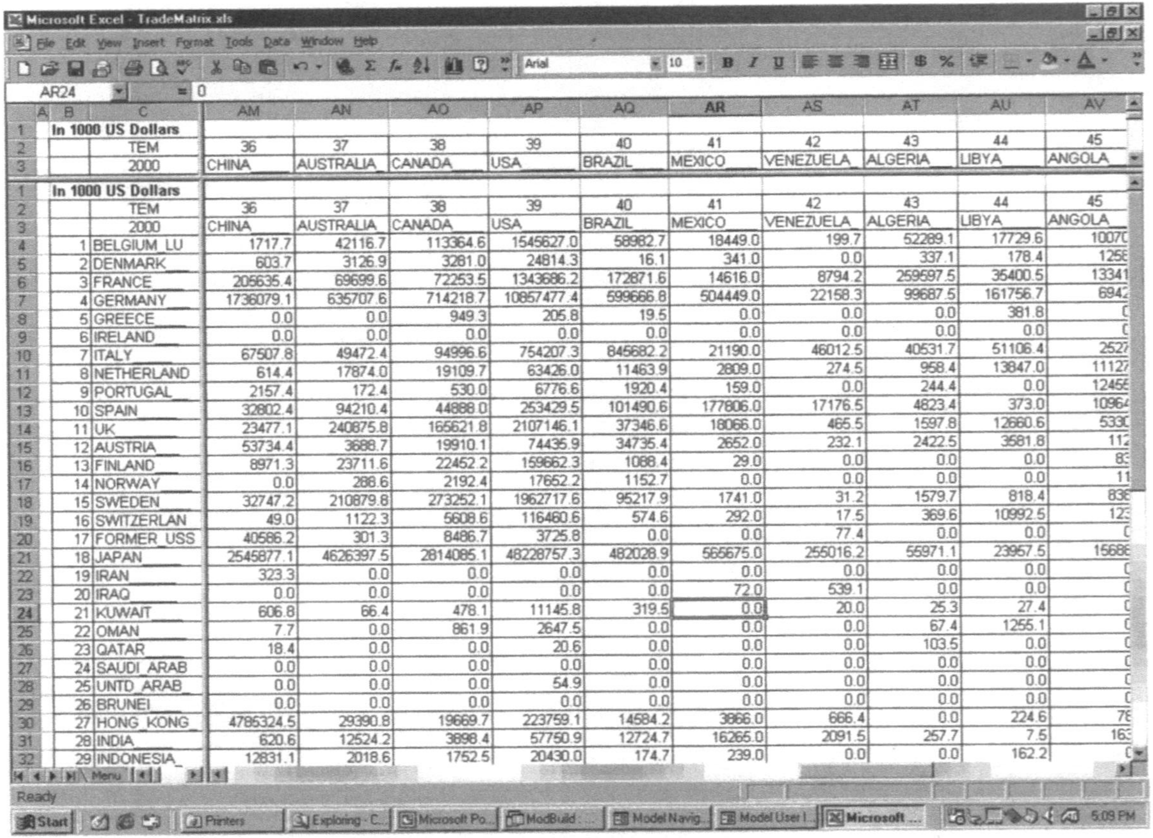

In 1000 US Dollars		AM	AN	AO	AP	AQ	AR	AS	AT	AU	AV
	TEM	36	37	38	39	40	41	42	43	44	45
	2000	CHINA	AUSTRALIA	CANADA	USA	BRAZIL	MEXICO	VENEZUELA	ALGERIA	LIBYA	ANGOLA
1	BELGIUM_LU	1717.7	42116.7	113364.6	1545627.0	58982.7	18449.0	199.7	52289.1	17729.6	1007
2	DENMARK	603.7	3126.9	3281.0	24814.3	16.1	341.0	0.0	337.1	178.4	125
3	FRANCE	205635.4	69699.6	72253.5	1343686.2	172871.6	14616.0	8794.2	259597.5	35400.5	1334
4	GERMANY	1736079.1	635707.6	714218.7	10857477.4	599666.8	504449.0	22158.3	99687.5	161756.7	694
5	GREECE	0.0	0.0	949.3	205.8	19.5	0.0	0.0	0.0	381.8	0
6	IRELAND	0.0	0.0	0.0	0.0	0.0	0.0	0.0	0.0	0.0	0
7	ITALY	67507.8	49472.4	94996.6	754207.3	845682.2	21190.0	46012.5	40531.7	51106.4	252
8	NETHERLAND	614.4	17874.0	19109.7	63426.0	11463.9	2809.0	274.5	958.4	13847.0	1112
9	PORTUGAL	2157.4	172.4	530.0	6776.6	1920.4	159.0	0.0	244.4	0.0	1245
10	SPAIN	32802.4	94210.4	44888.0	253429.5	101490.6	177806.0	17176.5	4823.4	373.0	1096
11	UK	23477.1	240875.8	165621.8	2107146.1	37346.6	18066.0	465.5	1597.8	12660.6	533
12	AUSTRIA	53734.4	3688.7	19910.1	74435.9	34735.4	2652.0	232.1	2422.5	3581.8	11
13	FINLAND	8971.3	23711.6	22452.2	159662.3	1088.4	29.0	0.0	0.0	0.0	8
14	NORWAY	0.0	288.6	2192.4	17652.2	1152.7	0.0	0.0	0.0	0.0	11
15	SWEDEN	32747.2	210879.8	273252.1	1962717.6	95217.9	1741.0	31.2	1579.7	818.4	83
16	SWITZERLAN	49.0	1122.3	5608.6	116460.6	574.6	292.0	17.5	369.6	10992.5	12
17	FORMER_USS	40586.2	301.3	8486.7	3725.8	0.0	0.0	77.4	0.0	0.0	0
18	JAPAN	2545877.1	4626397.5	2814085.1	48228757.3	482028.9	565675.0	255016.2	55971.1	23957.5	1568
19	IRAN	323.3	0.0	0.0	0.0	0.0	0.0	0.0	0.0	0.0	0
20	IRAQ	0.0	0.0	0.0	0.0	0.0	72.0	539.1	0.0	0.0	0
21	KUWAIT	606.8	66.4	478.1	11145.8	319.5	0.0	20.0	25.3	27.4	0
22	OMAN	7.7	0.0	861.9	2647.5	0.0	0.0	0.0	67.4	1255.1	0
23	QATAR	18.4	0.0	0.0	20.6	0.0	0.0	0.0	103.5	0.0	0
24	SAUDI_ARAB	0.0	0.0	0.0	0.0	0.0	0.0	0.0	0.0	0.0	0
25	UNTD_ARAB	0.0	0.0	0.0	54.9	0.0	0.0	0.0	0.0	0.0	0
26	BRUNEI	0.0	0.0	0.0	0.0	0.0	0.0	0.0	0.0	0.0	0
27	HONG_KONG	4785324.5	29390.8	19669.7	223759.1	14584.2	3866.0	666.4	0.0	224.6	7
28	INDIA	620.6	12524.2	3898.4	57750.9	12724.7	16265.0	2091.5	257.7	7.5	16
29	INDONESIA	12831.1	2018.6	1752.5	20430.0	174.7	239.0	0.0	0.0	162.2	0

Figure 2-6. Sample output of the COMPASS database: trade matrix simulation for motor vehicles in 2000.

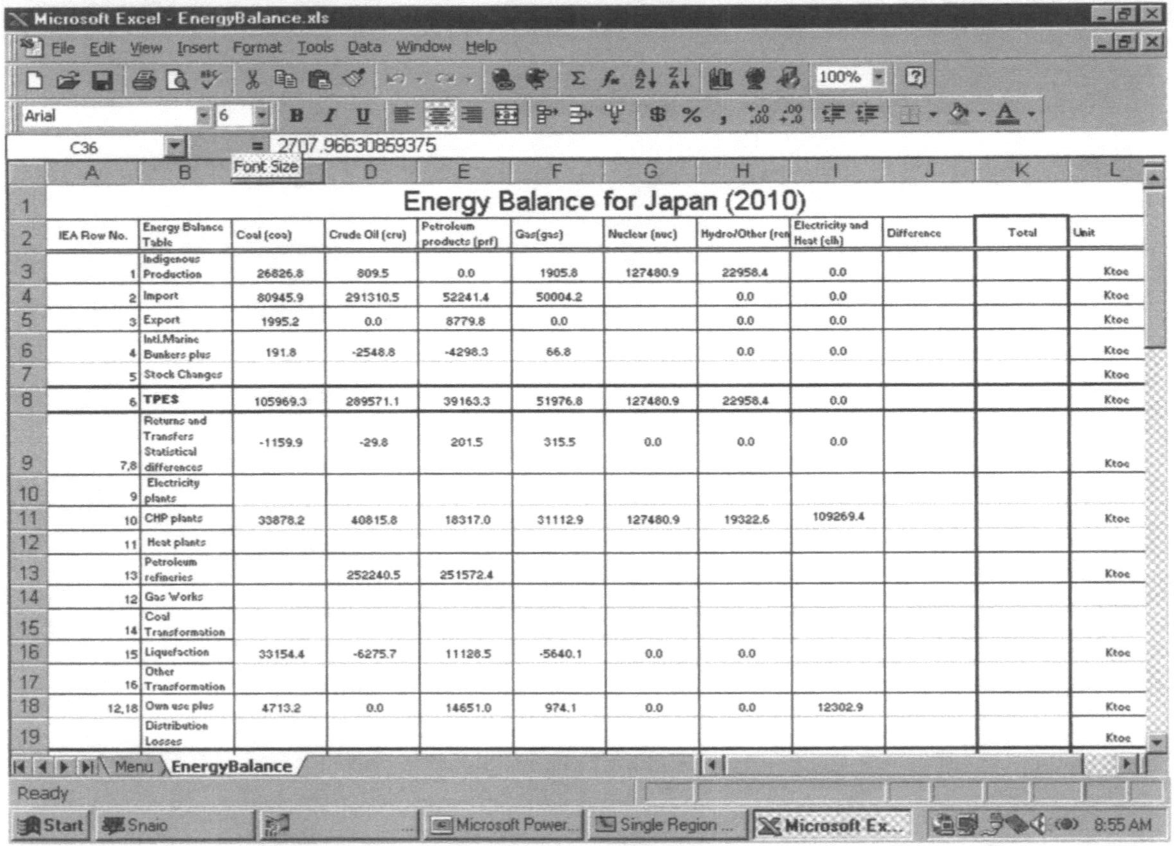

Energy Balance for Japan (2010)

IEA Row No.	Energy Balance Table	Coal (coa)	Crude Oil (cru)	Petroleum products (prf)	Gas(gas)	Nuclear (nuc)	Hydro/Other (ren	Electricity and Heat (elh)	Difference	Total	Unit
1	Indigenous Production	26826.8	809.5	0.0	1905.8	127480.9	22958.4	0.0			Ktoe
2	Import	80945.9	291310.5	52241.4	50004.2		0.0	0.0			Ktoe
3	Export	1995.2	0.0	8779.8	0.0		0.0	0.0			Ktoe
4	Intl.Marine Bunkers plus	191.8	-2548.8	-4298.3	66.8		0.0	0.0			Ktoe
5	Stock Changes										Ktoe
6	**TPES**	105969.3	289571.1	39163.3	51976.8	127480.9	22958.4	0.0			Ktoe
7,8	Returns and Transfers Statistical differences	-1159.9	-29.8	201.5	315.5	0.0	0.0	0.0			Ktoe
9	Electricity plants										
10	CHP plants	33878.2	40815.8	18317.0	31112.9	127480.9	19322.6	109269.4			Ktoe
11	Heat plants										
13	Petroleum refineries		252240.5	251572.4							Ktoe
12	Gas Works										
14	Coal Transformation										
15	Liquefaction	33154.4	-6275.7	11128.5	-5640.1	0.0	0.0				Ktoe
16	Other Transformation										
12,18	Own use plus	4713.2	0.0	14651.0	974.1	0.0	0.0	12302.9			Ktoe
	Distribution Losses										Ktoe

Figure 2-7. Sample output of the COMPASS database: energy balance simulation for Japan in 2010.

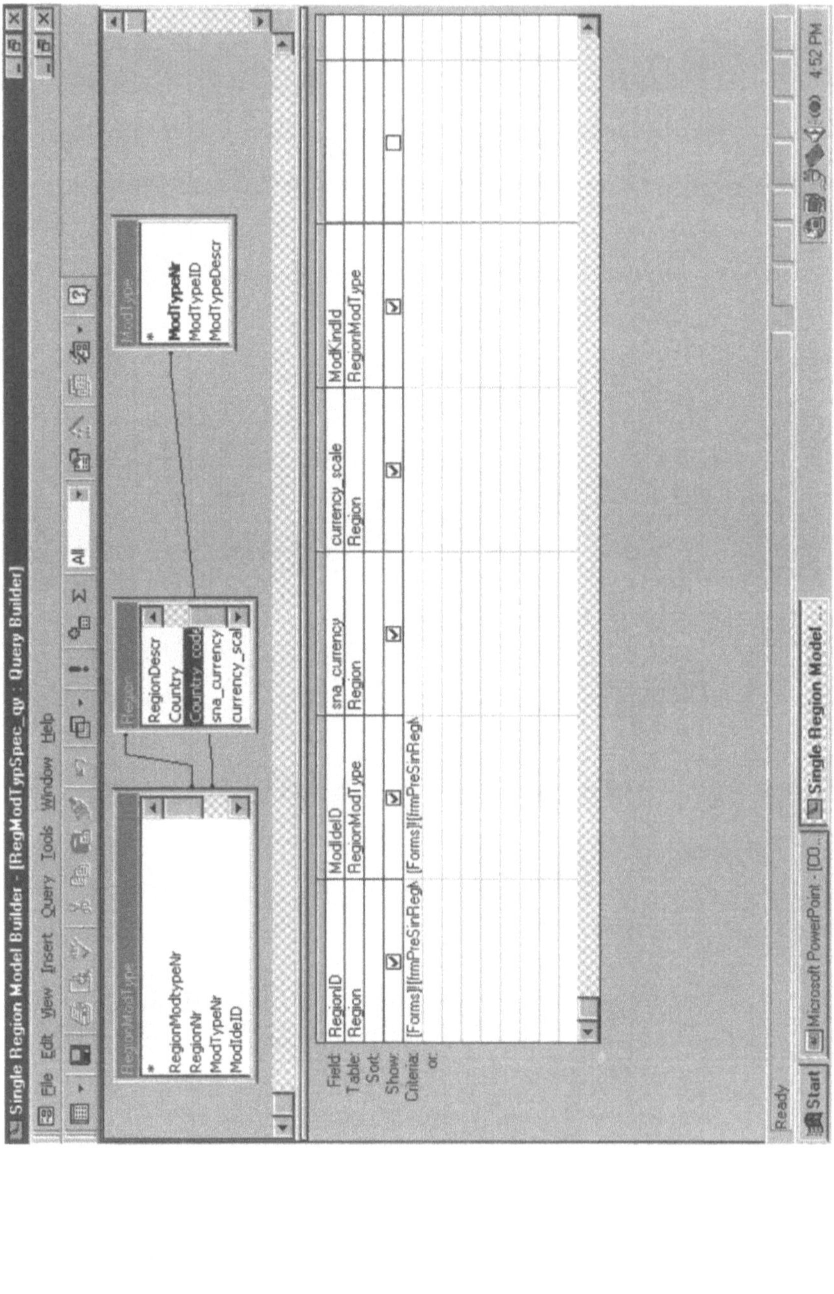

Figure 2-8. Database query.

7. Final words

It is demonstrated in this chapter that an object-oriented database and modelling system for COMPASS serves as:

(1) Model-building tool covering from a single country/region model to a globally linked model.
(2) Scenario description tool allowing changes in exogenous/policy variables, including overriding results from simulation runs.
(3) Simulation tool by which projection results are obtained and compared taking international and inter-industrial repercussions into account.
(4) Accounting tool where original data and/or simulation results can be put into an accounting framework such as System of National Accounts (SNA) and balance of payments, System of Input-Output, Energy Balances, and Trade Matrices where internal consistency can be examined. In other words, the iteration process is repeated until accounting consistency is obtained throughout the global system.

The database and simulation infrastructure such as this is indispensable to make a research project of the scope and magnitude of COMPASS. The advantages of the system are demonstrated in the following chapters.

3
IO, macro-finance, and trade model specification

BERND MEYER and CHRISTIAN LUTZ

1. Introduction

This chapter provides an overview of the model-structure of the economic part of the system focusing on the regional linkages. COMPASS is a world system, but it has a special focus on the APEC region. COMPASS is the only disaggregated model with double international linkage consisting of a multisectoral trade model and a linkage of the capital flows. The input-output structure will easily allow its detailed extension to energy and environmental modelling. This will be the third international linkage. This will be described in a later chapter.

We need information concerning the policy options that can avoid environmental catastrophes as well as conflict between nations. There seems to be no doubt that in the 21st century Asia will be the region with leading economic growth rates. This will have severe consequences for worldwide energy demand and global warming. COMPASS is modelling the economy–energy–environmental interdependencies worldwide, but with focus on the APEC countries. COMPASS will be an instrument that can be used to answer any questions which arise.

The main difference between COMPASS and other global economic-environmental models (see for example Edmonds *et al.*, 1994; Burniaux *et al.*, 1992) derives from the specific historical question which COMPASS addresses itself. The specific structure of production and consumption in the individual countries has to be described based on actual performance of the countries in focus, which needs an econometric analysis of the behaviour of the agents. This forbids open parametric modelling as is usual with CGE modelling.

Another important difference between COMPASS and other global models is the relationship between the macro-models and the input-output models. The usual way is a top-down modelling as in the GIO model (Duchin and Smyshlyaev, 1995). In this case the components of final demand are determined by the macro-systems and the input-output model gives only the sectoral structure. This means that structural change has no influence on final demand and it further produces inconsistency, because the solution of the input-output system implies value added, which may differ from the macro estimated GNP (see Grassini, 1995). COMPASS has a bottom-up structure, in which the generation of income in the sectors of the economy adds up to macroeconomic income figures.

K. Uno (ed.), Economy–Energy–Environment Simulation, 55–68.

A third difference concerns the linkage between the regions. The trade linkage is usually done in the following way: imports for good *i* are summed up to a world import of good *i*, and a region's export is a constant or variable share of world import (Nyhus, 1991; Duchin and Smyshlyaev, 1995). Instead of this procedure we will establish a multisectoral bilateral trade model with endogenized shares such as the one constructed by Ma (1996), which was used for the first time as part of the INFORUM International System (Nyhus *et al.*, 1996).

The linkage of capital flows is neglected in most models. There is a big variety of modelling approaches (Klein and Krelle, 1983) because with the full representations of the balance of payments of the regions the question of endogenization of exchange rates arises. For COMPASS the link of the capital flows is very important, because the restrictions of a clearing global capital market have to underlie the forecast for the rapidly growing APEC countries. We will try a modelling that endogenizes the exchange raes and uses the US interest rate to equalize worldwide capital demand and supply.

2. The general architecture

A first view of the system from a distance, as shown in Figure 3-1, gives the impression of a wheel: its axis is given by the multisectoral bilateral trade model.

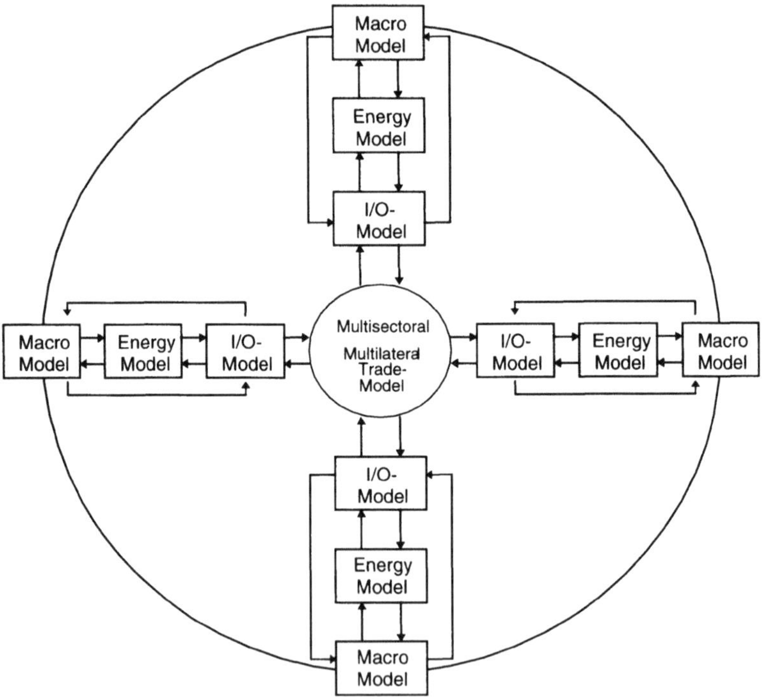

Figure 3-1. The Compass wheel.

The spokes are the country models consisting of an I/O model, an SNA (system of national accounts) system, and a financial model and their linkages between each other and the trade model. The tread is the linkage through the capital flows.

The trade model is based on trade matrices describing bilateral trade flows at the commodity levels. It obtains from each of the country models a vector of export prices and vector of import volumes, and it gives back to the country models a vector of export volumes and a vector of import prices.

The input-output model describes the production technology as reflected in the input structure of individual industrial activities at the disaggregated level. It gives the components of primary inputs such as labour income, operating surplus, depreciation, and indirect taxes and the vector of imports to the SNA system, where these streams are aggregated.

The SNA system depicts the redistribution of income between enterprises, government, households, and the rest of the world. It further calculates the aggregates of the components of final demand and gives them back to the input-output models. The SNA system receives the exports from the trade model and imports from the input-output model. It also obtains value added from the input-output model and computes disposable income and net lending figures, among others. The SNA system contains the balance of payments with exports and imports of goods and services and an explicitly modelled net flow of capital.

The financial model depicts the capital flows, the balance of the Central Bank, and the interest rates. The US interest rate is calculated by the condition that all capital flows sum up to zero. This is the condition which links all country models on the macro level, expressed by the 'tread' of the wheel in Figure 3-1.

3. The input-output models

The real import of good i $fm_i(t)$ (we omitted the country index) is depending on its relative price $p_i(t)/q_i(t)$ measured in the countries' currency and final demand $f_i(t)$ of good i:

$$fm_i(t) = fm_i[p_i(t)/q_i(t), f_i(t)] \qquad (1)$$

By definition we obtain the vector of real final demand minus imports of finished goods. With D as the matrix of domestic input coefficients and y as the vector of real gross production we get:

$$y(t) = [E-D(t)]^{-1}\,[f(t)-fm(t)]; \quad E\text{: unit matrix} \qquad (2)$$

The input coefficients are assumed to be exogenous variables, without losing generality to endogenize them as needed. The domestic input coefficient d_{ij} is defined as:

$$d_{ij}(t) = a_{ij}(t) * [1-m_{ij}(t)] \qquad (3)$$

a_{ij} is the technical coefficient and m_{ij} is its share of imports. The matrix AM of the coefficients for imported inputs has the elements:

$$am_{ij}(t) = a_{ij}(t) * m_{ij}(t) \qquad (4)$$

The relation between indirect taxes minus subsidies and real output may be constant:

$$T_j \cdot (\mathrm{t}) = \alpha_j * y_j \cdot (\mathrm{t}); \quad \alpha_j > 0 \tag{5}$$

The same hypothesis is assumed for depreciation

$$D_j \cdot (\mathrm{t}) = \beta * y_j \cdot (t); \quad \beta_j > 0 \tag{6}$$

Wages w_j are depending on a macro-productivity term of the preceding period and the consumer price index of the last period.

$$w_j(\mathrm{t}) = w_j[gnp(\mathrm{t}-1)/H(\mathrm{t}-1);\ p_c(\mathrm{t}-1)] \tag{7}$$

H: total working hours
p_c: consumer price index
gnp: gross national product

The labour input (measured in hours) h_j is explained by the real wage rate, a time trend and gross production.

$$h_j(\mathrm{t}) = h_j[w_j(\mathrm{t})/p_j(\mathrm{t}), y_j(\mathrm{t}), \mathrm{t}] \tag{8}$$

The labour input coefficient is given by:

$$z_j(\mathrm{t}) = h_j(\mathrm{t})/y_j(\mathrm{t}) \tag{9}$$

and the salaries with:

$$l_j(\mathrm{t}) = w_j(\mathrm{t}) * h_j(\mathrm{t}) \tag{10}$$

The vector of unit cost u is defined as:

$$u(\mathrm{t}) = D'(\mathrm{t}) * p(\mathrm{t}) + AM'(\mathrm{t}) * q(\mathrm{t}) + Z(\mathrm{t}) * w(\mathrm{t}) + \alpha + \beta \tag{11}$$

D: matrix of domestic input coefficients
AM: matrix of input coefficients of imported goods
Z: diagonal matrix of labour input coefficients
w: vector of wages
α: vector of indirect taxes minus subsidies per unit of output
β: vector of depreciation per unit of output

The price of sector j is now dependent on its unit cost:

$$p_j(\mathrm{t}) = p_j[u_j(\mathrm{t})] \tag{12}$$

This approach allows us to estimate different prices for the row of the input-output table. If the data allow, we could estimate different prices for final consumption and intermediate inputs, all explained in a regression by unit cost.

Now the balance of the columns of the input-output table can be used to calculate operating surplus g as a residual.

$$g_j(\mathrm{t}) = p_j(\mathrm{t}) * y_j(\mathrm{t}) - u_j(\mathrm{t}) * y_j(\mathrm{t}) \tag{13}$$

Most of the industrialized countries in Asia, Europe, and North America are represented by an input-output model. The global economy can be considered to consist of layers of different economic activities, with different resource requirements and environmental implications. This is particularly important in that the model is capable of tracing the effect of structural change of the economies. It is widely known that industrial branches are characterized by widely diverse energy requirement per unit output, implying that environmental impact may be mitigated through structural change.

The dynamic input-output models have its inputs final demand items such as consumption and investment generated in the macro models. The export vector is given by the trade model. The import vector is determined by the final demand of good (*i*) and the relative price measured in the country's currency. The sector disaggregation of final demand vector is treated as share variables. The input coefficients in the basic model are treated as exogenous variables, but are subject to various policy experiments including treatment as endogenously determined variables (e.g., depending on energy prices) or exogenous scenario variables (e.g., diffusion of a particular technology in the future or international technology transfer affecting input coefficients).

4. The macro-financial models

The macro models refer to a combined SNA-financial framework. The SNA system receives generated value added from the input-output system, distributing it to private households, enterprises, government, and the rest of the world. Disposable income is then calculated, which in turn determines the levels of consumption. The model has investment functions. The cost component of GDP such as indirect taxes, depreciation, wages and operating surplus are obtained from the input-output model. Where detailed information is lacking, a simplified model framework (macro simulator) has been prepared, largely based on trend information.

The financial model converts the balance of current account *SUR* into dollars dividing it by the exchange rate of the countries' currency per dollar:

$$CUR^l(t) = SUR^l(t)/e^l(t) \tag{14}$$

The balance of capital account is dependent on the balance of current account and the interest rate and the rate of inflation of country *l* and of other countries, which are linked by autonomous capital flows with country *l*.

$$CAP_l(t) = CAP^l[CUR^l(t), rl(t), r^w(t), \pi^l(t), \pi^w(t)] \tag{15}$$

r^l = interest rate of country *l*
r^w = interest rate, country *w*
π^l = rate of inflation, country *l*
π^w = rate of inflation, country *w*

The difference of inflation rates shall be a proxy for exchange rate expectations. The standard specification for the exchange rate e of country l is:

$$e^l(t) = e[r^l(t), r^w(t), \pi^l(t), \pi^w(t)] \tag{16}$$

In some countries additionally the exchange rate of a dominating currency such as DM or the yen are important variables. The relation between the net foreign assets of the country and its reserve money may also influence the market valuation of the currency. In very rapidly inflating countries the relation between the gdp deflator and the US gdp deflator gives a very good explanation of the exchange rate.

The overall balance in dollars is given by definition:

$$ALL^l(t) = CUR^l(t) + CAP^l(t) \tag{17}$$

Multiplying by the exchange rate yields the change of the central bank's net foreign assets in the currency of country l.

$$\Delta NFA^l(t) = ALL^l(t) * e^l(t), \tag{18}$$

The stock of net foreign assets follows.

$$NFA^l(t) = \Delta NFA^l(t) + NFA^l(t-1). \tag{19}$$

The stock of net domestic assets NDA is explained by a monetary/policy-reaction function, which describes the efforts of the central bank to let the monetary base grow parallel to needs coming from the real and the price development:

$$NDA^l(t) = NDA^l[gdp(t), p(t), NFA(t)] \tag{20}$$

gdp = gross domestic product
p = deflator of gdp

We expect a negative sign for the coefficient of NFA, because we assume that the central bank in the case of a monetary expansion caused by foreign influence will try to stabilize the monetary base.

We then get, by definition, the reserve money:

$$RM^l(t) = NDA^l(t) + NFA^l(t) \tag{21}$$

The country interest rates can be calculated solving a simple Keynesian money market:

$$MD^l(t) = p^l(t) \cdot MD^l[gdp^l(t), r^l(t)] \tag{22}$$

$$MS^l(t) = m^l(t) * [RM(t)] \tag{23}$$

$$MD^l(t) = MS^l(t) \tag{24}$$

MD = money demand
MS = money supply
RM = reserve money

m = money multiplier
gdp = gross domestic product
p = deflator of gdp
r = rate of interest

The equilibrium condition is that the sum of all capital flows is zero. From iteration to iteration the interest rate is rising (falling), if there is a demand surplus (deficit) in the international capital market.

The clearing of the international capital market formally can be written as follows:

$$DIFF(t) = \sum_{l=1}^{n} NFC^{l}(t) \tag{25}$$

$$CAL(\tau) = \begin{cases} -ETA & \text{if } DIFF > ALPHA \\ 0 & \text{if } -ALPHA < DIFF < ALPHA \\ +ETA & \text{if } DIFF < -ALPHA \end{cases} \tag{26}$$

$$ETA, ALPHA > 0$$

$$r^{w}(t, \tau) = r^{w}(t, \tau - 1) + CAL(\tau) \tag{27}$$

$$\tau = \text{iteration index}$$

DIFF is the demand suplus of the world capital flows. If *DIFF* is bigger than a very small number *ALPHA*, which is an acceptable deviation from a cleared capital market, the US interest rate is raised by an *ETA*, and in the next iteration the capital import of the countries will be reduced.

The balance of payments bridges the SNA to the financial model. The balance of capital account is explained by interest rate and inflation rates, while the exchange rates remain constant. The country interest rate can be calculated by solving a set of money market equations representing demand and supply of money. In future versions the interest rate in the United States is obtained by equalizing global demand and supply of the capital flows.

Explicit inclusion of saving-investment balance at the global level, as well as within individual countries, is deemed essential in a long-term growth model such as the COMPASS. The model does not have the explicit production function in the framework, largely due to the unavailability of reliable capital stock series for many of the countries in question. If the model is driven by the demand side alone, without proper consideration to the supply side, growth would continue almost indefinitely in a developing economy context where actual constraints lie with supply capacity which can only be enhanced through capital formation. The COMPASS examines whether a country's desire to invest is actually feasible under its own capacity to save and its ability to use someone else's savings channelled through global capital flow.

5. The trade model

The trade model will have the following structure:

$$x_l^i(t) = \sum_{k=1}^{n} S_{lk}^i(t) * m_k^i(t) \tag{28}$$

i: goods index
l, k: country index
n: number of countries
x: exports in constant dollars
S_{lk}^i : trade share of country l in the imports of country k for product i
m: imports in constant dollars

with

$$S_{lk}^i(t) = S_{lk}^i\left[\tilde{p}_l^i(t)/\tilde{p}(t), t\right] \tag{29}$$

$\tilde{p}_l^i$: price of good i in the exporting country l in dollars
$\tilde{p}^i(t)$: world market price of good i in dollars
t: time trend

The import prices for good $q^i{}_k$ in country k are compiled as weighted averages of the export prices of the different exporting countries:

$$\tilde{q}_k^i(t) = \sum_{l=1}^{m} S_{lk}^i(t) * \tilde{p}_l^i(t) \tag{30}$$

q_k^i: import price of good i in country k in dollars

The COMPASS is designed to capture the interdependencies among regions of the world through merchandise trade, service trade, primary income flows, financial flows, energy, and the environment. The trade model is intended to focus on the linkages of the global community through trade. The trade flow is disaggregated to 25 commodity classes. Input-output models and macro models are specified in order to facilitate such linkage.

A single-region model offers the values of import demand and export prices (as represented by the domestic output price) to the trade model. Exchange rates are also generated by the country macro models. The trade model in turn picks up the import demand and allocates the demand among exporters according to trade share. The trade shares and world prices will be determined by an iteration process based on the relative prices and a time-trend variable, subject to two constraints:

(1) The sum of shares from all sources in a given commodity class must add up to 1.
(2) Total imports in the world equals total exports in the world.

Key variable at play is the export prices supplied by the country models, adjusted for exchange rate to obtain prices expressed in dollar terms. World import price (i.e. in terms of US$) is generated based on the export prices and trade shares, which is then converted to import prices in terms of national currency. The national macro models use the results of the trade model concerning allocated exports and world prices (i.e. import prices) to obtain new import levels and export prices. Iteration between the country models and the trade model is continued until convergence is obtained with desired margin.

6. Details of the SNA system

This chapter gives a listing of the equations of a SNA system, that takes generated income from an input-output system, transforms it by distribution between enterprises, households, government and rest of the world into disposable income of private households and the government, which are determinants of private and public consumption. It further shows capital formation, the balance of payments with trade and capital flows, and the financial system with the balance of the central bank. The interdependent internationally linked SNA system will allow estimation of a world interest rate.

This section gives the numbers of the tables from the UN statistics which are needed to calculate the entire system. The variable names are based on the naming conventions explained in an earlier chapter. The reader will find all definitions and the main and typical specifications of the regressions functions of the system, although it is not possible to show here all of the alternative specifications. A regression function always has the name of the dependent variable as the name of the function and its arguments in brackets. For brevity the country codes are omitted for all variables. The equations are organized following the structure of the UN statistics.

fin_dem.rel: final demand

Table 1.2. Expenditure on the GDP in constant prices

1. Government Expenditure (1)
   $gttsna(t) = gttsna[ydisep(t), \ldots]$
2. Private Consumption
   $cttsna(t) = cttsna[ydiseh(t), INTLEN(t), \ldots]$ (2)
3. Gross Capital Formation
   $kttsna(t) = kttsna[(GCFSEP(t) + GCFSEC(t))/pkttsna(t), \ldots]$ (3)
4. Exports
   $xttsna(t) = rxttsnasio * xttsio(t)$ (4)
5. Imports
   $mttsna(t) = mttsna[mttsio(t), \ldots]$ (5)
   Real GDP:
   $gdp(t) = gttsna(t) + cttsna(t) + kttsna(t) + xttsna(t) - mttsna(t)$ (6)

Prices for Components of Final Demand

Deflator of Government Expenditures
$pgttsna(t) = pgttsna[p\_outsio(t), \ldots]$ (7)

Table 1.2. continued

Deflator of Private Consumption	
$pcttsna(t) = pcttsna[p\_outsio(t), \ldots]$	(8)
Deflator of Investment	
$pkttsna = pkttsna[p\_outsio(t), \ldots]$	(9)
Deflator of Exports	
$pxttsna(t) = pxttsna[p\_outsio(t), \ldots]$	(10)
Deflator of Imports	
$pmttsna(t) = pmttsna[p\_outsio(t), \ldots]$	(11)
Rate of Iflation (Private Consumption)	
$INFLCTT(t) = 100.0 * (pcttsna(t) - pcttsna(t-1))/pcttsna(t-1)$	(12)
Expenditure on the GDP in Current Prices	
$GTTSNA(t) = gttsna(t) * pgttsna(t)$	(13)
$CTTSNA(t) = cttsna(t) * pcttsna(t)$	(14)
$KTTSNA(t) = kttsna(t) * pkttsna(t)$	(15)
$XTTSNA(t) = xttsna(t) * pxttsna(t)$	(16)
$MTTSNA(t) = mttsna(t) * pmttsna(t)$	(17)
Nominal GDP:	
$GDP(t) = GTTSNA(t) + CTTSNA(t) + KTTSNA(t) + XTTSNA(t) - MTTSNA(t)$	(18)
Price of GDP:	
$pvttsna(t) = GDP(t)/gdp(t)$	(19)

dis_inc.rel: income

Table 1.3. Cost components of the GDP

1. Indirect Taxes, Net	
$TSTSNA(t) = RTSTSNASIO * TSTSIO(t)$	(20)
2. Consumption of Fixed Capital	
$DETSNA(t) = RDETSNASIO * DETSIO(t)$	(21)
3. Compensation of Employees, Paid by Resident Producers	
$WATSNA(t) = RWATSNASIO * WATSIO(t)$	(22)
4. Operating Surplus	
$URTSNA(t) = GDP(t) - TSTSNA(t) - DETSNA(t) - WATSNA(t)$	(23)

Table 1.4. General government current receipts and disbursements

Receipts	
1. Operating Surplus	
$URTSEP(t) = URTSEP[URTSNA(t), \ldots]$	(24)
2. Property and Entrepreneurial Income	
$YPRSEP(t) = YPRSEP[YPPSEP(t) + YPPSEC(t) + YPPSEH(t) + YPPSEF(t)]$	(25)
3. Taxes	
A. Indirect Taxes	
$TTTSNA(t) = TTTSNA[GDP(t), \ldots]$	(26)
B. Direct Taxes	
$DITSEP(t) = DITSEH(t) + DITSEC(t)$	(27)
C. Social Security Contributions	
$SORSEP(t) = SORSEP[SOPSEH(t), \ldots]$	(28)
4. Other Current Transfers + 3.D Compulsory Fees, Penalties	
$OTRSEP(t) = OTRSEP[OTPSEH(t), \ldots]$	(29)
Disbursements	
1. Government Final Consumption Expenditure	
$GTTSNA(t)$: see equation (13)	

Table 1.4. continued

2. Property Income
   *YPPSEP(t) = YPPSEP[INTGOV(t) * NLESEP(t), YPPSEP(t−1), . . .]* (30)
3. Subsidies
   SUTSNA(t) = TTTSNA(t) − TSTSNA(t) (31)
4. Other Current Transfers
   A. Social Security Benefits and B. Social Assistance Grants
   SOPSEP(t) = SOPSEP[SOPSEH(t), YDISEP(t), . . .] (32)
   C. Other
   OTPSEP(t) = OTPSEP[SOPSEH(t), OTPSEH(t), . . .] (33)
   Net Saving
   SAVSEP(t) = URTSEP(t) + YPRSEP(t) + DITSEP(t) + TTTSNA(t) + SORSEP(t) + OTRSEP(t) − GTTSNA(t) − YPPSEP(t) − SUTSNA(t) − SOPSEP(t) − OTPSEP(t) (34)
   Disposable Income in Current Prices
   YDISEP(t) = GTTSNA(t) + SAVSEP(t) (35)
   Disposable Income in Constant Prices
   ydisep(t) = YDISEP(t)/pgttsna(t) (36)

Table 3.13. General government: capital accumulation account, total finance of accumulation

Finance of Gross Accumulation
1. Gross Saving
   A. Consumption of Fixed Capital
   DETSEP(t) = DETSEP[DETSNA(t), . . .] (37)
   B. Net Saving
   SAVSEP(t): see equation (34)
2. Capital Transfers
   CTRSEP(t) = CTRSEP[CTPSEH(t) + CTPSEC(t) + CTPSEF(t), . . .] (38)

Gross Accumulation
1. Gross Capital Formation
   GCFSEP(t) = GCFSEP[(INTGOV(t) − INFLCTT(t), YDISEP(t), . . .] (39)
2. Purchases of Land
   PULSEP(t) = PULSEP(t) (40)
3. Purchases of Intangible Assets
   PUASEP(t) = PUASEP(t) (41)
4. Capital Transfers
   CTPSEP(t) = CTPSEP[GDP(t), . . .] (42)
   Net Lending
   NLESEP(t) = SAVSEP(t) + CTRSEP(t) + DETSEP(t) − GCFSEP(t) − CTPSEP(t) − PULSEP(t) − PUASEP(t) (43)

Table 1.5. Current income and outlay of corporate and quasi-corporate enterprises

Receipts
1. Operating surplus
   URTSEC(t) = URTSEC[URTSNA(t), . . .] (44)
2. Property and Entrepreneurial Income Received
   YPRSEC(t) = YPRSEC[YPPSEP(t) + YPPSEC(t) + YPPSEH(t) + YPPSEF(t), . . .] (45)
3. Current Transfers
   OTRSEC(t) = OTRSEC[OTPSEP(t) + OTPSEC(t) + OTPSEH(t) + OTPSEP(t), . . .] (46)

Table 1.5. continued

Disbursements

1. Property and Entrepreneurial Income
   $$YPPSEC(t) = YPPSEC[YPPSEC(t-1), INTGOV(t) * NLESEC(t), URTSNA(t), \ldots] \quad (47)$$
2. Direct Taxes
   $$DITSEC(t) = DITSEC[URTSEC(t), YPRSEC(t), \ldots] \quad (48)$$
3. Other Current Transfers
   $$OTPSEC(t) = OTPSEC[GDP(t), \ldots] \quad (49)$$
4. Net Saving
   $$SAVSEC(t) = URTSEC(t) + OTRSEC(t) + YPRSEC(t) - YPPSEC(t) - DITSEC(t) - OTPSEC(t) \quad (50)$$

Table 3.2.3. Corporate and quasi corporate enterprise capital account finance

Finance of Gross Accumulation

1. Gross Saving
   A. Consumption of Fixed Capital
   $$DETSEC(t) = DETSEC[DETSNA(t), \ldots] \quad (51)$$
   B. Net Saving
   SAVSEC(t): see equation (50)
2. Capital Transfers
   $$CTRSEC(t) = CTRSEC[CTPSEH(t) + CTPSEC(t) + CTPSEP(t) + CTPSEF(t), \ldots] \quad (52)$$

Accumulation

1. Gross Capital Formation
   $$GCFSEC(t) = GCTSEC[URTSEC(t), INTGOV(t) - INFLCTT(t), \ldots] \quad (53)$$
2. Purchases of Land
   $$PULSEC(t) = PULSEC(t) \quad (54)$$
3. Purchases of Intangible Assets
   $$PUASEC(t) = PUASEC(t) \quad (55)$$
4. Capital Transfers
   $$CTPSEC(t) = CTPSEC[GDP(t), \ldots] \quad (56)$$

Net Lending
$$NLESEC(t) = SAVSEC(t) + CTRSEC(t) + DETSEC(t) - GCFSEC(t) - PULSEC(t) - PUASEC(t) - CTPSEC(t) \quad (57)$$

Table 1.6. Current income and outlay of households and non-profit institutions

Receipts

1. Compensation of Employees, Received
   A. From Resident Producers
   $$WDRSEH(t) = WDRSEH[WATSNA(t), \ldots] \quad (58)$$
   B. From Rest of the World
   $$WMRSEH(t) = CERSEF(t) \quad (59)$$
   Total
   $$WTRSEH(t) = WDRSEH(t) + WMRSEH(t) \quad (60)$$
2. Operating Surplus
   $$URTSEH(t) = URTSEH[URTSNA(t), \ldots] \quad (61)$$
3. Property and Entrepreneurial Income
   $$YPRSEH(t) = YPRSEH[YPPSEP(t) + YPPSEC(t) + YPPSEH(t) + YPPSEF(t), \ldots] \quad (62)$$
4. Current Transfers
   A. Social Security Benefits and B. Social Assistance Grants
   $$SORSEH(t) = SORSEH[SOPSEP(t), \ldots] \quad (63)$$

Table 1.6. continued

C. Other

$$OTRSEH(t) = OTRSEH[OTPSEP(t) + OTPSEC(t) + OTPSEH(t) + OTPSEF(t), \ldots] \tag{64}$$

Disbursements

1. Private Consumption

   CTTSNA: see equation (14)

2. Property Income

$$YPPSEH(t) = YPPSEH[INTGOV(t) * CTTSNA(t), YPPSEH(t-1), \ldots] \tag{65}$$

3. Direct Taxes and other Transfers

   A. Social Security Contributions

$$SOPSEH(t) = SOPSEH[WATSNA(t), \ldots] \tag{66}$$

   B. Direct Taxes

$$DITSEH(t) = DITSEH[WATSNA(t) + YPRSEH(t), \ldots] \tag{67}$$

4. Other Current Transfers + 3.C. Fees, Fines and Penalties

$$OTPSEH(t) = OTPSEH[WATSNA(t), \ldots] \tag{68}$$

5. Net Saving

$$SAVSEH(t) = WDRSEH(t) + YPRSEH(t) + URTSEH(t) + SORSEH(t) + OTRSEH(t) - CTTSNA(t) - YPPSEH(t) - SOPSEH(t) - DITSEH(t) - OTPSEH(t) \tag{69}$$

   Disposable Income in Current Prices

$$YDISEH(t) = CTTSNA(t) + SAVSEH(t) \tag{70}$$

   Disposable Income in Constant Prices

$$ydiseh(t) + YDISEH(t)/pcttsna(t) \tag{71}$$

bal_pay.re1: balance of payments

Table 1.7. External transactions on current account

Payments to the Rest of the World

1. Imports of Goods and Services

   MTTSNA(t): see equation (17)

2. Factor Income to the Rest of the World

   A. Compensation of Employees

$$CEPSEF(t) = CEPSEF[WATSNA(t), \ldots] \tag{72}$$

   B. Property and Entrepreneurial Income

$$YPPSEF(t) = YPPSEF[YPPSEP(t) + YPPSEC(t) + YPPSEH(t), \ldots] \tag{73}$$

3. Current Transfers to the Rest of the World

$$OTPSEF(t) = OTPSEF[OTPSEP(t) + OTPSEC(t) + OTPSEH(t), \ldots] \tag{74}$$

4. Surplus of the Nation on current Transactions

$$SURSEF(t) = RECSEF(t) - MTTSNA(t) - CEPSEF(t) - YPPSEF(t) - OTPSEF(t) \tag{75}$$

Receipts from the Rest of the World

1. Exports

   XTTSNA(t): see equation (16)

2. Factor Income from Rest of the World

   A. Compensation of Employees

$$CERSEF(t) = CERSEF[WATSNA(t), \ldots] \tag{76}$$

   B. Property and Entrepreneurial Income

$$YPRSEF(t) = YPPSEP(t) + YPPSEC(t) + YPPSEH(t) + YPPSEF(t) - YPRSEP(t) - YPRSEC(t) - YPRSEH(t) \tag{77}$$

Table 1.7. continued

3. Current Transfers from the Rest of the World

$$OTRSEF(t) = OTPSEP(t) + OTPSEC(t) + OTPSEH(t) + OTPSEF(t) - OTRSEP(t) - OTRSEC(t) - OTRSEH(t) \quad (78)$$

Payments to the Rest of the World Plus
Surplus of the Nation and Current Transactions

$$PAYSEF(t) = MTTSNA(t) + YPPSEF(t) + CEPSEF(t) + OTPSEF(t) + SURSEF(t) \quad (79)$$

Receipts from the Rest of the World

$$RECSEF(t) = TTSNA(t) + CERSEF(t) + YPRSEF(t) + OTRSEF(t) \quad (80)$$

Current Account

$$CURBAL(t) = CURBAL[SURSEF(t)/NACUSD(t), \ldots] \quad (81)$$

Net Capital Exports

$$CAPBAL(t) = CAPBAL[(INTGOV(t) - INFLCTT(t)), INFLCTTus(t), INTGOVj(t), \ldots] \quad (82)$$

mon_mar.rel: financial system

Change of Foreign Assets of the Central Bank

$$ALLBAL(t) = CURBAL(t) + CAPBAL(t) \quad (83)$$

Exchange Rate

$$NACUSD(t) = NACUSD[INTGOV(t), INTGOVj(t), INFLCTT, \ldots] \quad (84)$$

Foreign Assets, Net of the Central Bank in National Currency

$$ASS(t) = [FORASS(t-1), ALLBAL(t) * NACUSD(t), \ldots] \quad (85)$$

Claims on Central Government

$$CLAGOC(t) = CLAGOC[RESMON(t), \ldots] \quad (86)$$

Claims on Deposit Money Banks

$$CLAMBK(t) = CLAMBK[RESMON(t), \ldots] \quad (87)$$

Central Government Deposits

$$GOVDEP(t) = GOVDEP[RESMON(t), \ldots] \quad (88)$$

Other Deposits

$$OTHDEP(t) = OTHDEP[RESMON(t), \ldots] \quad (89)$$

Reserve Money

$$RESMON(t) = [gdp(t), pvttsna(t), TTREND(t), \ldots] \quad (90)$$

Money M1

$$MONEY1(t) = MONEY1[RESMON(t), INTDIS(t), \ldots] \quad (91)$$

Interest on Government Yields

$$INTGOV(t) = INTGOV[MONEY1(t), GDP(t), \ldots] \quad (92)$$

Discount Rate

$$INTDIS(t) = INTDIS[INFLCTT(t), \ldots] \quad (93)$$

Lending Rate

$$INTLEN(t) = INTLEN[INTDIS(t), INTLEN(t-1), \ldots] \quad (94)$$

4
Endogenized trade shares in a global model

BERND MEYER and CHRISTIAN LUTZ

1. Introduction

Globally integrated modelling is achieved by employing trade matrices for linking country models. In the case of COMPASS this is achieved by having time-series for trade matrices of 25 commodity groups linking 53 countries and regions. In addition to country models that are estimated econometrically based on empirical data, trade matrices themselves are available over time in response to changing relative prices (endogenous in the model) and other factors. This chapter focuses on an attempt to empirically explain the trade shares.

Prices play an important role for energy demand and supply. In the first place the shares of the different energy carriers are dependent on their relative prices. Secondly, the energy intensities in factor demand of the industries and in household consumption are following price signals. Finally, energy prices influence production costs of all intermediate and final products and have an impact on product prices and demand for the different intermediate and final products. So a shock of energy prices will not only have a direct impact on energy demand, but also have indirect effects resulting from changes in the structure of the economy.

In a global context this means that the competition of the different economies is hit by energy price changes, because countries have a very different usage of energy carriers, energy intensities in production, and very different structures of intermediate and final demand.

In general energy forecasting needs an exact forecast of the world economic developments, and in times of globalization this can be achieved only by a realistic anticipation of structural change in international trade.

Global energy problems should be discussed in a multisector-multicountry context. The multisector approach is necessary, because energy influences structural change of the economies and is on the other hand dependent on it reflecting different energy intensities among industrial sectors. A regional modelling concept is useful, because of the big differences in the economic structures and the energy markets. From the view of modelling policy decisions the regions should be politically defined as countries. This makes sense, because policy decisions reflect the structures of the economies in question.

The global model COMPASS (Meyer and Uno, 1999; Uno *et al.*, 1999) fulfils these requirements. It depicts the interdependencies of prices and volumes on the

K. Uno (ed.), Economy–Energy–Environment Simulation, 69–80.

different stages of production and consumption for 53 countries and links the economies by a multisector global trade model. This chapter focuses on the latter part of the model.

In section 2 we first give a short overview of the role of the trade modelin the system COMPASS and continue with the alternatives of modelling trade in section 3. In the literature three approaches for modelling bilateral trade are discussed: the first is the hypothesis of given structures in international trade, which can be rejected with a short review of the facts. The second is modelling import functions based on the Armington hypothesis. This approach is used – for example – in the general equilibrium model GTAP (Hertel and Tsigas, 1997). The third approach is modelling export shares. In global multisector models this approach was first realized in the INFORUM system (Ma, 1996; Nyhus *et al.*, 1996). In section 3 we discuss the different approaches and prefer the third.

In section 4 the logical structure of the trade model of COMPASS is presented. Results of the econometric estimations of the functions of the trade model are discussed in section 5. Section 6 shows that trade matters: we present the results of two simulations with the whole COMPASS system with constant *versus* price-dependent trade shares. We compare the impacts of a price shock for primary energy on GDP and CO_2 emissions for the G7 countries. The results show that modelling of trade plays an important role in depicting the interdependency of energy, economy and environment.

2. The role of the trade model in the global 3E system COMPASS

The structure of the global energy-economy-environment model COMPASS is depicted in Figure 3-1 in the previous chapter: it shows a wheel, in which the bilateral trade model is the axis. The spokes are the country models, which always consist of a macro model and for most OECD and APEC countries of an input-output model and an energy model. The tyre represents the linkage of the countries via the global equality of savings and investments.

A deeper insight into the linkage of the trade model with the other parts of the system is shown in Figure 4-1 for one specific country. The global trade model receives a vector of import volumes and export prices from every country model. For every country it calculates a vector of export volumes and import prices.

The input-output models consist of 36 sectors. They obtain the vector of export volumes and import prices from the trade model and aggregated investment and private and public consumption figures from the macro models, and distribute these to 36 sectors. From the energy models they receive prices for the energy carriers. The input-output models calculate the vectors by industry of gross production, intermediate demand, the vector of imports and the vectors for the different components of primary inputs. The input-output models further estimate the vector of unit costs and the vector of prices.

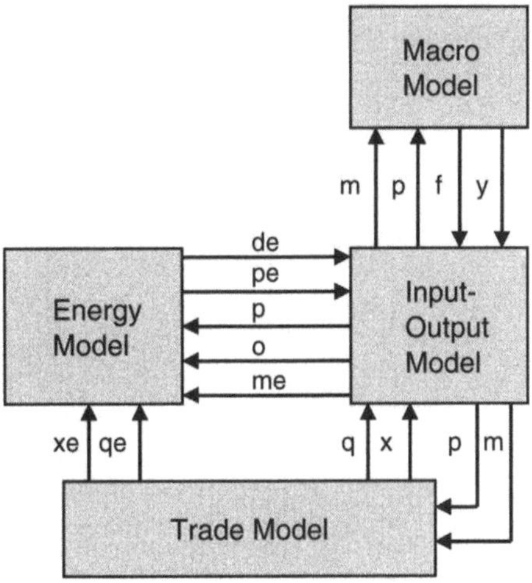

f: aggregate components of final demand
p: pice vector
q: vector of import prices
x: export vector
m: import vector
o: output vector
de: vector of energy demand
pe: vector of energy prices
qe: vector of energy import prices
xe: vector of energy exports
me: vector of energy imports

Figure 4-1. The linkage of the trade model and the other models for one specific country.

The macro models aggregate primary income, import volumes and prices coming from the input-output models. They show the redistribution of income between the government, households, enterprises and foreign countries and calculate the disposable income for these institutions, which are important determinants of private and public consumption. The macro models further depict monetary markets and calculate the interest rate and other determinants of investment. The accounting system of the macro models further contains the balance of payments.

The energy models receive the vector of gross production by industry and final demand by branches as well as industry prices, energy import volumes from the input-output models. The trade model delivers energy import prices and energy export volumes to the energy models. The energy models calculate primary and secondary energy demand for seven carriers in detail, the conversion of energy and CO_2 emissions of the different fossil energy carriers. Based on energy import prices the energy models further determine wholesale and retail prices for the energy carriers, which are delivered to the input-output models.

The 53 countries of the model include all OECD countries, all APEC countries, all OPEC countries, Russia and India (see Table 4-1). For the moment the

Table 4-1. Countries and regions in the trade matrices

1	Belgium and Luxemburg	19	Iran	37	Australia
2	Denmark	20	Iraq	38	Canada
3	France	21	Kuwait	39	USA
4	Germany	22	Oman	40	Brazil
5	Greece	23	Qatar	41	Mexico
6	Ireland	24	Saudi Arabia	42	Venezuela
7	Italy	25	United Arab Emirates	43	Algeria
8	Netherlands	26	Brunei	44	Libya
9	Portugal	27	Hong Kong	45	Angola
10	Spain	28	India	46	Nigeria
11	United Kingdom	29	Indonesia	47	Rest of Europe
12	Austria	31	Korea	48	Rest of Asia
13	Finland	31	Malaysia	49	Rest of Oceania
14	Norway	32	Philippines	50	Rest of North America
15	Sweden	33	Singapore	51	Rest of South America
16	Switzerland	34	Thailand	52	Rest of Africa
17	Former USSR	35	Taiwan	53	Rest of Continents
18	Japan	36	China		

structure discussed above is realized for the most important OECD and APEC countries. The others are represented only in the trade model and with macro models, from which 18 models are simple macro simulators.

All oil-, gas- and coal-producing countries, the bigger developing countries and the industrialized countries are explicitly part of the system. Additionally seven regions are defined, which include the countries not explicitly mentioned, so that the system closes on the global level. It is guaranteed that energy demand and supply can be depicted in a global dimension simultaneously with goods and capital markets.

3. Alternatives of modelling bilateral trade

In multisector systems the trade model transforms the vector of import volumes of the different countries into the export volumes of delivering countries. Of course a simple hypothesis is that the regional structure of the imports of the different countries is constant. In that case we would only need information concerning the regional structure of the imports for every country for one year. Given this information the exports of every country could be calculated by definition.

A look at the dynamics of international trade raises some doubts. Table 4-2 compares the shares of some selected countries in total world exports. We see, that China, Thailand and Hong Kong have rapidly growing shares coming from a very low level. But most of the G7 countries also have rising trade shares.

Table 4-3 shows that on the industry level the dynamics of international trade are even stronger. Here we see for the strongly expanding product group 'office and computing machinery' the same picture as for the totals concerning China, Thailand and Hong Kong, but now the shares of some G7 countries such as

Table 4-2. World export demand in US$: shares of selected countries (in %)

	1980	1996
China	1.083	3.080
Thailand	0.342	1.039
Hong Kong	1.064	3.332
France	5.081	5.401
Germany	8.594	10.107
Italy	3.543	4.880
UK	5.952	5.109
Japan	6.232	6.969
Canada	3.505	4.313
USA	11.281	11.825
Rest	53.322	43.944

Table 4-3. World export demand of office and computing machinery in US$: shares of selected countries (in %)

	1980	1996
China	0.041	2.176
Thailand	0.024	2.491
Hong Kong	3.870	6.497
France	6.321	2.479
Germany	10.592	3.848
Italy	4.027	2.239
UK	8.785	5.672
Japan	14.427	14.431
Canada	1.008	3.235
USA	31.378	18.708
Rest	19.526	38.222

USA, Germany, UK and Italy are shrinking. We are convinced that this dynamic cannot be depicted by a trade matrix with constant shares, where every change in the regional structure of exports has to be explained by the regional structure of the import vector.

A more elaborate approach was presented by Armington (1969). He postulated that the share of country l in imports of country k of good i depends on the price for good i, that country l is taking in country k relative to the average of the import prices of good i over all delivering countries. He further set the hypothesis that the corresponding price elasticity might be independent from the delivering country.

$$s_{kl}^{i}(t) = b_{kl}^{i} \cdot \left(q_{kl}^{i}(t) / q_{kl}^{i}(t)\right)^{\mu_k^i} \quad (1)$$

s_{kl}^{i} = share of imports of good i in country k delivered from country l
b_{kl}^{i} = positive constant parameter
q_{kl}^{i} = import price of good i in country k from country l in US$
q_{k}^{i} = average of import prices in country k over all delivering countries in US$
μ_{k}^{i} = price elasticity of imports of good i in country k.

The hypothesis is a critical one: it supposes that, for example, the price elasticity of imported cars in Germany would be the same for Japanese, Korean, American, French and Italian cars. But for many goods consumers have preferences with respect to the delivering country.

The advantage of the hypothesis is that price elasticities can be estimated without having time-series of the shares of the delivering countries. A time-series of the total imports of good i in country k is enough.

For an econometric estimation of the Armington hypothesis see for example Brenton (1989). Often model-builders do not estimate on their own, but take estimated price elasticities from literature. They need only one observation of the shares and the relative import prices for one year. With the predetermined price elasticity they can calculate the parameters b of equation (1). This procedure is called calibration. With a time-series for the import prices they can use equation (1) to forecast the import shares. In this way bilateral trade relations are modelled in the GTAP system (Hertel, 1997; Hertel and Tsigas, 1997) and in many other CGE models.

If we want to achieve a better representation of the preferences of the agents we have to estimate specific price elasticities for the delivering countries. To execute the regressions we need a data set, which is at least by factor 15 bigger than in the Armington case. We reject the Armington hypothesis, because we want to have a more adequate analysis and an explicit econometric test of the used equations. This is not possible in the described calibration procedure of CGE modelling.

If we change our perspective from the importing to the exporting country, the share of the imports of good i in country k coming from country l obviously is the share of the exports of country l of good i to country k in the imports of good i in country k:

$$x_l^i(t) = \sum_{k=1}^{n} s_{lk}^i(t) \cdot m_k^i(t) \tag{2}$$

s_{lk}^i = share of exports of country l in the imports of good i of country k
x_l^i = nominal exports of good i in country l
m_k^i = nominal imports of good i in country k.

Following earlier work by Klein and Van Petersson (1973), Moriguchi (1973), Hickmann and Lau (1973) and Gana *et al.* (1979), Ma (1996) specified the following function for the endogenization of the trade shares:

$$\log\left(s_{lk}^i(t)\right) = a_{lk}^i + b_{lk}^i \cdot \log\left[\left(p_l^i(t)/q_k^i(t)\right)\right] + c_{lk}^i \cdot \log\left(K_l^i(t)/K_w^i(t)\right) + d_{lk}^i \cdot t \tag{3}$$

p_l^i = export price of good i in country l in US\$
q_k^i = import price of good i in country k in US\$
K_l^i = capital stock invested in production of good i in country l
K_w^i = capital stock invested in production of good i in the world.

Using this approach Ma estimated price elasticities that change with the exporting and the importing country. The capital stocks are indicator variables, which will measure the influence of changes in the quality of the production in the delivering country and its competitors. The time trend has to catch other continuously developing influences. His good results encourage us to follow his approach.

4. The logical structure of the trade model in the COMPASS system

As stated above in equation (2) the exports of the model are calculated as follows:

$$x_l^i(t) = \sum_{k=1}^{n} s_{lk}^i(t) \cdot m_k^i(t) \tag{4}$$

x_l^i = nominal exports of country l of good i
s_{lk}^i = share of exports of country l in the imports of good i of country k
n = number of countries (53)
m_k^i = nominal imports of country k of good i.

Here s_{lk}^i is the nominal share of the export of country l in the imports of country k of good i depending on the export price of country l relative to the average of the import prices of good i in country k and a time trend.

$$s_{lk}^i(t) = s_{lk}^i\left[p_l^i(t)/q_k^i(t), t\right] \tag{5}$$

s_{lk}^i = share of exports of country l in the imports of good i of country k
p_l^i = price of good i in the exporting country l in US$
q_k^i = import price of good i in country k in US$
t = time trend.

Since capital stocks by industries were not available for the different countries, we had to reduce Ma's specification.

The import prices of good i in country k are calculated as weighted averages of the export prices of the different countries. The weights are given with the trade shares

$$q_k^i(t) = \sum_{l=1}^{n} s_{lk}^i(t) \cdot p_l^i(t) \tag{6}$$

q_k^i = import price of good i in country k
s_{lk}^i = share of exports of country l in the imports of good i of country k
p_l^i = price of good i in the exporting country l.

For a better understanding of the role of the trade model in energy modelling we take a look at the logical flow in the case of a rise in oil prices, for example. In equation (6) rising export prices in oil-producing countries increase the import prices for oil in the other countries. This will influence wholesale prices and retail prices in the energy models, which will induce substitution and changes in energy intensities in the energy models of the different countries. The input-output models transform the price changes for energy carriers in domestic price changes for the different goods and calculate reactions on product demand including import demand. From here we come back to the trade model, where in equation (4) the world exports for the different goods are calculated, using trade shares that have been changed by the adjusted relative prices as equation (5) of the trade model shows. The change of the exports then feeds back into the country models and induces further reactions on macro variables, production of industries, demand for commodities and energy demand.

Ma's trade model is part of the INFORUM system (Nyhus *et al.*, 1996). This consists of about 100 product groups compared to only 25 in COMPASS, but only 13 countries compared to 53 in COMPASS. The INFORUM system is limited to only a part of global trade.

5. Empirical results of the estimation of trade shares

The database of the empirical work comprises time-series of trade matrices of Statistics Canada with about 6000 product groups of the producing sectors from 1980 to 1996 in US dollars. Services are not included. We aggregated these raw data to the 25 product groups of the two-digit classification scheme of the OECD input-output tables. The data were also adjusted, so that the identities of the trade matrix in equation (2) are not violated.

We estimated the nominal trade shares by the logarithmic function of (5) with the ordinary least squares (OLS) estimation. The estimated equation can be written as:

$$\log\left(s_{lk}^{i}(t)\right) = \beta_{1,lk}^{i} + \beta_{2,lk}^{i} \cdot \log\left(p_{l}^{i}(t)/q_{k}^{i}(t)\right) + \beta_{3,lk}^{i} \cdot \log(t) \tag{7}$$

The trade shares were estimated automatically using a C++ program. With 25 product groups and 53 times 53 trade shares for each group not less than 70,225 trade shares had to be analysed. In the program equation (7) and also its alternatives without a time trend (8) or without price dependency (9) were estimated:

$$\log\left(s_{lk}^{i}(t)\right) = \beta_{1,lk}^{i} + \beta_{2,lk}^{i} \cdot \log\left(p_{l}^{i}(t)/q_{k}^{i}(t)\right) \tag{8}$$

$$\log\left(s_{lk}^{i}(t)\right) = \beta_{1,lk}^{i} + \beta_{3,lk}^{i} \cdot \log(t) \tag{9}$$

If at least one value of the time-series was zero, the trade share would be set constant at the level of the last observation. Afterwards, the three equations (7), (8) and (9) were estimated for the remaining variable trade shares. For each trade share the selection procedure was the following: first, the estimated value of the price elasticities $\beta_{2,lk}^{i}$ was restricted to economic plausibility. Because the trade shares are nominal this price elasticity had to be less than 1. On the other hand extreme price elasticities lower than −10 were rejected:

$$-10 \leq \beta_{2,lk}^{i} \leq 1 \tag{10}$$

Secondly, the selection followed econometric criteria: the *t*-values for the estimated coefficients had to be above 2 or below −2. The Durbin-Watson statistic had to be between 1.1 and 2.9. If more than one of the three equations above accomplished these restrictions, the program would choose the estimation with the highest R^2. If all specifications failed, trade share in question would be kept constant in the simulation period.

The algorithm above is applied for 25 product groups and 53 countries; this means for 70,225 time-series. As 42,588 of the time-series have one or more zero values, 27,637 trade shares were estimated. A total of 12,564 equations (45.5% of the non-zero trade shares) hold the restrictions discussed above and the trade shares can be endogenized. Altogether 9300 trade shares (one-third of the non-zero trade shares) are explained only with a time trend, 8.0% depend only on prices and 3.8% are determined by prices and trends. So price dependency is given for 3264 of the trade shares.

Table 4-4 shows these results in detail. The most price-dependent commodity groups are drugs and medicines (14.3%), food beverages and tobacco (14.0%) and wood and furniture products with 13.6%. In mining and quarrying only 6.7% of the trade shares depend on relative prices.

6. The relevance of endogenized trade shares: results of simulations with the whole system

The econometric analysis of price dependency of trade shares has shown that over all sectors not more than 12% of the non-zero trade shares are price-dependent. A look at the sector details shows further that the energy products do not belong to the group with the strongest price dependency. Does this mean that prices do not matter?

Without information regarding the importance of the price-dependent shares for the solution of the system the answer is difficult. Of course, there is the expectation that bigger shares are more price-dependent than smaller ones. In this regard the price-dependent trade volume might be much bigger than 12%. However, the importance of a trade share is not only given with its number; it also depends on the import total of that good in the observed country, with which the share is multiplied to obtain the export of the delivering country. Importance of the trade shares in energy analysis further depends on the product group and its direct and indirect relevance for the energy markets.

It seems that the identification of important trade shares is rather difficult. Therefore, another approach is chosen, which uses all information that is hidden in the whole COMPASS system: first, a simulation is run with an energy price shock with constant nominal trade shares and then the same simulation is repeated with endogenized trade shares. A comparison will then give

Table 4-4. Results of the estimation of the import shares

		Number of Coefficients	At Least One Value of the Series is Zero	Estimated Coefficients				Constant Coefficients
				Only with Trend	Only with Prices	Prices and Trend	Total	
1	Agriculture, forestry and fishery	2809	1445	498	118	56	672	692
2	Mining and quarrying	2809	1753	429	48	23	500	556
3	Food, beverages and tobacco	2809	1198	534	155	70	759	852
4	Textile, apparel and leather	2809	1251	485	123	82	690	868
5	Wood and furniture products	2809	1733	320	97	49	466	610
6	Paper, paper pro. and printing	2809	1473	451	100	61	612	724
7	Industrial Chemicals	2809	1342	546	124	52	722	745
8	Drugs and medicine	2809	1609	433	128	43	604	596
9	Petroleum and coal products	2809	2047	252	72	23	347	415
10	Rubber and plastic products	2809	1755	409	70	38	517	537
11	Non metallic mineral products	2809	1628	396	98	39	533	648
12	Iron and steel	2809	1706	390	76	48	514	589
13	Non ferrous metals	2809	1820	333	68	37	438	551
14	Metal products	2809	1415	442	127	36	605	789
15	Non electrical machinery	2809	1345	519	110	71	700	764
16	Office and comp. machinery	2809	1879	289	80	37	406	524
17	Electrical apparatus nec	2809	1490	453	106	50	609	710
18	Radio, TV and com.Equipment	2809	1695	373	96	33	502	612
19	Shipbuilding and repairing	2809	2327	106	28	8	142	340
20	Other transportation means	2809	2214	168	43	18	229	366
21	Motor vehicles	2809	1687	381	75	57	513	609
22	Aircraft	2809	2310	130	35	16	181	318
23	Professional goods	2809	1433	452	129	54	635	741
24	Other manufacturing	2809	1258	501	104	49	654	897
25	Electricity, gas and water	2809	2775	10	3	1	14	20
	Total	70,225	42,588	9300	2213	1051	12,564	15,073

an answer. Of course, the case of constant nominal shares also implicitly assumes price dependency, because the real shares have a price elasticity of -1. So we actually compare two model versions with price-dependent trade shares.

In the business-as-usual (bau) scenario the price for crude oil is fixed at the level of $22 and the price for coal is also constant for the period 2000-2010. In the alternative scenario the oil price grows by 4% per year, an increase of about 50% against the bau in 2010, and the price for coal in every country follows the path of the price for refined petroleum. We choose these two prices because in our trade matrix petroleum and coal are a combined product group.

Which effects can be expected? On the energy markets the price shock induces substitution between energy carriers: in final energy demand gas and renewable energy carriers, and in electricity production nuclear energy and renewables will expand their shares. Rising energy prices will reduce the energy intensities of the sectors. Both effects diminish CO_2 emissions. GDP will be reduced, because higher energy prices will raise the output prices of the sectors and this diminishes the demand for goods. This is the reason for a further reduction of CO_2 emissions.

Table 4-5 shows that for all countries our expectations are fulfilled. However, the results are very different in the countries. Let us first consider the results for constant nominal trade shares: in the US we observe the strongest reduction of GDP and compared to this a relatively low reduction of CO_2. This is plausible, since a low reduction of CO_2 indicates a low reduction of energy demand and a stronger impact on costs, prices of goods and GDP. In general this relation also holds for the other countries. Only for Japan is the picture different. The reaction of CO_2 emissions is very low. The reason is that, in Japan, energy efficiency follows strong long-run trends and is not price-elastic. The low impact on GDP in Japan can be explained by low reactions of the countries' goods prices on changes in energy costs.

In the case of endogenized trade shares the effects on CO_2 emissions do not change drastically, but the impacts on GDP are very different for the US, France and Canada. In the US and Canada we have a 30% and a 26% lower reduction of GDP, whereas in France the reduction of GDP is 40% higher than

Table 4-5. Effects of higher energy prices in G7 countries: deviations of GDP and CO_2 emissions from the BAU scenario in the year 2010 (in percent)

	Constant Trade Shares		Endogenous Trade Shares	
	GDP	CO_2	GDP	CO_2
USA	−1.74	−3.01	−1.25	−2.88
Japan	−0.63	−1.36	−0.58	−1.29
Germany	−1.08	−8.90	−0.99	−8.42
France	−1.14	−6.62	−1.60	−6.82
Italy	−0.58	−11.70	−0.43	−11.83
Great Britain	−0.97	−9.27	−1.00	−9.58
Canada	−0.93	−4.68	−0.69	−4.38

in the case of constant nominal shares. One interpretation may be that in the US and Canada the price-elasticity of the exports is on average less than one in absolute terms, and in France on the average more than one. However, there are many other effects, such as changes in the sectoral and regional import structures, which are responsible for the result.

7. Conclusions

Global 3E (energy-economy-environment) studies have to analyse questions in a multicountry and multisector approach. The multisector approach is necessary, as changes in energy demand and supply have drastic impacts on the structure of the economy. The multicountry approach is useful because global policy is done by the governments of actually existing countries and not of artificial country groups in regions. Policy decisions are always predisposed by the structure of the economy, that has to be modelled. In consequence we get a huge system with a large number of countries. This chapter shows that, also in the global context of 53 explicitly modelled countries, the endogenization of trade shares is possible without the restrictions of the Armington hypothesis, typically used for CGE modelling.

As we could show, endogenized trade shares have an influence on the impacts of energy prices on the economy. The deviations from the bau scenario in the case of constant shares and in the case of endogenized shares are different. Of course we are presenting results from work in progress. Further work will show how a variation of the economic and econometric restrictions on the estimation of the trade share functions influences the results.

The impact of the different modelling approaches (constant *versus* endogenized shares) on the levels of exports and GDP in both scenarios is not discussed. There are very strong differences in respect to the sector and the regional structure of the world economy, but this is beyond the aim of this chapter.

5
Developing an energy balance simulation model

YUMIKO UMEHARA

1. The need for an energy balance model

The COMPASS energy balance model is constructed on the basis of the 36 sectors of an input-output and macro econometric model with the financial and trade balances. The economic models of COMPASS cover 60 countries/regions in the world. They are all constructed on the basis of the original object-oriented database system that stores all the data sets required for modelling such as IO, SNA, financial, trade and energy statistics. The energy balance model, as one of the modules in COMPASS, retrieves the IEA (International Energy Agency) energy balance table from the COMPASS database system. Then it simulates the energy balance and compiles the energy balance table for the future.

Policies to combat climate change require country- and sector-specific analysis of the greenhouse gas mitigation opportunities. Nevertheless recent research focuses on international aspects, such as the impacts of the Kyoto Mechanism. Although there are many energy-economics models in the world, the engineering models lack economic disaggregation, whereas economic models tend to lack energy end-use and process descriptions. Assessment of both policy measures geared towards specific activities together with international environmental framework calls for description of economy and energy in detailed disaggregation levels. The energy balance model contains seven energy sources for primary, transformation, utility and three end-use sectors, namely the (1) industry, (2) transportation, and (3) residential/commercial sector including agriculture. It can be used as the energy material flow model, and as the energy-environmental policy simulation tool both at aggregate and disaggregate levels. The important aspects embedded in energy balance modelling are the potential of technological change and the fuel switching towards de-carbonized society, that are likely to be induced by any policy to increase energy costs. These are econometrically estimated and employed in the simulation. This model is operated together with the COMPASS international team. COMPASS (Comprehensive Model for Policy Assessment) is a dynamic modelling system to capture multi-country and multi-sector interactions. This chapter presents the development of the energy balance model.

In literature on energy modelling the observed correlation between the economic development stage and energy intensity trends often leads to the

K. Uno (ed.), Economy–Energy–Environment Simulation, 81–98.

assumption of AEEI (autonomous energy efficiency improvement). AEEI is entirely apart from any conservation induced by rising energy prices. Recent comparison of six widely used models produced AEEI ranging from 0% to 1% per year. This implies a difference of 2.5 billion tons of carbon by 2020. The difference widens to 42 billion tons of carbon by 2100, a figure 7 times higher than the 1990 emissions level. The numerical value of the AEEI is highly controversial and remains problematical.

It is our view that modelling attempts based on assumed AEEI, or those based on AEEI measured at highly aggregate level, such as GDP and total energy consumption, are highly inadequate.

In contrast, the work presented here is based on detailed energy balance tables that are compiled annually by the IEA for a large number of countries. Energy balance tables consist of three parts: primary energy supply, energy conversion, and final use in industry, transport, and residential and other sectors. Of the primary energy sources, coal, crude oil, petroleum, natural gas, nuclear, hydro and others are distinguished. Electricity is generated by converting primary energy. The COMPASS examines economy-energy interactions at three points in the energy balance tables; namely (1) energy intensity change, defined as the energy consumption per output in respective industrial branches; (2) energy conversion efficiency, depicting the ratio between input of primary energy sources into electricity generation and the obtained electricity; and (3) the substitution among primary energy sources. Of these, (1) and (3) are econometrically examined, while (2) is an exogenous technological parameter.

In the model, relative energy price and output determine the energy intensity in three end-use sectors. Energy intensity is interpreted to represent the technology levels of a pertinent industrial branch at a given time.

The demand for electricity, in turn, determines the amount of transformation needed. The electricity generation is modelled with the conversion efficiency explained by time trend.

The share of each fuel in total primary energy supply, as well as in electricity generation, is also dependent on the relative price of the fuels in each sector. Shares of nuclear and new and renewables (hydro and others) are exogenous. The shares of remaining fuels are rescaled in the model so that the total share adds up to unity.

The economic variables such as relative prices of energy and production level of individual industrial branches are provided by the input-output/macro/financial/trade models. The link between input-output and commodity trade of energy-related products is important since this link enables endogenization of world oil prices. This becomes possible by utilizing a model incorporating the main OECD members, APEC members, Russia, oil-exporting countries, and other developing countries.

2. Objectives and features

2.1. Objectives

Considering that there are a number of existing energy-economics models operational today, it is imperative that the advantage of the COMPASS modelling approach, particularly focusing on the energy balance model, should be clarified before describing its structure. There are three major objectives in developing an energy balance model.

First, COMPASS aims at depicting a dynamic time path of interaction between energy production (seven energy sources) and consumption (three end-use sectors) in a disaggregated way. It gives both monetary and physical magnitudes of energy supply/demand in combination with the input-output model. Such materials/energy balance modelling is partly based on accounting frameworks represented by a system of integrated environmental and economic accounts (SEEA), known as green GNP. This framework consists of material inputs delivered by the natural environment, transformations within the economy, i.e. extraction, conversion, manufacturing, consumption, and material outputs (residuals) such as wastes etc. (Uno, 1995). However, such an accounting is only a snapshot of the interaction between economy and environment. The dynamic change of environmental and related activities can be described by a time-series comparison of such a framework. The idea of the negative outputs returning to the natural environment is not specifically treated in monetary value in our framework. Currently, only the CO_2 emissions are modelled, the cost of which is an externality for the whole globe to be internalized by social economic systems. The model is built on a country-by-country basis, and gives a global dynamic picture in linkage with the commodities trade of fossil coal, oil, and gas.

Secondly, COMPASS includes the technology aspect explicitly. The energy-input structure of industrial productions, household activities, transportation or commercial services, describes the level of energy production and transformation technology together with the end-use technology. For example, energy production technology can be observed as conversion efficiency. Similarly, technology in the iron and steel sector, for example, can be represented by energy intensity. However, unit consumption of energy in households, or in transportation services, cannot be calculated as the energy required to cook or light, or a passenger to travel, which represent the end-use technologies in a real sense. The development of internationally comparable physical data of durable stock representing 'lifestyle', is needed to study technologies in a modelling framework.

Thirdly, and most importantly, COMPASS captures dynamic interactions between technology, market mechanisms, and government policy on a sectoral basis. The energy balance model, combined with the IO/SNA, financial and trade balance models, distinguishes indirect taxes and subsidies for individual sectors in addition to inputs of material and labour in production processes. It is particularly important that, in practice, prospect policy measures to reduce CO_2 emissions are in most cases sector-specific. Indeed, little study has been done on linking the disaggregated analysis of driving force of the emissions with the policy options at domestic level, while keeping international comparability in mind. One of the few studies conducted to look into the existing low-cost mitigation opportunities on a country basis is the disaggregated decomposition analysis by Unander and Schipper (1999). They distinguished the activity effect, energy service effect, intensity effect, and fuel mix effect as causes of the domestic CO_2 emissions. This enables one to draw a picture of where the energy is still used inefficiently, and where the policies can be most effectively adopted, but there is no feedback regarding the policy impact of economic activities on energy aspect within this framework itself. The energy balance model, through the full disaggregation of the sectors that are linked internationally, is able to test the policy scenarios in the industry, transportation, residential/commercial (agriculture, public sectors are included), and energy sectors.

Let us try to clarify the features of the energy balance model as a tool for assessing policy measures to combat climate change. Repetto and Austin (1997) provided a list of focal points of the available models as follows:

(1) The extent to which substitution among energy sources, energy technologies, products, and production methods is possible.
(2) The extent to which market and policy distortions create opportunities for low-cost (or no-cost) improvements in energy efficiency.
(3) The likely rate of technological innovation and the responsiveness of such change to price signals.
(4) The availability and likely future cost of non-fossil, backstop energy sources.
(5) The potential for international 'joint implementation' of emissions reductions.
(6) The possibility that carbon tax revenues would be recycled through the reduction of economically burdensome tax rates.

Currently the COMPASS is designed to capture all of these except (4). The main features of the model are that it explicitly estimates the (1) substitution among energy sources, and (3) the price responsiveness of technological change. The model is also able to generate world and domestic energy prices when it is fully integrated with the energy sectors in an IO model. This is a great advantage since energy prices for non-OECD countries are not fully available from the IEA database. The removals of price distortions, a part of (2) that arises mostly from subsidies, specifically play significant roles in reducing CO_2 emissions in EITs (economies in transition) and developing countries. The assessment of joint implementation, which is item (5), can be described as the technology (input coefficients/energy intensity) transplanted to EITs. Either energy intensity or the input coefficient can be easily transferred across countries/sectors. Finally, item (6) is important in assessment of the market-based policies such as carbon tax or permits trading. The impacts of a rise in energy price will lower carbon intensity, and a change in mark-up price will affect all energy-using sectors through cost functions. The impact can also be observed as the second-round effects in the monetary policy response to control inflation in linkage with the macro-financial model.

2.2. *Theoretical standpoint*

(a) Type of model

The model that we present can be classified as a multi-sector econometric energy balance model. This type of model cannot be mapped into a conventional model categorization.

The most widely known categorization of energy-economics models draws a distinction between an *economic* and an *engineering* model. A *top-down* model has a detailed description of the economy depending on specific parameters. The other type, called *bottom-up*, entails the individual consumption pattern of energy and related technologies (IPCC, 1996b). Repetto and Austin (1997) point out that the top-down models typically assume that all cost-effective improvements in energy efficiency have already been realized, an assumption contradicted by actual experience. On the other hand the bottom-up models have found inefficiencies in energy use to the degree that around 20% of current carbon emissions could be eliminated at no cost.

The two modelling approaches have often been cited as alternatives, but recent developments show both approaches are supplementary. The MARKAL (Loulou and Kanudia, 2000) and the AIM (National Institute for Environmental Studies) are examples of such models integrating both types. Such integrated models allow us to consider the possibility of choices among a menu of technologies, under a certain constraint on economic activity. The parameters for the economic part can be modified if the result from the technology model diverges from assumptions at the macroeconomic levels. This approach serves to increase the consistency of potential technological progress in the economic context, thus the results may not be too optimistic. Weyant (1999) categorized the models presented at the energy modelling forum in their economic model types and energy/carbon model types (see Table 5-1). In the presented categories, the COMPASS energy balance model may be mapped into the last row of an economic model type; however, an energy model type may lie between energy demand/supply sectors and technology detail. We will discuss our modelling approach in economy and energy together with a theoretical background.

(b) Multi-sector econometric model

For convenience of describing this modelling approach let us compare it with CGE (computable general equilibrium model). In COMPASS the energy

Table 5-1. Model types

	Energy/Carbon Model		
Economy Model	Energy Supplies & Demands by Sector	Energy Technology Detail	Carbon Coefficients
Aggregate Production/Cost function		CETA MERGE3 GRAPE	FUND RICE
Multi-sector General Equilibrium	MIT-EPPA WorldScan G-Cubed	ABARE-GTEM AIM MS-MRT SGM	
Multi-sector Macroeconometric	Oxford		

Source: Weyant *et al.* (1999).

supply/demand is expressed in monetary value in an IO framework, and in physical volume in an energy balance model. Together with IO, SNA-financial, and trade balance they comprise a multi-sector economic and energy flow model. Let us first clarify the type of economic model of the COMPASS. The system includes employment, financial markets, international capital flows, and monetary policy. According to West (1995), the model can be termed as the IOE (input-output + econometric model). CGE provides the optimal solution by assuming consumers and producers allocate their resources in order to maximize their welfare or profits. Although it is convenient in analytical terms, IOE does not assume the equilibrium in the market (Meyer and Uno, 1999a; Ayres, 1994), which does not reflect the real world. It does not necessarily assume perfect knowledge in the market place, thus it describes economy in 'as is base' and reflecting the reality of the economic behaviour, without *a priori* assumptions not based on empirical evidence. What mostly makes IOE differ from CGE is that IOE is a dynamic modelling approach. Therefore, IOE is capable of depicting time paths of the endogenous variables reacting to the exogenous impacts on economic systems. Thus, with the energy balance model, not only can we test the impacts of certain policy measures, but we can also explore how the objective of such a policy is being met with the passage of time.

(c) Energy model and technological change

Many energy-economics models employ production functions, which assume the factors of production are continuously variable and substitutable (Allen, 1968). However, whether substitutability exists between energy and capital is controversial. Some observers claim that energy and capital are complementary in the short run, and substitutable in the long run; others conclude there is no such evidence. Slade *et al.* (1993) summarize the reasons for such divergence in research outcomes. They note that the important difference in the observation lies in the estimation of capital stock. The substitution between energy and other factors, commonly presented as ESUB, lies between 0.3 and 0.7 across the models (Azar, 1996). The higher this value is, the easier it is to reduce CO_2 emissions. Austin and Repetto (1997) argue that, despite being more or less disaggregated, models differ in the assumed ease of substitution among products and technologies in response to cost change, and it is difficult for modellers to get it right.

Let us briefly look at the organization of the energy-models employing production functions at an aggregated/disaggregated level. In the DICE model (Nordhaus, 1994) there is only one product: world output, and two factors of production: capital and labour inputs. In the Global 2100 (Manne and Richels, 1992) that encompasses the subdivided regions of the world, the electrical energy and non-electrical energy, capital, and labour have the constant elasticity of substitution. The MARKAL is an engineering model in which the production function allows substitution between the pair capital-labour and energy. The two-level CES (constant elasticity of substitution) function (Kosaka and Yano, 1999) separates labour from capital for the substitution for energy inputs. The nested-CES functions are employed by a multi-sector general equilibrium model such as OECD GREEN (Burniaux *et al.*, 1992), GemWTrap (Bernard and Vielle, 1999), among others. In any of these models the production factors substitution plays a central role in the formulation of energy demand functions and consequently determines the costs required for reducing emissions.

The well-established production functions are unquestionably theoretically satisfying. Nevertheless, we do not use such a function in linking between energy and economy in the model. We consider, though, that marginal, but econometrically estimated technological change (shifts in isoquants), rather than constant substitution (movements alongside an isoquant), need to be modelled. Currently in COMPASS, the input coefficients (*aij*) are yet to be modelled to shift according to the change in labour and material costs. Higher productivity presents lower labour cost in producing the same value, and mark-up pricing is being implemented. In the energy balance model the technological change is considered to occur by the change in relative prices of energy to labour and material costs.

In addition, there is no internationally comparable capital stock data. We will keep exploring the quality and availability of R&D data from individual country sources. It is an important step to discuss in enhancing the model, since it relates to the discussion of endogenous technological change. Rather, at this stage, we try to observe the price responsiveness of energy intensity – a change in mixed form of the technologies and production processes – that is not radical, but surely more than marginal in nature.

2.3. Energy intensity

In the model we use estimated price elasticity of energy intensities at end-use sectors, testing the hypothesis of price-induced technological change. It is probable that the rate of technological change and its price responsiveness varies across countries and sectors. As for energy efficiency improvement, such variations may be due to differences in the existing energy tax, share of carbon-intensive fuel, and the energy intensity level. If the rise in relative price induces the inefficient technologies to retire and investments are made towards installing more energy-efficient ones, the emissions can be reduced efficiently. This approach allows the model to look into the low-cost opportunities spreading across the economy in the long run in response to the initial market signals, whereas socioeconomic adjustment requires some time lags. The reasons we employ such an approach are as follows.

Energy intensity can be interpreted as an *ex-post-facto* realization of selected technology. In reality, the causes of the fall/rise in energy intensities are complex. They include technological improvements in individual energy use, structural shifts in energy systems, inter-fuel substitutions, economic structural change from more to less energy-intensive industries, and changing lifestyles. Not necessarily all, but many, of these factors affecting energy efficiency improvement form the long-term trend in energy intensity. Therefore, whether the economy can lower energy intensities constitutes an important key in mitigating any greenhouse gas emissions. Most of today's domestic climate change measures/policies are directed towards efficiency improvement.

Nakicenovic *et al.* (1998) found that the energy intensities are lower at higher per-capita GDP, except for the former USSR. They attribute the result, first, to more efficient energy conversions and end-use technologies that become affordable, and second, to the higher quality of energy carriers made available by the energy sector. The observed correlation between the economic development stage and energy intensity trends leads to the assumption of the AEEI in conventional energy-economic modelling. The AEEI of 0.5% per year results in an energy conservation growth at 0.5% less than GDP, and is entirely apart from any conservation induced by rising energy prices.

The OECD model comparison project (Dean and Hoeller, 1992) conducted sensitivity analyses of the model outcomes by standardizing essential assumptions. The six widely used models were compared including CRTM (Rutherford, 1992), ERM by Edmonds-Reilly (Barns *et al.*, 1992), OECD GREEN (Burniaux *et al.*, 1992), IEA model (Vouyoukas, 1992), Global 2100 (Manne-Richels, 1992) and WW (Whalle-Wigley, 1992). They found that AEEI ranges from 0% to 1%, resulting in a gap of 2.5 billion tons of carbon by 2020, which increases to 42 billion tons of carbon by 2100. This is nearly 7 times larger than the emissions level in 1990. The emission abatement cost is crucially sensitive to the baseline emission level where highly controversial AEEI assumption plays a key role.

Jorgenson and Wilcoxen (1993) recognized in the neoclassical approach that the economy could be more energy-using, contrary to the conventional wisdom of autonomous conservation, if the relative price stays constant. Azar and Dowlatabadi (1999) pointed out that AEEI is not an appropriate measure of technological change since the energy efficiency improvement is not likely to be captured unless specific policies are adopted. They claimed that, for heavy industries, energy costs are the major concerns, and they must figure prominently in economic activity for technological change to reduce energy intensity. Such discussions arise in the wake of the evolutionary theory of economics, in which the technological change is considered to occur in the economic disequilibrium, which creates incentives for new technological development and thus drives economic growth (Ayres, 1994).

As will be discussed in the next section the estimated elasticity is constant, by taking the double-log specification, and is applied to the energy balance simulation for the future. In this approach the change in elasticity itself has not yet been studied in the current framework of our energy balance model.

2.4. Energy substitution

The term energy substitution needs to be clearly defined here; it refers to the inter-energy substitution induced by price change. There are few descriptions of the model specification regarding the constant elasticity for inter-energy substitution. The GREEN model distinguishes the elasticity for old vintages and new vintages. The parameter for the inter-energy elasticity is 0.25 for old and 2.0 for new vintages. It is assumed to be 10.0, quite large, between conventional and backstop technologies. The parameter is uniform across countries and energy use sectors. The Oxford model (Cooper *et al.*, 1999), which is similar in sector disaggregation to COMPASS, presents the elasticity for electricity generation. The parameters are modified by technology models so that the 'coal, oil, and gas total' and 'nuclear/non-fossil fuel' elasticity of 0.5 includes a 0.2 technological gain, which is not likely to be reversed.

If elasticity is high or it is easier for an economy to de-carbonize, and whether the models treat non-fossil or backstop energies, matters greatly in cost evaluation. For example models that exclude backstop technologies overstate the economic impact of carbon tax because the economy is assumed to rely indefinitely on conventional fuels even after their costs were higher than the costs at which alternative energy sources would be profitable (Repetto and Austin, 1997). The major backstop technologies for production of electricity include: solar photovoltaics and thermal, fluidized-bed combustion, integrated gasification/combined cycle, geothermal, wind, and fuel cells. The most important drawback is their cost relative to a conventional combustion system, i.e. the capital costs of coal combustion are between \$1260 (1983 US\$/kW) and \$1580, whereas the solar thermal system cost is about \$2000 to \$3000. The costs are expected to decline in the earlier stage of diffusion, alongside the curve of learning by experience. There are also possibilities that other relatively inexpensive technologies can also substitute for heavy fuels. Fuel switching to new energies can be prompted by the future prospective of returns to investment, supported by government subsidies. However, entering into new technologies is highly uncertain and may not be very promising for the medium-term contribution to CO_2 abatement. In most of the energy markets, fossil fuel 'lock-in' could be persistent, in particular with relatively low oil prices. In such a case it is possible that the renewable energy technologies remain restricted to niche markets.

3. The model design

3.1. Energy balance and sectors

The development of an energy model is based on the time-series of the IEA energy balance table that are reworked and stored in the database system of COMPASS. The structure of the energy table in the model is shown in Table 5-2 together with standard variables names used in the database and modelling. This consists of seven energy sources for primary, secondary, and final consumption. We added the line 'Additional*' for convenience of modelling.

Let us first describe the construction of the table. TPES (total primary energy supply) is equal to the sum of domestic production, import, and stock change minus export (IEA Row no. 1-5). It is also equal to TFC (total final consumption) plus all the inputs for energy transformation and own use and losses (expressed as a negative) plus all the inputs for energy transformation and own use and losses (expressed as a negative) and $\pm$ statistical difference (no. 7-18). The transformation includes electricity generation, etc. (no. 9-12), petroleum refinery (no. 13), coal transformation (no. 14), liquefaction of gas (no. 15), and other transformation (no. 16). TFC consists of energy use in Industry, Transportation, and Residential/Commercial sectors plus Non-energy use.

Industry, Transportation, and Residential/Commercial sectors contain consumption of each fuel, represented with the prefix *ind*, *tra*, and *pro*, respectively. These end-use sectors are grouped with the IO 36 sectors according to the International Industrial Classification Rev. 2 and Rev. 3 as is shown in Table 5-3 (United Nations, 1990). We need to combine Rev. 2 and Rev. 3 because the recent IEA data are classified according to Rev. 3, but the COMPASS classification has not yet been updated. This grouping enables us to link the whole energy balance table with the IO table. In the near future, the supply of seven fuels will be grouped with the energy sectors in IO, such as petroleum and coal products, mining and quarrying, and electricity and gas.

3.2. Logical structure

The flow chart of the energy balance model is drawn in Figure 5-1. The economic variables coming from IO/macro/financial and trade models drive energy-related materials consumption, as is indicated in the figure as IO model. The arrow shows the dependence of the variables; the arrows with double lines show that variables are used in regressions, and arrows with thick lines indicate

Table 5-2. Energy balance table in the model

IEA No.	Energy Balance Table	Coa	Crude Oil	Petroleum	Gas	Nuclear	Hydro/ Other	Electricity and Heat	Difference	Total	Unit
1	Indigenous production	Outcoa	outcru	outprf	outgas	outnuc	outren	outelh			Ktoe
2	Import	Mttcoa	mttcru	mttprf	mttgas	mttnuc	mttren	mttelh			Ktoe
3	Export	Xttcoa	xttcru	xttprf	xttgas	xttnuc	xttren	xttelh			Ktoe
4	Intl. marine bunkers plus	Nttcoa	nttcru	nttprf	nttgas	nttnuc	nttren	nttelh			Ktoe
5	Stock changes										Ktoe
6	*Total Primary Energy Supply*	Petcoa	petcru	petprf	petgas	petnuc	petren	petelh			Ktoe
7,8	Returns and transfers	Sdtcoa	sdtcru	sdtprf	sdtgas	sdtnuc	sdtren	sdtelh			Ktoe
	Statistical differences										
9	Electricity plants							ougelh =			
10	CHP plants	Inecoa	inecru	ineprf	inegas	inenuc	ineren	(ougele) + (oughea)			Ktoe
11	Heat plants										
13	Petroleum refineries		inpcru	ourprf							Ktoe
12	Gas works										
14	Coal transformation										
15	Liquefaction	Otfcoa	otfcru	otfprf	otfgas	otfnuc	otfren				Ktoe
16	Other transformation										
12,18	Own use plus	Loocoa	loocru	looprf	loogas	loonuc	looren	looelh			Ktoe
	Distribution losses										Ktoe
19	*Total Final Consumption*	Ftecoa	ftecru	fteprf	ftegas	ftenuc	fteren	fteelh		fteorg = ftedes	Ktoe
Aditional*	***Energy use***	Feecoa	feecru	feeprf	feegas		feeren	feeelh			Ktoe
20	***Industry sector***	Indcoa	indcru	indprf	indgas		indren	indelh		feeind	Ktoe
21	Iron and steel (ias)										Ktoe
22,23	Chemical (che)										Ktoe
24	Non-ferrous metals (nfm)										Ktoe
25	Non-metalic minerals (nmp)										Ktoe
26	Transport equipment (tre)										Ktoe
27	Machinery (mac)										Ktoe
28	Mining and quarrying (maq)										Ktoe
29	Food and tobacco (fbt)										Ktoe
30	Paper, pulp and printing (pap)										Ktoe

Table 5-2. continued

IEA No.	Energy Balance Table	Coa	Crude Oil	Petroleum	Gas	Nuclear	Hydro/ Other	Electricity and Heat	Difference	Total	Unit
31	Wood and wood products (waf)										Ktoe
32	Construction (con)										Ktoe
33	Textile and leather (tal)										Ktoe
34	Non-specified Industry (oth)										Ktoe
35,36,37, 38,39,40, 41,42	***Transport sector*** Air, road, rail, Internal navigation and non-specified transport	Tracoa	tracru	traprf	treagas		traren	traelh		feetra	Ktoe
43	***Tertiary sector***	Procoa	procru	proprf	progas		proren	proelh		feepro	Ktoe
44	Agriculture										Ktoe
45,46,47	Public/commerce, residential										Ktoe
48,49, 50,51	***Non-energy use***	Fnecoa	fnecru	fneprf	fnegas					fnetot	Ktoe
52	Electricity generated	Geecoa		geeprf	geegas	geenuc	geeren			geetot	Gwh
55	Heat generated	Gehcoa		gehprf	gehgas	gehnuc	gehren	(gehele) + (gehhea)		gehtot	TJ
Aditional*	Electricity and heat generated	Gencoa		genprf	gengas	gennuc	genren	genelh	gendif	gentot	Ktoe

Source: COMPASS working paper Umehara (1999).

Note: GWh*0.0086 is equal to Mtoe. TJ*0.002388 is equal to Mtoe.

Table 5-3. Sector integration of input-output and energy balance (end-use level)

Total Final Energy Consumption	IEA ISIC (Rev. 2)	IEA ISIC (Rev. 3)	COMPASS IO Branches Description
Industry sector			
1 Iron and steel	371		12 Iron and steel
2 Chemical	351, 352, 354, 355, 356		7 Industrial chemicals 8 Drugs and medicine 9 Petroleum and coal products 10 Rubber and plastic products
3 Non-ferrous metals	372		13 Non-ferrous metals
4 Non-metallic minerals	36		11 Non-metallic mineral products
5 Transport equipment	384		19 Ship building and repairing 20 Other transport 21 Motor vehicles 22 Aircraft
6 Machinery	38		14 Metal products 15 Non-electrical machinery 16 Office and computing machinery 17 Electrical apparatus, 18 Radio, TV and communication equipment 23 Professional goods
7 Mining	23, 29		2 Mining and quarrying
8 Food processing, beverages and tobacco	31		3 Food, beverages and tobacco
9 Pulp paper and printing	34		6 Paper, paper products and printing
10 Wood and wood products	33		5 Wood products and furniture
11 Construction	50		26 Construction
12 Textiles and leather	32		4 Textiles, apparel and leather
13 Non-specified	****	25, 33, 36, 37	24 Other manufacturing
Transport sector Air Road Rail Internal navigation Non-specified transport	****	60, 61, 62	29 Transport and storage
Other sectors Agriculture Public/commerce Residential Non-specified other		01, 02, 05 41, 50, 51, 52, 55, 63, 64, 65, 66, 70, 71, 72, 73, 74, 75, 80, 85, 90, 91, 92, 93, 95 and 99	1 Agriculture, forestry and fishing 25 Electricity and gas and water (Water Only) 27 Whole sale and retail trade 28 Restaurants and hotels 30 Communication 31 Finance and insurance 32 Real estate and business services 33 Community, social and personal services 34 Producers of government services
Non-energy use	****	****	****

Sources: Compiled by the present author based on IEA (various year a,b); OECD (1995a); and United Nations (1990).

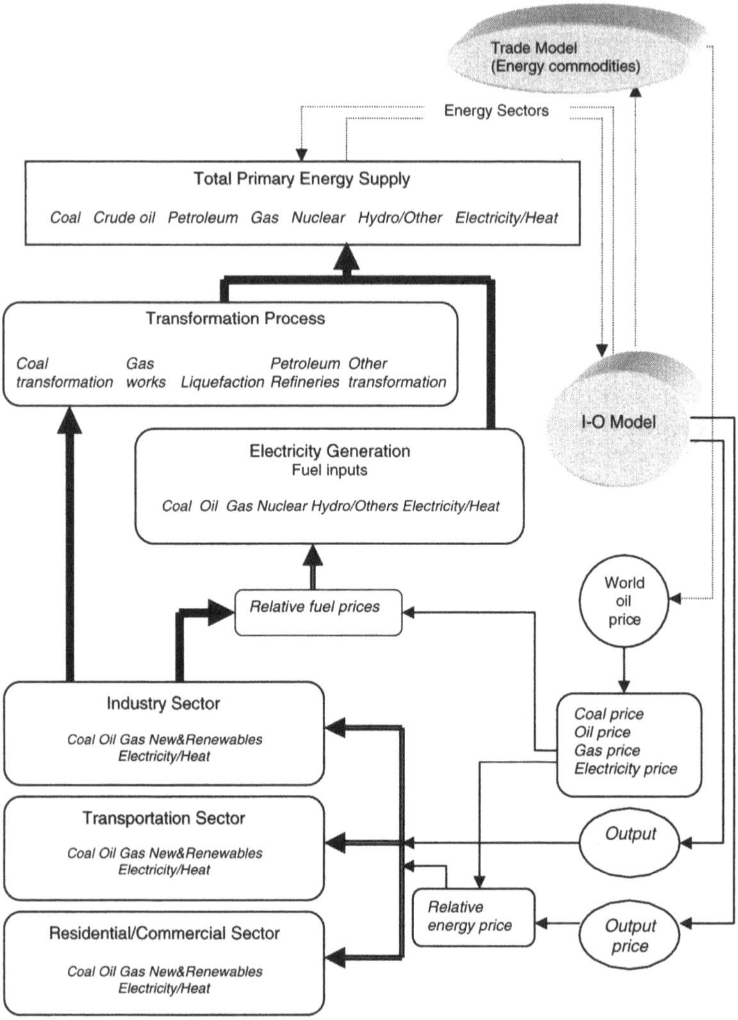

Figure 5-1. Structure of energy balance model.

that the aggregations are done by energy sources. A dotted line indicates the links with economic spheres. This is important for future development since this link enables us to endogenize the world oil price when other regions of the world, including oil-exporting countries, are incorporated in an energy balance model. When this is achieved this will be the third global closure within the COMPASS framework, including commodity trade, international financial transactions, and world energy flow.

In the model the relative energy price and *output* determine the energy intensity levels in three end-use sectors, which represent the technology defined here as the energy consumption per unit sector *output*. Energy consumption in end-use sectors is calculated as the estimated energy intensity multiplied by the *output*. Then final energy demand/supply is determined through transformation and electricity generation.

Shares of each fuel in the end-use sectors, as well as in electricity generation, are also dependent on the relative price of the fuels in each sector. Shares of nuclear and new and renewables (hydro/others) are exogenous. Electricity generation is also modelled with conversion efficiency depending on time trend. With the share of nuclear and new and renewables given by policy, the remaining shares of the rest of the fuel inputs are recalculated so the total share adds up to unity.

3.3. *Equations*

(a) Price equations

Price indexes for coal, oil, gas are used to calculate the weighted energy price index in three end-use sectors and utility sector.

$$pe_{ij(t)} = p_{j(t)} \times s_{ij(t)} \tag{1}$$

$$pe_{i(t)} = \sum_j pe_{ij(t)} \tag{2}$$

Pe = Energy price index (1990 = 100)
i = End-use sectors; Industry, Transportation, Residential
j = Energy carriers: Coal, Oil, Gas, Electricity

The weighted output price indexes are obtained in the input-output models (modelled in I-O).

Unit cost of production is determined by the unit price of domestic and imported products, labor, plus indirect tax minus subsidies minus depreciation per unit of output.

$$u_{(t)} = D'_{(t)} \times p_{(t)} + AM_{(t)} \times q_{(t)} + Z_{(t)} \times w_{(t)} + a + b \tag{3}$$

D = matrix of domestic input coefficients
AM = matrix of input coefficients of imported goods
Z = diagonal matrix of labour input coefficients
w = vector of wages
a = vector of indirect taxes minus subsidies per unit of output
b = vector of depreciation per unit of output
p = vector of domestic price
q = vector of import price.

The price of sector *i* is now dependent on the unit cost of sector *i*.

$$p_{i(t)} = p_i[u_{i(t)}] \tag{4}$$

(b) Energy intensity equation

$$\log \frac{E_{i(t)}}{X_{i(t)}} = a + \beta \log\left(\frac{Pe_{i(t)}}{P_{i(t)}}\right) + t \tag{5}$$

E_i = Energy consumption in sector *i* (unit: Ktoe)
X_{it} = Output in sector *i*
t = Time trend.

Now we have the estimated energy intensity, then multiply it by the *production output* is equal to the energy demand in sector *i*.

$$E_{i(t)} = X_{i(t)} \times \frac{E_{i(t)}}{X_{i(t)}} \tag{6}$$

(c) Energy substitution equation

$$s_{ij} = f\left[\log\left(s_{ij(t-1)}\right), \log\left(\frac{pe_{ij(t)}}{pe_{i(t)}}\right)\right] \tag{7}$$

(d) Final energy consumption equations

$$E_{c(t)} = \sum_i (s_{ic(t)} E_{i(t)}) \tag{8}$$

$$E_{o(t)} = \sum_i (s_{io(t)} E_{i(t)}) \tag{9}$$

$$E_{g(t)} = \sum_i (s_{ig(t)} E_{i(t)}) \tag{10}$$

$$E_{e(t)} = \sum_i (s_{ie(t)} E_{i(t)}) \tag{11}$$

(e) Transformation and conversion equations

$$e_{tc(t)} = r_c E_{c(t)} \tag{12}$$

$$e_{to(t)} = r_o E_{o(t)} \tag{13}$$

$$e_{tg(t)} = r_g E_{g(t)} \tag{14}$$

$$s_{uj_{(t)}} = \log\left(\frac{p_{uj}}{p_u}\right) \tag{15}$$

$$e_{uc(t)} = S_{uc}E_{e(t)} \quad (16)$$

$$e_{uo(t)} = S_{uo}E_{e(t)} \quad (17)$$

$$e_{ug(t)} = S_{ug}E_{e(t)} \quad (18)$$

(f) Fuel inputs equations for electricity generation

Fuel j input is determined by the relative price of the fuel over weighted price of fuels in the electricity sector.

(g) Total primary energy supply equation

$$F_c = e_{tc} + e_{uc} \quad (19)$$

$$F_o = e_{to} + e_{uo} \quad (20)$$

$$F_g = e_{tg} + e_{ug} \quad (21)$$

$$F_c = E_e \quad (22)$$

(h) Carbon emission equations

The carbon coefficient (MMTC/Mtoe) of coal is 1.08, oil is 0.84 and gas is 0.64.

$$C_{ic(t)} = c_c E_{ic(t)} \quad (23)$$

$$C_{io(t)} = c_o E_{io(t)} \quad (24)$$

$$C_{ig(t)} = c_g E_{ig(t)} \quad (25)$$

4. Energy balance table and the model

The logical flow of the energy model is as follows. Economic agents (utilities, industry, and consumers) react to changes in relative prices of energy to the price of the branch output (i.e., elasticity of energy intensity in terms of $pei_i = pene/pout$ i) and relative prices among primary energy sources (i.e. elasticity of energy substitution in terms of coal prices *vs*. average energy price, or crude oil price *vs*. average energy price, etc.) is explained in the model. It is not treated as exogenous assumptions. In a simulation experiment these can be made policy variables, and can be forcefully made to take particular values coming from a scenario. The same thing can be said about technology (input coefficients are endogenous or can be policy variables through R&D or technology transfers), trade pattern (country shares are endogenous depending on prices or R&D, but can be determined by policy through liberalization), industrial structure (final demand = lifestyle is endogenous but can be changed by environmental policy, and output of each industrial branch can be endogenous or imposed for the sake of a policy experiment), and cost structure (production cost is endogenously explained, but, with the introduction of emission trade, new cost components can be added).

The emission of CO_2 can be calculated based on technically given parameters for different fossil fuels (APERC, 1998).

Coal: 1.08 tC/toe
Gas: 0.64 tC/toe
Oil: 0.84 tC/toe

The price for tradable permits is obtained by the market-clearing process. Define the demand surplus of world CO_2 emission rights. If the demand for the permit is bigger than the supply the price of the permit is raised by a very small fraction. The demand will be reduced by a small margin in the next iteration. The iterative process determines the clearing of the international energy market. In the input-output framework carbon tax can be treated as indirect tax which is represented in the value-added sphere.

In summary, the logical flow of the energy block is from final consumption through conversion to primary energy demand. Econometrically estimated variables are energy consumption by different industrial sectors and the selection among different energy carriers. Technological parameters are the conversion efficiency from primary energy to secondary. The economic block (see Chapter 3) generates prices of output in individual branches (*p* or *pout*) whereas energy prices in respective branches are derived from exogenously given prices for primary energy carriers and their shares in energy use in the branch in question.

Part III

Economy–energy–environment, 2010

6
Policy agenda

YUMIKO UMEHARA

1. Kyoto and beyond

The global climate change issue is gaining ever-growing attention as the seriousness of the greenhouse effect becomes widely recognized. This poses a serious challenge to the global community. For one, global warming itself is a serious threat, in addition posing a shift of our mindset from economic growth to sustainability of the way of life and the resource base. Second, for lack of a global policy agent and a global market mechanism encompassing environmental spheres, we are left with a situation in which we are faced with a threat on a global scale. We are not equipped with an effective mechanism for decision making and policy implementation except for national governments that are, at best, responsible for domestic constituencies of the current generation.

Since the beginning of the industrialization in the late 19th to early 20th century, a series of new innovative technologies has contributed to the high productivity growth. The appearance of new technologies has eliminated environmental problems while creating new ones. However, activities with heavy reliance on the burning of fossil fuels outpaced sequestration of GHGs (greenhouse gases) by the eco-system. This caused the dramatic rise in CO_2 (carbon dioxide) emissions. The concentration of CO_2 reached 340 ppmv (1/million in volume) in 1980. This is expected to double to 560 ppmv possibly as early as 2030, and could lead to a rise in global mean temperatures unprecedented in human history. The industrialized countries have disproportionately increased the accumulation of CO_2. Nevertheless, LDCs (less developed countries) are vital in contemplating any long-term solution. The most commonly used reference scenario of CO_2 emissions trajectory, IPCC's IS92a, indicates that LDCs' total emission level will reach the level of industrialized nations by the year 2037. Therefore, the methods by which the global society addresses the climate change issue involve issues such as the division of responsibilities and differing capabilities to respond. Climate change is unique in the human history of pollution in that it is caused by all human activities, and the consequences are uniform over the globe as a whole.

The Framework Convention to Climate Change (UNFCCC) was signed in 1992 at the United Nations Conference on Environment and Development

K. Uno (ed.), Economy–Energy–Environment Simulation, 101–110.

(UNCED). The objective of the Convention is:

> to achieve stabilization of greenhouse gas concentrations in the atmosphere at a level that would prevent dangerous interference induced by human activities in a time-frame sufficient for ecosystems to adopt to climate change, and secure food production so the economic development proceed in a sustainable manner (Article 2).

It requires the parties to take precautionary measures to anticipate, prevent, or minimize the causes of climate change and mitigate its adverse outcomes cost-effectively. The parties should also take into account the benefit of present and future generations, on the basis of equity and in accordance with their common but differentiated responsibilities.

After the Earth's Summit conference in 1992, a series of COP (Conferences of the Parties) took place to materialize the UNFCCC. The Kyoto Protocol, adapted in 1997 at COP-3, was the first stage of the international framework that enforces the emission limitations on Annex-I parties – most of the industrialized countries (UNFCC, 1997). Table 6-1 shows the target reduction from the 1990 emission levels for the major Annex I parties. At the time of writing this chapter, the Protocol has been signed by 84 countries and is subject to ratification. However, only a few of the small island countries have so far ratified. The negotiations are likely to become more difficult, as discussions of the implementation of the Protocol are sensitive to a number of complex interest groups.

When environmental ministers from G8 countries met in Otsu City in Japan, in April 2000, one of the events leading to the Summit in Okinawa in July 2000, the most heavily debated point was whether to set a concrete date for putting the Kyoto Protocol into effect. The Protocol, which was adopted at the Third Conference of the Parties (COP-3) in Kyoto in December 1997, obliges industrialized countries to reduce greenhouse gas emissions in order to mitigate global warming. The United States and Canada could not agree to include a specific deadline to ratify the Protocol because of domestic legislative procedures. All European Union countries, Japan, and Russia agreed that the Kyoto

Table 6-1. Kyoto targets and 1996 CO_2 emissions, G7 countries

	Target in 2010 (1990 = 100)	1996 Emission (1990 = 100)
Canada	94	110
USA	93	109
Japan	94	111
France	100	101
Germany	79	93
Italy	92	110
UK	87.5	100

Protocol should be ratified as soon as possible. The joint communiqué stated that 'We confirm our commitment to ensure that results achieved at COP-6 help promote the ratification and entry into force of the Kyoto Protocol as soon as possible. For most countries, this means no later than 2002.' The Sixth Conference of the Parties (COP-6) to the UN Framework Convention on Climate Change was held in the Hague, Netherlands, in November 2000, but had to re-convene in Bonn in July 2001 in an effort to come to an agreement on the details of implementing the Protocol in order to resolve major differences among participants. The US has shown considerable reluctance in following up the consensus achieved in Kyoto. Japan demonstrated unwillingness, recognizing that meeting the target would be extremely difficult given the current excess emissions by the target date, in addition to the fact that its energy efficiency is already highest in the world, leaving less room for future reduction. The EU as a group of countries strongly committed to the implementation phase, admitted carbon sequestration sinks for Japan and Canada, and contributed to keep the Protocol alive. The focus of COP-7 that met in Marrakesh, Morocco, in November 2001 was whether the participating countries could agree on the rules for compliance. Earlier in the year, a top official of the US said that the treaty would impose an excessive burden on US industries and that it is inappropriate to exempt developing countries, including China and India, from the mandatory emissions targets. Since then, the United States started to contemplate its own way of approaching the generally-agreed need to reduce greenhouse gas emissions, but not through a single international agreement. Japan opted to treat the emission targets as 'politically' binding, but not 'legally' binding, before the treaty is formally ratified next year. In the end, it has decided to ratify the Protocol even without waiting for US participation. Russia also agreed after obtaining a concession that doubles the credits it could claim for carbon absorption of its forests and agricultural land. The Rio Plus 10 meeting is scheduled to be held in 2002, marking the 10th anniversary of the 1992 Earth Summit in Rio De Janeiro and laying the groundwork for future environmental issues.

2. Policy issues

The innovative element of the Protocol is to introduce flexibility mechanisms such as 'Joint Implementation (Article 6)', 'Clean Development Mechanism (Article 12)', and 'Emissions Trading (Article 17)'. Such international cooperative modes of emission abatement contribute greatly to reducing the economic burden of the countries with high abatement costs under the set of specific targets. However, these mechanism are supplemental and major reductions of GHGs will have to be achieved domestically by each party (Figure 6-1). Therefore, the selection of appropriate policy measures is currently being sought by governments. Despite the fact that the measures should be based on our understanding on the greenhouse effect, and its potential consequences, economic considerations should play an important role in choosing the means of achieving the target. For instance, the

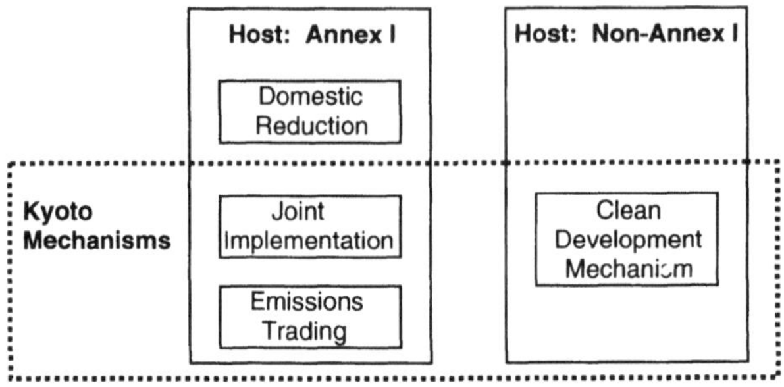

Figure 6-1. Kyoto mechanisms.
Source: Sijm *et al.*, 2000.

Best Practice already proposed by the UK government includes the Climate Change Levy, which will be imposed on industrial sectors, but not on transport of the household sector with consideration to social equity. Within industries, energy-intensive sectors will be given lower rates of levy so that they will not lose competitiveness. How would this affect economy and environment at the sectoral level? What would be the consequence if Japan, for instance, took the same step towards levying on industries? Are there any low-cost mitigation opportunities? The interaction between technology, market mechanism, and government policy needs to be studied in more detail. The energy balance model presented here is built in response to such needs coming both from the policy designing process and international policy coordination.

The specified limits on GHG emissions in the Kyoto Protocol would reduce the Annex I countries' average annual emissions between 2008 and 2012 to about 5% below the 1990 level. The basket of the six key GHGs are included – CO_2, CH_4, N_2O, halocarbons, perfluorocarbons, and sulphur hexafluoride, which are not controlled by the Montreal Protocol. The quantified reduction targets for Annex I countries and the required reductions from the reference case (IOE99) are shown in Table 6-2. All industrialized countries except Australia are required to reduce emissions, whereas economies in transition, such as Eastern Europe and the former Soviet Union, are projected to have excess emissions over target by 374 MMTC (million metric tons of carbon). The industrialized countries are allowed to enter the 'emission reduction units' among themselves, and of 'certified emission reduction' under the clean development mechanism. Such flexible mechanisms contribute greatly to reducing the global net cost of mitigating GHGs. However, since the differentiated targets are likely to result in heavy economic burdens if the mechanisms do not operate effectively, many studies focused on the cost-effectiveness of the different trading regimes.

Who will be the buyers and who will be the sellers in international emission trading? A variety of model results on the cost of Kyoto targets are presented and studied in the work by the EMF (Energy Modeling Forum) (Weyant, 1999). The costs differ in great extent across the variety of models presented.

Table 6-2. Projected 2010 emission and required reduction of Annex I parties

Region and Country	1990 Emission (MMTC)	2010 Reference Case (MMTC)	2010 Kyoto Target MMTC)	Change from Reference Case, 2010 (%)
Annex I Industrialized				
United States	1384	1790	1252	−30
Canada	126	162	118	−27
Western Europe	936	1021	862	−16
Japan	274	322	258	−20
Australia	90	133	97	−14
Total Annex I Industrialized	2772	3408	2586	−24
Transitional (EE/FSU)				
Former Soviet Union	991	666	990	49
Eastern Europe	299	270	320	18
Total EE/FSU	1290	935	1309	40
Total	4062	4344	3895	−10

Source: Energy Information Administration, US Department of Energy (1999).

Table 6-3. Model types

Economy Model	Energy/Carbon Model: Energy Supplies & Demands by Sector	Energy Technology Detail	Carbon Coefficients
Aggregate Production/Cost function		CETA MERGE3 GRAPE	FUND RICE
Multi-sector General Equilibrium	MIT-EPPA WorldScan G-Cubed	ABARE-GTEM AIM MS-MRT SGM	
Multi-sector Macroeconometric	Oxford		

Source: Weyant *et al.* (1999).

Comparing the results in all 13 models, Japan is faced with the highest carbon price, indicating the highest marginal abatement cost. The average carbon tax required to meet the target for Japan under a No Trading (NT) regime is around \$400/tc (1990 US\$/metric ton of carbon). The average for the USA and CANZ (Canada/Australia/New Zealand) lies between \$150 and \$200, and the EU average ranges from \$200 to \$300. The differences in carbon prices reflect pre-existing energy prices, share of coal, amount of energy produced domestically, and the required reduction in emissions given the baseline emission growth (Bernstein *et al.*, 1999). Russia, where actual emission is less than the target level, will be the only seller. All of the model results concluded that there is a great cost reduction under Annex I Trading (AI) than under NT. A rough overview shows the cost drop to 1/4 for Japan, 1/3 for EU, and 1/2 for USA and CANZ with the equalized carbon price for AI parties. Under Global Trading

(GT), the equalized global carbon price may be less than $50, in which case the costs can be halved compared to the AI case. This is due to additional sales of emission right by non-Annex I countries. This kind of model exercise gives excellent insights into policy making in broader perspective.

3. The players

Annex I countries refers to the UNFCC Climate Convention list and include all OECD members and countries with economies in transition. All other countries are referred to as non-Annex I countries. (Annex B countries, not touched upon here, refer to the Kyoto Protocol lists and include those developed countries that have accepted the target for their GHG emissions and include OECD members, Central and Eastern Europe, and Russia.)

Faced with the increased energy cost in meeting the Kyoto targets, energy-intensive industries may shift to non-Annex I countries where the energy costs are lower, in which case the reduction of GHGs can be readily offset. It is pointed out that GHG emissions in non-Annex I countries might increase in response to mitigating efforts in the Annex I countries, representing 'carbon leakage'. On the part of the industrialized countries, this means significant production reduction and employment loss. Full global trading eliminates leakage because all regions then face emissions caps. In this context participation by the developing countries to emissions trading is another critical issue to be addressed. This needs to be discussed in relation to the equity issue. Therefore the economic trends in countries that did not sign the Protocol, represented by South-East Asia and China in our analysis, are an important policy issue.

In the former Soviet Union and some East European countries, emissions in the target year are expected to be lower than in the 1990 base. The term 'hot air' points to the difference between the assigned amount of GHG emissions and the actual level in the 2008-2012 target year. This represents the emissions reduction caused by the decline in the economy, but not by energy efficiency improvements. If Russians can sell the difference to the international emission permit market they can realize a windfall gain. If this happens, Annex I countries would not be obliged to reduce GHG emissions domestically. The economic performance of Russia in the future is to be carefully looked at in this context.

3.1. *The scenarios for industrialized countries*

Compared to the 1990 level the United States is required to reduce emissions by 7% in 2010. Of this, the United States expects to achieve a 4% cut through domestic measures, sink, and reduction of GHGs other than CO_2, and the remaining 3% through emission trading, JI (joint implementation), and CDM (clean development mechanism). It is estimated that actual emissions would far exceed the target, necessitating a reduction of 30%. In order to secure low-cost measures the United States requires the participation of developing countries in the scheme.

The European Union aims at reducing the emission of global warming gas by 8% from the 1990 level. In reality, however, the emission in 2010 will be 7% larger than the 1990 figure. In other words, a 15% reduction will be required around the year 2010 in order to honour the commitment made at Kyoto. Energy requirement per unit of economic activity would be reduced at an annual rate of 1.5% or more till the year 2020, reflecting the structural change of the market and technological progress, whereas on the supply side, nuclear plants will be closed successively. Reflecting divergent reduction responsibilities within Europe, the EU's response will be based on national policies within individual members, and common measures are regarded only as supplementary. The EU will initiate consultation to individual members regarding emission trading, and expects to come up with a framework for emission trading within the European Union by 2005. Already, in some countries, a domestic scheme is being drawn up, despite the fact that this is the first attempt at a market solution to emission reduction.

Japan is committed to a 6% reduction. Actually, it is projected that CO_2 emission will be considerably higher than the target, reflecting a slower pace of energy saving under lower energy prices. If the economy recovers from the current slump an 18% reduction will be required from the projected level in 2010. Originally, it was envisaged that 1.8% of the 6% reduction would be achieved through emission trading, and the remaining portion by utilizing CO_2 sink and voluntary action plans drawn up by industries such as automobile and household appliances. Further effort will be required in order to meet the commitment, such as early implementation of emission trading, and expansion of JI with developing countries, among others.

In models assuming identical AEEI (autonomous energy efficiency improvement), or ESUB (substitution between energy and other sectors) for that matter, across the countries, only comparable differences can be found deductively in simple economic assumptions such as GDP or population. On the contrary, with the estimated price/output elasticity of energy intensity change, together with the inter-energy substitution, policy simulation can be done on the ground of observed underlying differences. Therefore, the approach at the disaggregated sector level based on empirically observed parameters is inductive, while not losing sight of an international overview.

3.2. South-East Asia

Asian economies exhibited high economic growth. This was attributable to favourable macroeconomic performance leading to a cycle with high saving leading to high investment ratio and high output growth, accompanied by high productivity and quality improvement, leading to successful penetration into the world market. Favourable international economic conditions also contributed. At an earlier stage, foreign economic assistance helped build up a social infrastructure (World Bank, 1993). More recently, foreign direct investment worked as the vehicle for financial resources as well as technological transfer.

In the mid-1990s, inflow of portfolio investment followed the economic success of the region, only to show that international exposure in this aspect could be reversed. The Asian economic crisis followed, shifting an overly optimistic outlook into excessively pessimistic views. As the economies entere the new millennium the situation looks more stable. 'Recovery is taking hold in much of Asia, and the upswing trend observed to date is likely to continue in the second half of 1999 through the year 2000. The acute phase of the Asian crisis is now over. However, many challenges posed by the Asian crisis remain' (APEC, 1999). The report points to two opposing views on the causes of the crisis. 'The panic camp' holds that the crisis was a self-fulfilling panic caused by the inherent instability of international financial markets. 'The home-grown camp' holds that the crisis was home-grown in the sense that crisis-affected economies had their own problems. Neither camp would contend that its view applies in a pure form, and there are views in between. A broad common understanding that emerged contends that 'financial instability, particularly, volatility of short-term capital flows play an important role. Economies had such problems as financial sector weaknesses and inflexible exchange policy that left them vulnerable to reversals of capital flows.' A recent report testifies on these accounts. 'The Asian crisis can be characterized as a capital account crisis, the origin of which was large inflows of private capital relative to the underlying current account deficit and of a largely short-term nature, followed by a sudden and massive reversal of capital flows. The magnitude of the swing in capital movement from massive inflows to sudden reversal accounted for about 11 percent of the DGP of crisis-hit Asian economies. In this capital account crisis, it was large swings in capital account balances that first caused the large current account deficits in 1995-1996, then the shift suddenly to large current account surpluses in 1998-1999, not the other way around as in previous conventional crisis.' (Asian Policy Forum, 2000; see also Yoshitomi and Shirai, 2000.)

The fact remains that the economies in South-East Asia represent the fastest-growing part of the world economy, implying, in turn, that energy demand in this region would continue to grow. One potential long-term constraint to growth, or indeed a cause of short-term disruption, would be insecurity in energy supply as reflected in large swings in international energy markets and capping of GHG emissions (including CO_2 emission, which is inseparably linked to energy consumption) imposed by the international community. We take up the economic structure and energy balance of Thailand, Malaysia, Singapore, the Philippines, Indonesia, Taiwan (Chinese Taipei), and Korea.

3.3. *China*

Currently, global warming gas emission is largely attributable to industrialized countries; within several decades, China, India, and other developing countries will emerge as the largest contributors. It is expected that Chinese growth will be somewhat lowered, and India's somewhat heightened, and both will continue to grow at an annual rate of 5% or so in the years to come. Chinese energy supply is heavily dependent on coal, and demand is largely for industrial use, but this pattern will quickly change. Energy supply will shift to oil and larger demand will originate in transportation and household sectors. India will become the world's most populous nation in the first half of the 21st century. Coupled with an increase in income, India will witness a rapid shift to commercial energy. Both China and India are cautious not to turn emission reduction to growth constraint. This points to the room for policy coordination such as technology transfer between industrialized countries and the big late-comers.

Literature suggests that past economic growth rates record in China had been biased upward, not because of intentional manipulation but because of statistical peculiarities reflecting price structure under central planning (Wu, 1998; Maddison, 1998a, State Statistical Bureau and Hitotsubashi University, 1997; see also OECD, 1996). The difference in view will also affect the trajectory of this enormous economy. A particularly important factor in deciding the future growth of the Chinese economy is the availability of financial resources. As Chinese official views hold, there would not be a shortage of final demand in terms of capital formation, household consumption, or government spending. Rather, the constraint on continued expansion is likely to be posed by the lack of investment funds. The international flow of financial resources would be crucial.

The chapter on China (Chapter 9) discusses some of the alternative estimates of the Chinese economy in the past, present, and future, and prepares a basis for simulation experiments at the global level that takes into account alternative paths of Chinese development. The main author of the chapter on China is an experienced multi-sector econometric model-builder and in a position to provide long-term structural analysis to the Chinese government. As for the macroeconomic performance, the author adheres to the official position that the economy will continue to grow at an annual rate of 10% in the foreseeable future. The structural change of the Chinese economy will be discussed and simulation results presented.

3.4. The case of Russia

The former Soviet Union and Eastern Europe are included in Annex I but are not subject to reduction obligations. Due to economic slumps, and a shift from coal to cleaner natural gas, these countries have increased the sellable emission right. When emission trading is conducted with Annex I they will obtain $45 billion income in 2010 according to one calculation. In drawing up an economic recovery plan, therefore, revenues from emission trading will be a positive factor to be considered in addition to crude oil exports.

Russia shifted to SNA in the early 1990s, at the time of the economic reform (Kuboniwa, 1997). The analysis contained in Chapter 10 employs input-output tables and macroeconomic data from the post-reform period. Russia is also a major producer of crude oil and gas, the price of which is a major factor determining the Russian balance of payments. A low international energy price has had an adverse effect on Russian economic reconstruction. Moreover, the economic downturn resulted in less than sufficient investment in the energy sector in Russia.

In this context the recovery scenario is examined. Institutional characteristics of the Russian economy have improved drastically through the measures taken during the past decade (Kuboniwa and Gavrilenkov, 1997; for the past growth record, see Maddison, 1995, 1998b). The situation in 1997 showed some positive development in terms of the macroeconomic aggregate. The 1998 financial crisis inflicted a severe blow to the Russian economy, resulting in a decline in GDP. Due to higher oil and gas prices, and promotion of exports, partly contributed by import-substitution in manufacturing, the economic growth rate was 3.2% in 1999, followed by more than 8% growth in 2000.

Looking into the future, there can be optimistic views that anticipate acceleration. This version of the prospect is based on the smooth progress of structural reform and the promotion of business activities and capital formation. The devaluation of the ruble in financial crisis gave an impetus for import-competing manufacturing industries. In addition, one may envisage that the sudden drop in economic activities in the wake of the financial crisis in 1998, on top of the continued decline in the 1990s, provides room for quick recovery. The inflow of direct foreign investment and advanced technology is an additional factor that can be counted on in the medium to long term. In contrast, pessimistic views stem from the prospect of lagging structural reforms. Distorted economic structure and low international competitiveness, the lack of an effective legal infrastructure, underdeveloped financial institutions, bureaucratic and corrupt administration, among others, are often listed as the factors adversely affecting the course of economic reform. This version of the view also anticipates continued inflation due to lack of hard budget constraints and explicit/implicit subsidies to enterprises.

It is noted that the dividing line between the optimistic views and pessimistic ones hinges on the prospect of economic reforms, which is a domestic factor, and not on an external factor such as the drop in international oil and gas prices or an economic crisis due to more financial turbulence. In other words, both optimists and pessimists assume that current improvement in the international economic environment represented by higher energy prices is expected to continue. The fact remains that the future of the economic performance of Russia is contingent upon the international economic environment.

Against this backdrop in Chapter 10, two economists from Russia provide an assessment of the reform process in the 1990s, describe the economic records at the turn of the century to 2004, and describe the official long-term social and economic program until 2010.

7
Price-induced energy intensity and inter-energy substitution of G7 countries

YUMIKO UMEHARA

1. Price-induced energy intensity

In this chapter, first the estimation results of energy intensity and energy substitution of G7 countries are described. Secondly, the assumptions embedded into energy balance simulation are explained. Then the chapter shows the results of the future energy balance table and CO_2 emission simulation up to 2010.

Let us look at the energy intensity of the Industry, Transportation and Tertiary sectors shown in Tables 7-1a, b, and c. The general trend in all sectors shows that the intensities have been declining during the period from 1980 to 1995. A slight increase observed from 1980 to 1985 was mostly due to the relatively low oil price after oil shocks in the 1970s. Amongst the seven countries, Canada and the USA show relatively high energy intensity, indicating a lower energy efficiency level compared to other countries. The reasons for this must be multiple; however, a major part of the inefficient use of energy could be attributable to abundant domestic production of energy, and hence the lower price of energy. Intensities in Japan are the lowest throughout the observation here, followed by European countries.

We now compare the observed energy intensity levels and trends (Tables 7-1a, b, and c) with estimated price elasticities (Tables 7-2a, b, and c) for respective sectors. In Table 7-1a, the Industry sector from 1980 to 1995, Germany and the USA have lowered the intensity by 0.5% per year and Canada has reduced by 0.4% per year. Japan has decreased 0.3% per year, and the rest show negligible or no improvement.

Let us look in parallel at the estimated price and output elasticity of intensity shown in Table 7-2a. An output elasticity of -0.5 indicates that the rise in production output by 1% will result in a 0.5% decline in energy intensity. During a course of economic development, such an energy efficiency improvement is a general incidence, as already discussed in the previous chapter. Our results show that the parameters of USA, France, and Germany are bigger than -1. The rest are between -0.5 and -1 except Japan which is -0.43. These figures do not contradict the conventional assumption of AEEI. However, the price effects on intensity are not taken into account in such an assumption. In

K. Uno (ed.), Economy–Energy–Environment Simulation, 111–134.

Table 7-1a. Energy intensity change of G7 countries: Industry sector (Mtoe/billion US$ of 1990)

	1980	1985	1990	1995
Canada	0.26	0.27	0.22	0.19
France	0.08	0.14	0.07	0.06
Germany	0.13	0.19	0.08	0.05
Italy	0.07	0.11	0.06	0.07
Japan	0.09	0.09	0.05	0.04
UK	0.08	0.12	0.08	0.07
USA	0.20	0.17	0.14	0.12

Table 7-1b. Energy intensity change of G7 countries: Transportation sector (Mtoe/billion US$ of 1990)

	1980	1985	1990	1995
Canada	1.73	1.73	1.43	1.57
France	0.67	1.09	0.53	0.48
Germany	1.73	2.25	0.91	0.54
Italy	0.48	0.79	0.49	0.54
Japan	0.74	0.81	0.38	0.25
UK	0.59	1.05	0.44	0.41
USA	1.75	1.31	0.97	0.88

Table 7-1c. Energy intensity change of G7 countries: Residential/Commercial sector (Mtoe/billion US$ of 1990)

	1980	1985	1990	1995
Canada	0.15	0.15	0.12	0.13
France	0.06	0.10	0.05	0.05
Germany	0.13	0.20	0.08	0.05
Italy	0.03	0.07	0.04	0.05
Japan	0.05	0.05	0.03	0.03
UK	0.08	0.12	0.06	0.06
USA	0.09	0.08	0.07	0.08

this analysis, most importantly, the price responsiveness of energy intensity is estimated separately.

We were able to observe significant results except for the USA. A price elasticity of −0.1 indicates that if the energy price rises by 1% with the price of other inputs staying constant, then the intensity will fall by 0.1%. The larger the parameter is, efficiency could be improved with relatively lower costs. The results interestingly indicate that the elasticity is high in Canada of −0.34, reflecting the low cost improvement possibility. Japan showed the lowest elasticity of −0.1, and the highest elasticity observed was in Italy, −0.53, followed by −0.47 in Germany.

Table 7-2a. Estimated parameters of energy intensity equations: Industry sector (figures in parentheses are *t*-values)

	Price Elasticity	Output Elasticity	Adjusted RSQ	DW
Canada	−0.34	−0.54	0.86	0.84
	(−3.20)	(−6.45)		
France	−0.22	−1.40	0.82	1.07
	(−1.70)	(−7.00)		
Germany	−0.47	−1.55	0.97	1.30
	(−2.26)	(−20.68)		
Italy	−0.56	−0.74	0.95	0.84
	(−3.58)	(−10.64)		
Japan	−0.10	−0.43	0.83	1.69
	(−2.25)	(−6.45)		
UK*	−0.15	−0.62	0.98	2.46
	(−2.1)	(−8.0)		
USA	−0.44	−1.70	0.85	1.33
	(−0.77)	(−5.17)		

* U.K is the results include the time trend of which parameter is −1.19 (−5.6).

Table 7-2b. Estimated parameters of energy intensity equations: Transportation sector (figures in parentheses are *t*-values)

	Price Elasticity	Output Elasticity	Adjusted RSQ	DW
Canada	−0.47	−0.53	0.88	1.71
	(−5.56)	(−6.59)		
France	−0.18	−0.13	0.82	2.18
	(−7.47)	(−4.92)		
Germany	−0.04	−0.42	0.94	1.63
	(−1.15)	(−13.12)		
Italy*	−0.09	−0.79	0.79	1.05
	(−1.22)	(−3.55)		
Japan	−0.09	−0.25	0.65	0.72
	(−1.95)	(−3.98)		
UK*	−0.26	−0.80	0.99	1.69
	(−7.98)	(−26.71)		
USA	−0.01	−0.58	0.97	1.04
	(−0.25)	(−13.60)		

* Italy, U.K, includes time trend of which parameter is 2.32 (3.91), 1.76 (10.42) respectively.

In the Transportation sector intensities are very high in Canada and the USA and in Germany, as shown in Table 7-1b. It should be remembered that the intensity in this sector does not represent energy consumption per physical service such as passengers/kilometre. Output in this sector is the monetary evaluated transportation services. Thus, a country with relatively low provision for public transportation service may end up having a high intensity level. USA

Table 7-2c. Estimated parameters of energy intensity equations: Residential/Commercial sector (figures in parentheses are *t*-values)

	Price Elasticity	Output Elasticity	Adjusted RSQ	DW
Canada*		−1.17 (−2.21)	0.83	1.28
France	−0.18 (−7.47)	−0.13 (−4.92)	0.82	2.18
Germany	−0.43 (−2.25)	−1.19 (−16.60)	0.97	1.33
Italy*		−1.00 (−2.14)	0.79	1.40
Japan*		−0.59 (−2.09)	0.94	2.23
UK*		−0.84 (−27.68)	0.98	1.32
USA	−0.21 (−1.70)	−0.89 (−6.53)	0.92	2.04

* Canada includes time trend of which parameter is 2.05 (1.20), Italy 1.15 (1.08), Japan 2.54 (5.03). UK does not include time trend.

intensity in this sector has declined by 0.5% per year, the same change as in the Industry sector, whereas Canada still shows the level achieving only 0.1% per year during the period from 1980 to 1995. EU countries' reductions in annual intensity are at about the same level as in 1995, an average of 0.5% per year. Again in Japan, the level is the lowest, and Japan has reduced the intensity level by 0.3% per year. In estimating the elasticity in the Transportation sector we tested equations with time trend, the results of which showed an overall trend of intensity rise in most of the countries without price consideration. Table 7-2b shows that the price elasticity in this sector is lower than in the Industry sector in France and Germany, but higher in Canada. The elasticity is at about the same level as that observed in the Industry sector for Japan. The result for the USA was not significant. The highest elasticity was observed in Canada, −0.47, followed by the UK with −0.26. France was −0.18, and Italy and Japan showed −0.09. The lowest elasticity was observed in Germany.

Now, let us look at Table 7-1c. In the Tertiary sector energy intensity is lowest in Japan, and again highest in Canada in 1995. Second highest is the USA, followed by the UK, and France, Germany, and Italy have the same intensity level. Germany has lowered intensity by 0.5% per year, and the rest of the countries by 1% per year, except Italy which showed 1% per year increase. The fall in intensity level was significant from 1985 to 1990 in France, the UK, and Germany. However, only Germany continued to reduce the level from 1990 to 1995. In this sector, elasticity of price was only observed in France, −0.18; Germany, −0.43 and the USA, −0.21 (Table 7-2c).

2. Price-induced inter-energy substitution

Tables 7-3a, b, and c show the estimated price elasticity of energy share. The figures show the point change in share; e.g. if the price of coal rises relative to the weighted price of energy the share would fall by the point shown in the 'energy total' of the group 'coal'. For example, in Table 7-3a, 0.06 in the first row for Japan means that if the share of coal was 30% in this sector, the relative price rise by 1% will push down the share to 29.4%. We also estimated the cross-energy price elasticity. For instance, coal price rise relative to oil price of Japan in the Industry sector will cut the share of coal by 0.04 point.

In the Industry sector, the share of coal in 1995 is: Canada 5%, France 10%, Germany 12%, Italy 6%, UK 15%, Japan 12%, USA 7%. Since coal is the most carbon-intensive energy the shift away from coal to less intensive energy

Table 7-3a. Estimated parameter of energy share equation: Industry sector: price elasticity of inter-energy substitution (percentage change in share)

	Canada	USA	Japan	France	Germany	Italy	UK
Coal							
Energy total	*	*	−0.06	−0.04	−0.10	*	*
Oil	*	*	−0.04	−0.03	−0.04	*	*
Gas	−0.04	*	−0.08	−0.03	−0.05	*	*
Electricity	*	*	−0.09	−0.09	−0.44	*	*
Oil							
Energy total	−0.06	*	−0.53	*	−0.29	−0.58	*
Gas	*	*	*	−0.30	*	−0.36	*
Electricity	*	*	−0.16	*	−0.03	*	*
Gas							
Energy total	−0.07	−0.26	*	−0.08	−0.03	*	*
Electricity	−0.03	−0.05	*	−0.04	*	*	*

* indicates that no significant results were observed.

Table 7-3b. Estimated parameter of energy share equation: Transportation sector: price elasticity of inter-energy substitution (percentage change in share)

	Canada	USA	Japan	France	Germany	Italy	UK
Coal							
Energy total	**	**	**	**	**	**	**
Oil	**	**	**	**	**	**	**
Gas	**	**	**	**	**	**	**
Electricity	**	**	**	**	**	**	**
Oil							
Energy total	*	*	−0.09	−0.15	−0.18	−0.16	*
Gas	−0.02	*	−0.003	*	−0.005	*	−0.008
Electricity	*	*	−0.002	*	−0.004	−0.005	*
Gas							
Energy total	−0.03	−0.01	**	**	**	−0.002	**
Electricity	−0.04	−0.01	**	**	**		**

** indicates that the fuel is not used in this sector.

Table 7-3c. Estimated parameter of energy share equation: Residential/Commercial sector: price elasticity of inter-energy substitution (percentage change in share)

	Canada	USA	Japan	France	Germany	Italy	UK
Coal to							
Energy total	*	*	−0.009	−0.01	*	*	*
Oil	*	*	−0.007	−0.01	*	*	*
Gas	*	*	*	*	*	*	*
Electricity	*	*	*	*	*	*	*
Oil to							
Energy total	−0.05	−0.02	−0.10	*	−0.25	*	*
Gas	*	−0.02	*	−0.05	*	*	*
Electricity	*	*	−0.15	*	*	*	−0.05
Gas to							
Energy total	*	−0.11	*	*	−0.08	−0.14	*
Electricity	*	−0.05	*	*	−0.02	−0.09	*

contributes greatly to CO_2 emission reduction. Germany had the biggest share of coal of 23% in 1990, and decreased by 2% per year to 1995. The USA has also decreased by 1% per year in the same period. Table 7-3a shows that in Japan, France, and Germany, the price elasticity of coal share was observed to a great extent. The price of coal relative to electricity was especially large in Germany, at 0.44. For the USA, Italy, and the UK, no significant results were observed. The share of oil is: France 33%, Canada 27%, Germany 31%, Italy 28%, Japan 50%, UK 33%, USA, 26% in 1995. The share is by far the highest in Japan. The elasticity of oil to energy price in this sector is high. Japan, Germany, and Italy have −0.53, −0.29, and −0.58, respectively. The share of gas in 1995 is: Canada and France 28%, Germany 29%, Italy 40%, Japan 5%, UK 32% and USA 35%. Italy has the largest share, followed by the USA and UK and other EU countries. Use of gas in Japan is only 5% of total energy in the Industry sector. The elasticity of gas price to energy price was large in the USA at −0.26. The biggest and smallest user of gas in the Industry sector is Italy, and Japan did not show significant results. Canada and USA showed −0.07 and −0.08, respectively.

In the Transportation sector the elasticity is generally smaller than in the Industry sector. There is no coal used in this sector for all seven countries. The oil is almost all the energy used in this sector, more than 90%. In the USA and UK, almost 99% is oil. The elasticity of oil to energy is large in Germany at −0.18, and France −0.15, Italy −0.16, and Japan −0.09, as shown in Table 7-3b. Gas is only a small portion in this sector; the largest share is 10% in Canada and the lowest is 0% in Japan, Germany, France, and the UK. Canada and the USA show −0.03 and −0.01, respectively; Italy shows a small elasticity of −0.002.

In the Residential/Commercial sector the coal share is 1% for Canada, France, and Japan; 3% in Italy; 4% in Germany and 5% in the UK. Table 7-3c shows that the elasticity of the relative price of coal to energy total is about −0.01 for France and Japan. No significant results were observed for other countries. The share of oil is: largest in Japan, 46%; 36% in Germany, 30% in France, 25% in Italy, 20% in Canada, and 14% and 12% in the UK and USA, respectively. Elasticity is large in Germany at −0.25, and Japan at −0.1. Canada and the USA show −0.05 and −0.02. One of the few observed results for the UK is the elasticity of oil price to electricity price at −0.05. The share of gas is very large in the UK at 56%, followed by 47% in Italy and 42% in the USA. France shows 20% and this is as small as 10% in Japan. The elasticity of gas price to energy price is −0.14 in Italy, −0.11 in the USA, and −0.08 in Germany.

We have also estimated the elasticity of electricity price to energy price in the three end-use sectors. However, we have only endogenized the share of electricity recently, and the results are almost as acceptable after briefly evaluating the parameters statistically. Careful tests on the level of significance are still necessary; therefore we do not include the results in this chapter. The share of 'New & Renewables' is calculated as the residual in the three sectors. The share of inputs in the electricity generation sector is also estimated for each fuel depending on relative price, except for policy variables such as 'Nuclear' and 'New & Renewables'. The results of the regressions are good in overall fit for every fuel share; nevertheless serial correlation may exist. In addition the significance of each estimated parameter also needs to be tested in this sector.

3. Major findings

Considering the difficulty of estimating the price elasticity of energy intensity change, we were quite successful in obtaining the results we expected. The parameters were observed at each of three different end-use sectors in respective countries, providing cross-country overview, and this helps to a great extent to understand the outcomes of the policy simulation studies to be done in the near future. In models assuming the same AEEI or ESUB across countries, only a comparable difference can be found deductively in simple economic assumptions such as GDP or population. In contrast, with an econometrically estimated price/output elasticity of energy intensity change based on empirical data, together with inter-energy substitution, policy simulation can be done on the grounds of already observed underlying differences. Therefore, our simulation approach at disaggregated sectors is inductive, while not losing sight of the international overview.

In the Industry sector the price responsiveness of energy intensity was quite a bit larger than we had expected. Autonomous technological change was also observed. The relative price-induced substitution of other energy for coal was observed in Japan, France and Germany, and the larger possibility of substitution of electricity for coal was especially indicated. In Canada a growing production of natural gas has resulted in substitution for coal. In France and Italy the substitution of natural gas for oil was observed. Germany and Japan also showed the potential substitution of electricity for oil. In the USA, the substitution of electricity for gas was observed. We need to look further into the UK as it showed no significant results. The pre-existing distorted energy market (such as coal subsidization) may be the reason that we were unable to observe price-induced energy substitution in this country.

In the Transportation sector the price elasticity showed that the fuel prices have induced energy efficiency improvements. The output elasticity also proved that increased output resulted in efficient use of energy. However, the positive elasticity observed with the time trend in all countries showed that, apart from economic factors, the change in quality of the service preferred by consumers resulted in more energy using. One of the reasons can be attributed to the increased purchases of recreation vehicles that are suitable for the lifestyle of modern-day people, but requiring more fuel to drive a kilometre. The substitution possibility in this sector is rather small, given the fact that a large part of the energy used is oil. Nevertheless, we have observed that greater use of gas and electricity is encouraged. For instance, the substitution of gas for oil was observed in Canada, Japan, Germany and the UK. Electricity has also substituted for oil in Japan, Germany, and Italy, and for gas in Canada and the USA. In France the parameters did not specify exactly from what fuel, but substitution for oil was also observed.

In the Tertiary sector the price elasticity of energy intensity was observed only in France, Germany, and the USA. We need to analyse further why other countries did not show significant results. The more disaggregation in this sector is needed because it is probable that inclusion of the Residential sector together with public, commerce, and other activities may explain the reasons, as it has always been speculated that this particular sector is the most price-unresponsive sector. It should be noted that the output elasticity was quite large in all countries. In the whole Tertiary sector, the expanding service activities accounted for most of the increasing output. Therefore, improved energy efficiency in, for instance, commercial buildings may have contributed towards lowering the intensity. In this sector substitution for oil was observed in all countries except Italy. Compared to the shift from coal to oil observed in Japan and France, this contributes greatly to CO_2 reduction. For example, substitution of gas for oil was observed in the USA and France. In Japan and the UK electricity substituted for oil and, though not specified in detail, other fuel substitution was observed in Canada and Germany. Further substitution of electricity for gas was observed in Italy, the USA, and Germany.

4. Energy balance and CO_2 emission simulation 2010

Before starting any policy simulation we need to test the performance of the model. To this end we have simulated a baseline energy balance and CO_2 emissions of G7 countries. By feeding a world oil price assumption of IEA projection – the most commonly used price in energy models – the model should be able to produce results similar to those presented in the IEA energy outlook. Up to the present we have simulated to the year 2010. The shares of 'Nuclear' and 'New & Renewables' in electricity production are also given from the policy review, also done by IEA. Estimated parameters described in the previous section are employed in the simulation model. Since this is a model of time-series balance table, we are able to obtain every year table for the simulated period; however, we show the results only for 2010.

4.1. Assumptions

The world oil price is projected to rise to \$25 in 2010 from \$12 in 1998. In the energy balance model the prices of domestic fuels are dependent on the world oil price, therefore the relative prices will also be forecast as the baseline case. Policy scenarios for share of 'Nuclear' and 'New & Renewables' are shown in Table 7-4. Share of Nuclear will stay constant from 1995 to 2010 in Canada. It is assumed to decline in Germany, the USA and most notably in the UK. France will continue to produce nuclear energy and the share in electricity production will rise by a small fraction. Japan is also assumed to expand nuclear generation, although the growth rate is set lower than originally proposed by the Japanese government. Share of 'New & Renewable' energy in generation will stay constant in Canada at 60%, Italy 17%, Japan about 10%, the USA about 10%, and the UK 32%. In France this share will decline to 10%, and Germany is assumed to increase the share to 15%, which is three times larger than in 1995.

4.2. Baseline energy balance simulation 2010

The COMPASS simulation results of the energy balance for G7 countries in 2010 are listed in Tables 7-5a to g. They are presented in the identical format of energy balance tables by the IEA. The baseline simulation experiment was conducted based on a fully linked COMPASS including macro/finance, IO, trade, and the energy balance models. They should be regarded as the test run

Table 7-4. Scenarios for the energy balance model: share of Nuclear and New & Renewables in electricity generation

	Nuclear 1995	Nuclear 2010	New & Renewables 1995	New & Renewables 2010
Canada	0.17	0.17	0.60	0.60
France	0.77	0.80	0.15	0.10
Germany	0.24	0.20	0.05	0.15
Italy	0.00	0.00	0.17	0.17
Japan	0.29	0.35	0.10	0.11
UK	0.26	0.13	0.32	0.32
USA	0.19	0.13	0.11	0.09

Table 7-5a. Simulated energy balance for Canada, 2010

IEA Row No.	Energy Balance Table	Coal (coa)	Crude Oil (cru)	Petroleum Products (prf)	Gas (gas)	Nuclear (nuc)	Hydro/Other (ren)	Electricity and Heat (elh)	Total	Unit
1	Indigenous production	49,215.4	113,207.9	0.0	149,702.6	33,001.4	47,871.0	0.0	392,998.3	Ktoe
2	Import	6956.5	49,898.5	6821.2	551.7	0.0	0.0	638.3	64,866.1	Ktoe
3	Export	23,588.7	60,516.2	12,413.8	64,707.8	0.0	0.0	3736.2	164,962.7	Ktoe
4	Intl. marine bunkers plus	1188.8	783.0	•68.0	1032.4	0.0	0.0	0.0	2936.2	Ktoe
5	Stock changes									Ktoe
6	**TPES**	33,772.0	103,373.1	•5660.6	86,578.8	33,001.4	47,871.0	•3097.9	295,837.8	Ktoe
7,8	Returns and transfers	663.4	0.0	2858.3	•1511.9	0.0	0.0	0.0		Ktoe
	Statistical differences									
9	Electricity plants									
10	CHP plants	26,137.1	1283.0	554.8	7417.3	33,001.4	37,748.6	62,899.0		Ktoe
11	Heat plants									
13	Petroleum refineries		102,090.1	101,618.3						Ktoe
12	Gas works									
14	Coal transformation									
15	Liquefaction	1655.0	0.0	0.0	3139.2	0.0	0.0			Ktoe
16	Other transformation									
12,18	Own use plus	233.2	0.0	8582.0	9979.2	0.0	0.0	9091.9		Ktoe
	Distribution losses									Ktoe
19	**TFC**	6410.1	0.0	89,679.2	64,531.2	0.0	10,122.4	50,709.2	221,452.1	Ktoe
20 through 34	**Industry sector (ind)**	5687.4	0.0	16,745.7	28,083.8	0.0	10,299.8	19,650.7	80,467.5	Ktoe
35 through 42	**Transport sector (tra)**	315.6	0.0	54,871.5	6691.7	0.0	313.2	453.6	62,645.7	Ktoe
43	**Other sectors (pro)**	62.1	0.0	11,511.8	29,755.7	0.0	2454.8	30,604.8	74,389.3	Ktoe
44	Agriculture									Ktoe
45,46,47	Public/commerce, residential, etc.									Ktoe
48,49,50,51	**Non-energy use**	344.9	0.0	6550.2	0.0				6895.1	Ktoe
52	Electricity generated	110,950.4		6094.6	41,052.7	126,324.3	438,830.0		725,082.6	Gwh
55	Heat generated	0.0		1177.2	18,141.7	3372.3	0.0	0.0	22,691.1	TJ
Additional	Electricity and heat generated	9541.7		552.2	3963.8	10,944.4	37,739.4	0.0	62,899.0	Ktoe

Table 7-5b. Simulated energy balance for France, 2010

IEA Row No.	Energy Balance Table	Coal (coa)	Crude Oil (cru)	Petroleum Products (prf)	Gas (gas)	Nuclear (nuc)	Hydro/Other (ren)	Electricity and Heat (elh)	Total	Unit
1	Indigenous production	48,884.6	2998.4	0.0	3731.6	26,585.6	7593.9	0.0	89,794.0	Ktoe
2	Import	9578.5	108,053.9	22,339.7	28,101.3	0.0	97.5	246.0	168,416.9	Ktoe
3	Export	465.2	445.6	15,537.7	616.2	0.0	0.0	6252.3	23316.9	Ktoe
4	Intl. marine bunkers plus	919.6	257.4	•3232.2	•708.9	0.0	9.1	0.0	•2755.0	Ktoe
5	Stock changes									Ktoe
6	**TPES**	58,917.6	110,864.1	3569.8	30,507.8	26,585.6	7700.5	•6006.3	232,139.1	Ktoe
7,8	Returns and transfers	103.8	0.0	5214.5	•788.7	0.0	•180.7	0.0		Ktoe
	Statistical differences									
9	Electricity plants									
10	CHP plants	50,711.9	0.0	20,260.1	•3860.8	26,585.6	0.0	43,866.2		Ktoe
11	Heat plants									
13	Petroleum refineries		110,864.2	111,566.0						Ktoe
12	Gas works									
14	Coal transformation									
15	Liquefaction	2798.4	0.0	2325.9	•34.3	0.0	0.0			Ktoe
16	Other transformation									
12,18	Own use plus	540.6	0.0	5787.9	447.5	0.0	0.0	6690.7		Ktoe
	Distribution losses									Ktoe
19	**TFC**	4970.6	0.0	91,976.4	33,166.7	0.0	7519.8	31,169.2	168,802.6	Ktoe
20 through 34	**Industry sector (ind)**	4591.3	0.0	16,712.2	13,003.6	0.0	347.9	8827.8	43,482.9	Ktoe
35 through 42	**Transport sector (tra)**	293.9	0.0	57,165.4	0.0	0.0	293.9	1029.0	58,782.2	Ktoe
43	**Other sectors (pro)**	85.4	0.0	13,013.8	20,163.0	0.0	12,635.5	21,312.3	67,210.1	Ktoe
44	Agriculture									Ktoe
45,46,47	Public/commerce, residential, etc.									Ktoe
48,49,50,51	**Non-energy use**	0.0	0.0	5084.9	0.0				5084.9	Ktoe
52	Electricity generated	197,885.3		237,074.1	−26,901.8	102,014.4	0.0		51,0072.1	Gwh
55	Heat generated	0.0		0.0	0.0	0.0	0.0	0.0	0.0	TJ
Additional	Electricity and heat generated	17,018.1		20,388.4	−2313.6	8773.2	0.0	0.0	43,866.2	Ktoe

Table 7-5c. Simulated energy balance for Germany, 2010

IEA Row No.	Energy Balance Table	Coal (coa)	Crude Oil (cru)	Petroleum Products (prf)	Gas (gas)	Nuclear (nuc)	Hydro/Other (ren)	Electricity and Heat (elh)	Total	Unit
1	Indigenous production	81,035.9	3609.1	0.0	23,572.7	34,527.9	8958.2	0.0	151,703.8	Ktoe
2	Import	12,228.3	125,637.8	44,809.9	55,311.9	0.0	0.0	3417.2	241,405.1	Ktoe
3	Export	1970.6	779.0	14,470.8	2395.3	0.0	0.0	3002.4	22,618.1	Ktoe
4	Intl. marine bunkers plus	2424.1	251.2	−710.7	−1323.3	0.0	0.0	0.0	641.3	Ktoe
5	Stock changes									Ktoe
6	**TPES**	93,717.7	128,719.0	29,628.4	75,166.1	34,527.9	8958.2	414.9	371,131.2	Ktoe
7,8	Returns and transfers	1545.0	0.0	−6704.6	−1293.6	0.0	0.0	0.0		Ktoe
	Statistical differences									
9	Electricity plants									
10	CHP plants	80,377.8	0.0	422.8	12,177.7	34,527.9	7644.8	57,017.7		Ktoe
11	Heat plants									
13	Petroleum refineries		128,719.0	129,281.1						Ktoe
12	Gas works									
14	Coal transformation									
15	Liquefaction	3675.3	0.0	6390.4	−718.8	0.0	0.0			Ktoe
16	Other transformation									
12,18	Own use plus	1244.4	0.0	7338.7	2801.5	0.0	0.0	8754.8		Ktoe
	Distribution losses									Ktoe
19	**TFC**	9965.2	0.0	138,052.9	59,612.1	0.0	1313.4	48,677.8	257,621.5	Ktoe
20 through 34	**Industry sector (ind)**	4085.2	0.0	24,363.7	20,952.5	0.0	134.6	17,768.4	67,304.4	Ktoe
35 through 42	**Transport sector (tra)**	404.0	0.0	78,481.9	3.5	0.0	405.5	1810.8	81,105.7	Ktoe
43	**Other sectors (pro)**	5271.5	0.0	31,138.3	38,656.1	0.0	1265.2	29,098.6	105,429.7	Ktoe
44	Agriculture									Ktoe
45,46,47	Public/commerce, residential, etc.									Ktoe
48,49,50,51	**Non-energy use**	204.5	0.0	4068.9	0.0				4273.5	Ktoe
52	Electricity generated	324,252.8		1906.0	41,634.5	132,437.5	55,577.1		555,807.9	Gwh
55	Heat generated	206,324.3		10,919.9	129,581.3	583.1	38,615.5	0.0	386,024.0	TJ
Additional	Electricity and heat generated	32,812.8		424.7	6675.0	11,403.5	5701.8	0.0	57,017.7	Ktoe

Table 7-5d. Simulated energy balance for Italy, 2010

IEA Row No.	Energy Balance Table	Coal (coa)	Crude Oil (cru)	Petroleum Products (prf)	Gas (gas)	Nuclear (nuc)	Hydro/Other (ren)	Electricity and Heat (elh)	Total	Unit
1	Indigenous production	−4398.8	5497.2	0.0	33,206.8	0.0	6724.6	0.0	41,029.8	Ktoe
2	Import	13,101.3	78,054.2	24,395.0	28,553.3	0.0	231.5	3324.9	147,660.2	Ktoe
3	Export	91.0	414.0	17,010.6	31.1	0.0	0.0	106.2	17,652.9	Ktoe
4	Intl. marine bunkers plus	−773.1	178.8	−2043.1	−224.6	0.0	0.0	0.0	−2861.9	Ktoe
5	Stock changes									Ktoe
6	**TPES**	7838.4	83,316.3	5341.4	61,504.4	0.0	6956.1	3218.7	168,175.3	Ktoe
7,8	Returns and transfers statistical differences	−25.3	0.0	108.5	−0.2	0.0	0.0	0.0		Ktoe
9	Electricity plants									
10	CHP plants	4241.9	0.0	8850.1	12,021.8	0.0	5747.4	19,504.6		Ktoe
11	Heat plants									
13	Petroleum refineries		83,315.6	82,984.3						Ktoe
12	Gas works									
14	Coal transformation									
15	Liquefaction	1487.3	0.0	5090.6	−42.9	0.0	162.4			Ktoe
16	Other transformation									
12,18	Own use plus	420.2	0.0	5425.9	668.3	0.0	0.0	2996.7		Ktoe
	Distribution losses									Ktoe
19	**TFC**	1663.7	0.0	69,067.6	48,857.0	0.0	1046.3	19,726.6	140,361.1	Ktoe
20 through 34	**Industry sector (ind)**	1192.6	0.0	8259.0	21,442.0	0.0	168.6	11,080.0	42,142.3	Ktoe
35 through 42	**Transport Sector (tra)**	246.5	0.0	47,625.1	295.8	0.0	246.3	836.5	49,250.2	Ktoe
43	**Other sectors (pro)** }	114.5	0.0 }	10,922.3 }	27,119.2 }	0.0 }	1034.0 }	7810.1 }	47,000.2	Ktoe
44	Agriculture									Ktoe
45,46,47	Public/commerce, residential and non-specified other									Ktoe
48,49,50,51	**Non-energy use**	110.1	0.0	2261.0	0.0				2371.1	Ktoe
52	Electricity generated	18,657.8		102,498.9	60,281.4	0.0	42,837.5		224,275.7	Gwh
55	Heat generated	0.0		0.0	0.0	0.0	9082.8	0.0	9082.8	TJ
Additional	Electricity and heat generated	1604.6		8814.9	5184.2	0.0	3900.9	0.0	19,504.6	Ktoe

Table 7-5e. Simulated energy balance for Japan, 2010

IEA Row No.	Energy Balance Table	Coal (coa)	Crude Oil (cru)	Petroleum Products (prf)	Gas (gas)	Nuclear (nuc)	Hydro/Other (ren)	Electricity and Heat (elh)	Total	Unit
1	Indigenous production	29,011.4	809.5	0.0	82.4	119,193.6	21,702.2	0.0	170,799.0	Ktoe
2	Import	80,945.9	292,669.4	52,241.4	50,004.2	0.0	0.0	0.0	475,860.8	Ktoe
3	Export	1995.2	0.0	8779.8	0.0	0.0	0.0	0.0	10,775.0	Ktoe
4	Intl. marine bunkers plus	191.8	−2548.8	−4298.3	66.8	0.0	0.0	0.0	−6588.5	Ktoe
5	Stock changes									Ktoe
6	**TPES**	108,153.9	290,930.1	39,163.3	50,153.4	119,193.6	21,702.2	0.0	629,296.1	Ktoe
7,8	Returns and transfers	−1230.9	−31.2	205.7	315.5	0.0	0.0	0.0		Ktoe
	Statistical differences									
9	Electricity plants									
10	CHP plants	31,649.2	38,000.0	17,053.3	29,287.8	119,193.6	18,066.5	102,166.0		Ktoe
11	Heat plants									
13	Petroleum refineries		256,585.1	255,905.5						Ktoe
12	Gas works									
14	Coal transformation									
15	Liquefaction	35,184.2	−6581.0	11,357.2	−5640.5	0.0	0.0			Ktoe
16	Other transformation									
12,18	Own use plus	5001.8	0.0	14,952.0	974.2	0.0	0.0	11,503.1		Ktoe
	Distribution losses									Ktoe
19	**TFC**	35,087.8	2894.3	251,912.0	25,847.5	0.0	3635.8	90,662.8	410,040.1	Ktoe
20 through 34	**Industry sector (ind)**	31,883.4	2894.3	79,725.2	10,941.8	0.0	3299.9	44,935.0	173,679.7	Ktoe
35 through 42	**Transport sector (tra)**	528.3	0.0	102,424.7	52.9	0.0	529.3	2319.1	105,854.3	Ktoe
43	**Other sectors (pro)**	} 2676.0	0.0 }	43,604.6 }	14,852.7	0.0 }	1548.9 }	43,408.7 }	106,091.0	Ktoe
44	Agriculture									Ktoe
45,46,47	Public/commerce, residential, etc.									Ktoe
48,49,50,51	**Non-energy use**	0.0	0.0	26,157.6	0.0				26,157.6	Ktoe
52	Electricity generated	158,041.2		197,118.7	152,734.4	457,370.8	142,557.1		1,183,174.6	Gwh
55	Heat generated	509.5		2341.8	8769.3	0.0	0.0	5671.6	17,292.1	TJ
Additional	Electricity and heat generated	13,603.7		17,008.1	13,344.6	39,333.9	12,259.9	135.4	102,166.0	Ktoe

Table 7-5f. Simulated energy balance for the United Kingdom, 2010

IEA Row No.	Energy Balance Table	Coal (coa)	Crude Oil (cru)	Petroleum Products (prf)	Gas (gas)	Nuclear (nuc)	Hydro/Other (ren)	Electricity and Heat (elh)	Total	Unit
1	Indigenous production	31,002.6	13,6481.3	0.0	76,376.1	22,877.2	2377.9	0.0	269,115.1	Ktoe
2	Import	11,007.8	44,348.0	8774.3	1505.3	0.0	0.0	1404.9	67,040.3	Ktoe
3	Export	862.7	88,308.1	22,298.3	869.0	0.0	0.0	2.0	112,340.1	Ktoe
4	Intl. marine bunkers plus	5507.9	920.1	−2359.0	768.0	0.0	0.0	0.0	4837.0	Ktoe
5	Stock changes									Ktoe
6	**TPES**	46,655.6	93,441.2	−15,883.0	77,780.4	22,877.2	2377.9	1402.9	228,652.2	Ktoe
7,8	Returns and transfers	−112.3	0.0	3715.5	−2915.8	0.0	0.0	0.0		Ktoe
	Statistical differences									
9	Electricity plants									
10	CHP plants	45,573.7	−14,866.5	−3892.6	14,717.1	22,877.2	1957.6	30,197.9		Ktoe
11	Heat plants									
13	Petroleum refineries		108,307.7	107,439.7						Ktoe
12	Gas works									
14	Coal transformation									
15	Liquefaction	262.2	0.0	1402.2	0.0	0.0	0.0			Ktoe
16	Other transformation									
12,18	Own use plus	46.6	0.0	7751.3	5174.3	0.0	0.0	5000.6		Ktoe
	Distribution losses									Ktoe
19	**TFC**	660.8	0.0	90,011.3	54,973.3	0.0	420.2	26,600.2	172,665.8	Ktoe
20 through 34	**Industry sector (ind)**	344.0	0.0	13,612.0	14,269.0	0.0	76.1	9739.5	38,040.5	Ktoe
35 through 42	**Transport sector (tra)**	316.8	0.0	61,699.1	0.0	0.0	316.8	1022.6	63,355.3	Ktoe
43	**Other sectors (pro)** }	0.0	0.0 }	6560.2 }	40,704.3	0.0 }	380.9 }	15,838.1 }	63,483.4	Ktoe
44	Agriculture									Ktoe
45,46,47	Public/commerce, residential, etc.									Ktoe
48,49,50,51	**Non-energy use**	0.0	0.0	8140.1	0.0				8140.1	Ktoe
52	Electricity generated	199,092.3		−44,899.9	82,350.6	87,784.5	17,556.9		351,138.2	Gwh
55	Heat generated	0.0		0.0	0.0	0.0	0.0	0.0	0.0	TJ
Additional	Electricity and heat generated	17,121.9		−3861.4	7082.2	7549.5	1509.9	0.0	30,197.9	Ktoe

Table 7-5g. Simulated Energy Balance for United States, 2010

IEA Row No.	Energy Balance Table	Coal (coa)	Crude Oil (cru)	Petroleum Products (prf)	Gas (gas)	Nuclear (nuc)	Hydro/Other (ren)	Electricity and Heat (elh)	Total	Unit
1	Indigenous production	629,338.6	399,990.1	0.0	454,728.3	201,225.3	116,812.5	0.0	1,802,094.8	Ktoe
2	Import	5300.3	386,926.9	41,954.8	65,771.0	0.0	31.9	4021.4	504,006.2	Ktoe
3	Export	54,746.3	2383.9	38,758.3	3533.0	0.0	0.0	786.6	100,208.0	Ktoe
4	Intl. marine bunkers plus	−6744.8	5019.6	−23,141.3	10,772.1	0.0	22.8	0.0	−14071.7	Ktoe
5	Stock changes									Ktoe
6	**TPES**	573,147.8	789,552.6	−19,944.8	527,738.4	201,225.3	116,867.1	3234.8	2,191,821.2	Ktoe
7,8	Returns and transfers	1775.0	0.0	43,901.9	−5956.8	0.0	−0.6	0.0		Ktoe
	Statistical differences									
9	Electricity plants									
10	CHP plants	536,909.1	0.0	−20,172.4	157,049.6	201,225.3	88,098.6	342,290.4		Ktoe
11	Heat plants									
13	Petroleum refineries		789,552.6	793,449.1						Ktoe
12	Gas works	11,125.5	0.0	0.0	0.0	0.0	0.0			Ktoe
14	Coal transformation									
15	Liquefaction									
16	Other transformation									
12,18	Own use plus	656.3	0.0	53,734.0	45,224.5	0.0	0.0	54,125.6		Ktoe
	Distribution losses									Ktoe
19	**TFC**	26,231.9	0.0	783,844.7	319,507.6	0.0	28,767.9	291,399.7	1,449,751.6	Ktoe
20 through 34	**Industry sector (ind)**	26,310.9	0.0	83,867.8	109,513.4	0.0	41,880.1	71,616.2	327,188.5	Ktoe
35 through 42	**Transport sector (tra)**	3048.8	0.0	585,714.5	18,228.7	0.0	2439.1	336.3	609,767.4	Ktoe
43	**Other sectors (pro)** }	2872.1	0.0 }	41,384.6 }	191,765.4	0.0 }	16,519.6 }	219,447.1 }	471,988.9	Ktoe
44	Agriculture									Ktoe
45,46,47	Public/commerce, residential, etc.									Ktoe
48,49,50,51	**Non-energy use**	0.0	0.0	72,877.8	0.0				72,877.8	Ktoe
52	Electricity generated	2,253,272.5		−206,953.3	638,703.1	772,143.6	429,223.9		3,886,389.5	Gwh
55	Heat generated	142,727.3		−103,601.3	267,498.7	0.0	30,933.5	0.0	337,558.2	TJ
Additional	Electricity and heat generated	197,189.8		−20,272.0	61,316.3	66,404.3	37,651.9	0.0	342,290.4	Ktoe

of the model rather than projection. Annual results are obtained, but we report here only on the terminal year of the simulation. The results are consistent with the IEA studies in magnitude and direction of energy demand/supply pattern change. The results proved the high performance and quality of the energy balance model.

Let us look at the simulation for Canada. The total TPES is 296 Mtoe compared to the IEA of 280 Mtoe. The difference comes from the larger supply of coal and nuclear results in our model. Breaking down on the sectors, total consumption in end-use sectors shows almost the same results as IEA, except that our model resulted in more energy use growth in the Tertiary sector, notably electricity consumption. As for France, the total TPES is 232 Mtoe, less than 281 Mtoe of IEA. The difference comes from nuclear, which is smaller in our model. Industry consumption of energy is also smaller. This is due to the lower electricity demand in our model, whereas consumption values in Transportation and the Tertiary sector are both larger compared to IEA. Simulation for Germany indicates that the total TPES is 371 Mtoe, larger than the 350 Mtoe of IEA. Our results show more oil and renewable supply, and less supply of nuclear than in IEA. Consumption in Industry is smaller in our results, mostly due to the smaller supply of coal. The other two sectors show more consumption, coming from a larger demand for gas and electricity, especially in the Tertiary sector. The total TPES of the UK is 229 Mtoe, which is less than the 249 Mtoe of IEA. This difference is due to the smaller supply of oil and gas, although coal supply is larger than IEA. Transportation and the Tertiary sector consumption values are almost the same as in IEA; however, they are less in Industry in our results. Italy's total TPES is 163 Mtoe, whereas IEA forecast is 188 Mtoe. Smaller supplies of coal and gas are reasons for the difference. The Transportation sector is larger than in IEA, while Industry and Tertiary are a little less. According to the results of Japan the total TPES is very large in our results of 629 Mtoe, compared to 562 Mtoe of IEA. This is due to the very high *output price* coming from the IO model of Japan. We need to look into the IO model and modify it if any inappropriate specification is adopted. The consumption values by the Industry and Tertiary sectors are almost the same as in IEA, although more electricity use in the Industry but less in the Tertiary sector is indicated. Transportation consumption reaches the level of the Tertiary sector in our model. Lastly, total TPES of the USA is 2191 Mtoe, compared to 2583 Mtoe of IEA. Supplies of gas and 'New & Renewables' are larger in our results, and the rest are smaller. Industry and Transportation consumption values of energy are greater in IEA, and the Tertiary sector shows almost the same results.

We have briefly looked at the energy balance simulation results and, except for the large difference in TPES in Japan and the USA, other forecasted countries showed almost the same magnitude as reported by IEA. However, the objective of this observation is not to compare the results with other studies, rather to check the consistency in general as already stated. It is surprisingly that the forecasts with the energy balance model and IEA show the results in great similarity to detail, if one looks at the two results in parallel. Therefore, we conclude that the energy balance simulation up to 2010 was very successfully done.

4.3. Baseline CO_2 emission simulation to 2010

Now we have simulated the energy balance, next comes the CO_2 emission simulation. The simulated Total emission, Utility emission (electricity production), Industry emission, Transportation emission, and Tertiary emission values are shown from Figure 7-1a to Figure 7-1g for respective G7 countries. The baseline emission is based on the baseline energy balance; therefore the assumptions and scenario for nuclear and renewables are also embedded into the simulation here. Therefore, in this baseline case, no additional energy or environmental policy is incorporated.

Let us look at Figure 7-1a. Canada is forecasted to increase total CO_2 emission as is indicated as Carbon Emission, by 2.2% per year from 1990 to 2010. Deducing other emissions shown in the figure from total gives the emission from other transformation processes. The annual increase of emission at each sector is 1.6% per year in the Utility sector, 3.8% per year in Industry, 1.9% per year

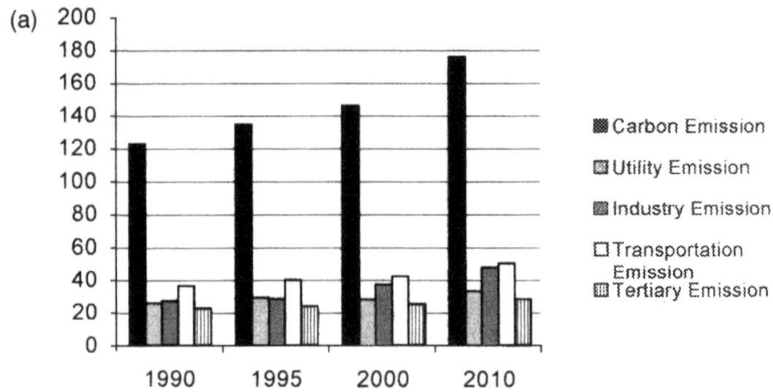

Figure 7-1a. Baseline CO_2 emission simulation: Canada (MMTC).

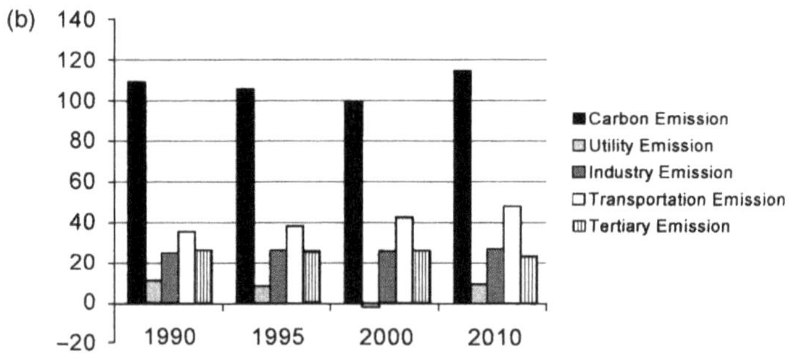

Figure 7-1b. Baseline CO_2 emission simulation: France (MMTC).

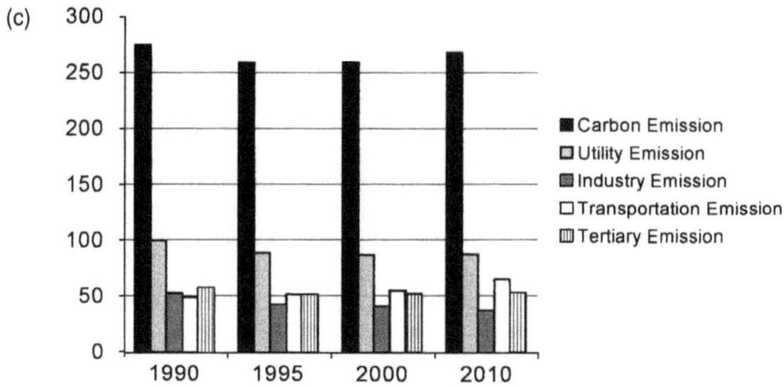

Figure 7-1c. Baseline CO_2 emission simulation: Germany (MMTC).

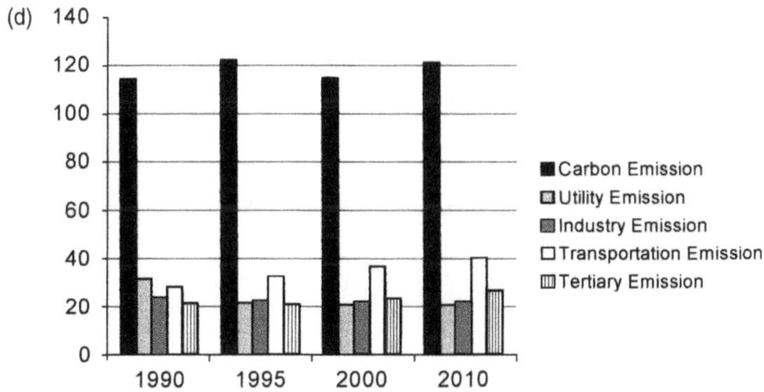

Figure 7-1d. Baseline CO_2 emission simulation: Italy (MMTC).

in Transportation, and 1.5% per year in the Tertiary sector. In order to achieve the Kyoto target of 6% reduction from 1990 level, Canada has to reduce almost 30% from this baseline emission in 2010. The reduction opportunities still lie in energy efficiency improvement and substitution. However, emission from the Transportation sector may be a big challenge because of the vast land area of this country.

Figure 7-1b is the simulation results of France. A large contribution by the extremely large share of nuclear energy can be seen in the very small total emission growth. From 1990 to 2010 total emission increased by 0.2% per year, and Industry sector emission grows at 0.5% per year. The largest increase is seen in the Transportation sector at 1.8% per year. Utility and Tertiary sector emission declines by −0.6% per year and −0.4% per year, respectively. The negative Utility emission in 2000 shows that a variable in the model may not be correctly calculated for this year, and we will have to investigate what is causing such results. As for the Kyoto target, France is obliged to reduce its future

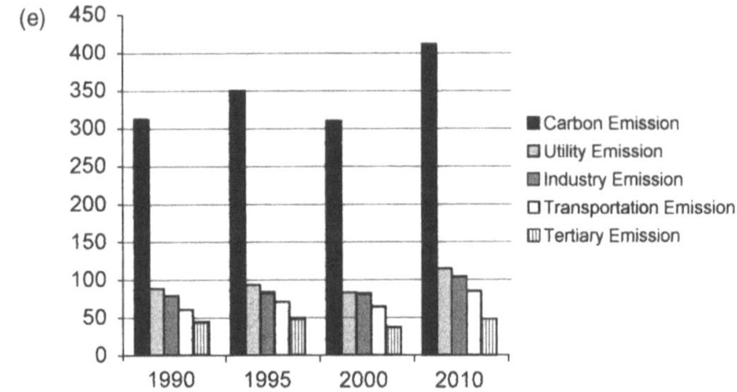

Figure 7-1e. Baseline CO_2 emission simulation: Japan (MMTC).

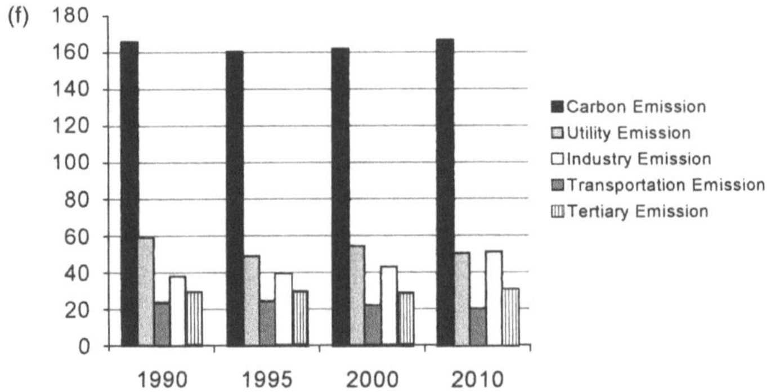

Figure 7-1f. Baseline CO_2 emission simulation: United Kingdom (MMTC).

emission level to the 1990 level as part of a burden-sharing agreement with other European Union member states. Therefore, the target will probably not cast a heavy economic burden on this country.

Figure 7-1c shows the CO_2 emission of Germany. In this country all sectors will reduce CO_2 emission during the period from 1990 to 2010 except the Transportation sector. Germany is required to reduce its emission by 21% from the 1990 level. This seems quite ambitious. The price responsiveness of technological change was quite large in this country, in all end-use sectors. In addition the heavy reliance on coal that is subsidized by the government indicates there are low cost mitigation opportunities. Whether Germany can enter the unregulated energy market and a make healthy policy mix, such as promotion of nuclear and renewables, is crucial. The efficiency improvement in office buildings and housing, and voluntary agreement, tax on petrol and diesel, promoted modal shift in transportation-related activities are expected to contribute greatly towards reducing emission. Nevertheless, whether Germany can meet the target is not clear, and government needs to coordinate the policies better.

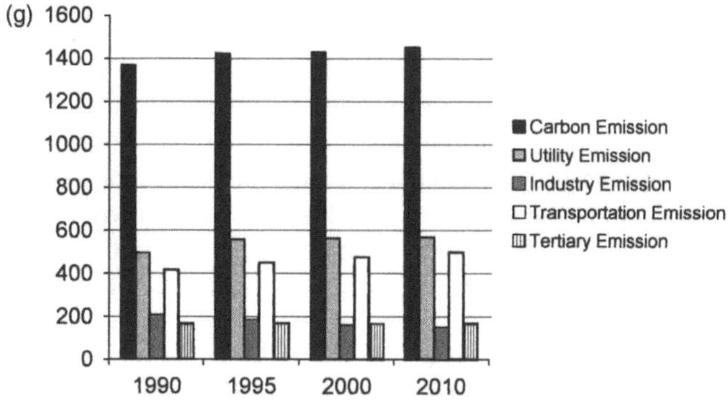

Figure 7-1g. Baseline CO_2 emission simulation: USA (MMTC).

The CO_2 emission of Italy is shown in Figure 7-1d. The total emission is forecast to increase at 0.4% per year from 1990 to 2010. The emission from Utility is expected to decline at 1.8% per year, and Industry is going to reduce emission at 0.4% per year. Increasing sectors are Transportation and Residential/Commercial, 2% and 1.2% respectively. Italy is targeting to reduce emission by 8% from the 1990 level. The high price elasticity of energy intensity, and price-induced substitution of gas for oil observed in Industry indicate that the low cost reduction opportunities are still to be found in this sector. At the same time, energy intensity in the Transportation sector has risen from 1990 to 1995, with the positive time trend indicating that the energy intensity is not declining regardless of relative price changes. At initial observations, the challenge seems to lie in emission reduction by the Transportation sector. In the Residential/Commercial sector the price-induced substitution of electricity to gas was observed, although technological change was only captured in output elasticity. Since Italy is not a nuclear-producing country, with growth of renewables supply probably remaining marginal, gas and electricity are going to play an important role both in terms of energy mix policy and climate change.

Figure 7-1e shows the emission forecast for Japan. Because of the rapid expansion of *output price* obtained from the IO model we have simulated energy balance with a larger figure for each of TPES, resulting in a higher level of emission in 2010. It is also questionable that the base year emission in 1990 is already higher in our simulation by more than 30 MMTC than the outlook by EIA, shown in Table 6-2 in Chapter 6. In Japan the emission from Utility is larger than in other sectors presented, despite the fact that the simulation results include the possibility of nuclear expansion, at a maximum 20% increase from present. As discussed in section 1 above, the potential low cost opportunities in this country seem hard to find, which observation is consistent with the outcomes from the model exercises introduced in Chapter 6. To meet the Kyoto target requires many efforts both on supply and demand sides. However, considering that 70% of the total energy supply depends on imports, and the shift to gas is limited due to there being no connection to pipelines across the surrounding seas, reduction by the supply side should not be viewed optimistically. In addition, whether the supply of renewables could augment as drastically as expected in the government plan remains questionable. The top-runner in end-use sectors and voluntary agreement in Industry are expected to contribute to CO_2 emission reduction, the effects of which are also unknown. In this respect, the government of Japan is faced with a big challenge in coordinating and institutionalizing measures to induce more energy-efficient technologies and cleaner energies to enter the market.

Figure 7-1f shows the emission of the UK. CO_2 emission in this country is forecast to stay almost at the same level as in 1990, except the rise in the Transportation sector at 1.7% per year during the period from 1990 to 2010. The UK has the emission reduction target at 12.5% below the 1990 level. The share of coal has declined to a great extent due to the privatization in 1994, and government considers reducing emissions through greater reliance on renewables and combined heat and power. In addition, government expects that nuclear energy proportion will decline steadily. How to meet the target and its own domestic aim of fuel diversity and energy security, without lowering the social equity figure, is one of the challenging tasks for this country.

Figure 7-1g shows the emission simulation of the USA. Total emission is forecast to increase by 0.3% per year. The Utility emission also increases by 0.75% per year, and Transportation increases by 1.0% per year. The Tertiary sector also increases by 0.17% per year, while the Industry sector emission declines by 1.2% per year. In section 3, we observed that the price elasticity of energy intensity was found only in the Tertiary sector in this country; whereas price-induced energy substitution, especially the shift from oil to gas, and gas to electricity, was observed in all sectors. In fact, fuel switching is an important factor in mitigating CO_2 emissions. Currently it is faced with trade-offs, whether to keep the energy prices low through competitive market while imperatives lie in climate change consideration.

5. Conclusion

Although the assessment of the domestic policy measures to tackle climate change was not an objective of this chapter, we were already able to draw a cross-country/sector picture of low cost mitigation opportunities and its challenges. This is one of the advantages of the energy model, as stated earlier, that it starts with disaggregated analysis of energy use and process, demand and supply. Now we are convinced of the validity of theories about price-induced technological change. The inter-energy substitution that lies behind the modelling was also proved to apply for the observed period. On top of that, we have tested the credibility of data, and the performance of the model, which were also proved good. We easily found from the simulated result whether something strange was occurring in variables, thanks to the very well-organized object-oriented database system and transparent modelling process. The assessment of domestic policy measures is to be conducted in the near future, and this will be able to provide great insights into how the country could/should meet the Kyoto target with its own domestic priorities in other policies. Especially, and most importantly, the policy such as a domestic carbon tax would induce technological change both in energy and end-use sectors through price signals, which are the important framework in our model.

It is already possible to conduct carbon tax simulation after baseline simulations were successfully done (see Chapter 11 for the effect of carbon tax on employment in G7 countries), The mechanism within the model of simulating such a policy is as follows. First, an imposed carbon tax rate will be added onto the energy price as mark-up percentages increase. The tax is imposed on each energy carrier according to their carbon content. The fraction of energy price increase is also added onto the price functions of energy sectors in the IO model (functions (3) and (4) in Chapter 5). The rise in energy costs will change the economic structure and international trade, and consequently change the *output* and *output prices*. The change in these two variables will affect the relative price of energy back into the energy balance model, the level of which induces the energy intensity change and inter-energy substitution. Therefore, the energy demand would decline as a result of the carbon tax. As described previously, the rate of changes in *output* (cost of CO_2 reduction) in relation to the rate of energy efficiency improvement and fuel switching are critically dependent on the estimated price elasticity of these elements. Therefore, a country with high price elasticity of energy intensity and energy substitution will probably result in a lower requirement level of carbon tax.

It should be noted here that it is crucially important to enable the model to handle international aspects of climate change policies, as discussed in Chapter 6. Despite the fact that the international mechanisms such as emissions trading should be supplemental to domestic policy measures, it is imperative that the parties design the mix of available policy options. There should be more insight into available opportunities and economic cost together with related impacts on socioeconomic trade-offs such as inflation or unemployment. We turn to this topic in Chapter 11.

It should be stressed as the concluding remark that, although it is not an easy task for economists to better capture the economic costs together with the economic benefits, the policy should be directed towards long-term benefit. If better technologies come into operation they will, in most cases, contribute greatly to operate and produce required services more efficiently than before. Such innovation augmentation could surely change the pattern of energy use in our society. However, such technological change is not likely to be continually induced unless the properly designed policy measures send price signals to users. The policies cannot be derived from models, but they should not be designed on human intuition. This is because the climate change issue is likely to affect not only the parties required to reduce GHGs, but also the whole globe, especially the developing countries within a very long time frame. We hope that the energy balance model will contribute to capture economy–energy–environment interaction and its policy implications, and thus to stimulate discussions on the economic analysis of policy measures to avoid climate change in a more in-depth way.

8
The case of South-East Asia

DIRK VANWYNSBERGHE, YUMIKO UMEHARA, and KIMIO UNO

1. Object-oriented approach to model building

Asia is diverse. The various countries differ in terms of per-capita income, size of the economy, and structure. In this chapter we take up countries in East and South-East Asia, namely Thailand, Malaysia, Singapore, Indonesia, the Philippines, Taiwan (Chinese Taipei), and Korea. They constitute a distinct group of countries in the global community despite all the differences among them; they have rapidly emerged in the world trade scene and proved to be the most dynamic engine of the world economy (Jung *et al.*, 2000; Pesaran *et al.*, 1998; World Bank, 1993; World Bank, various years b; ADB, various years b). Asian economic outlook is available in ADB, various years a; World Bank, various years a. Today, Korea is an OECD member. Taiwan is legally a part of China but its position as an economic entity is distinct. All of the countries that we take up are the core members of the APEC (Asia Pacific Economic Cooperation) (see Table 1-1 in Chapter 1).

The South-East Asian countries have succeeded in catching up with the West through their industrialization drive. This chapter looks into the energy and environment of the region. Despite its importance in the long run this issue tends to be neglected in the face of more urgent, short-term concerns. Because of their geographical proximity the relation with Japan is crucial. In other words the Asian economies are linked with the Japanese economy through trade, direct investment, and technology transfer, particularly after the 1985 realignment of the yen *vs*. the US dollar.

Energy analyses were conducted systematically for the G7 countries in Chapter 7. In this chapter it is intended to carry out a comparable work for Asian economies. Although physical data are available for those countries, price data are not.

In our object-oriented approach to database construction and model building, data set is an object and logic is an object. For energy analysis we need three main data. First is the physical quantity of energy, disaggregated by energy source and by final use. The empirical basis of the analysis is provided by the International Energy Agency's *Energy Statistics and Balances of Non-OECD Countries* (Korea is an OECD member) (IEA, various years b). This source provides energy balance

K. Uno (ed.), Economy–Energy–Environment Simulation, 135–144.

tables in an identical format as for OECD members, with slight differences in the order of listing. After adjusting for this difference, the data set for non-OECD countries can be treated in the same manner as the data set for OECD members.

Second is the output price of economic branches so that relative price can be derived. This is obtained from the economy model of COMPASS.

Third is the energy prices corresponding to the final use categories, which are not readily available. The problem is the energy prices. In the case of the OECD countries data are available from the IEA *Energy Prices and Taxes* (IEA, various years c). For Asian countries (and other developing economies) this is not the case. This chapter starts by describing the process of preparing energy price data, an indispensable step in attempting an econometric analysis of responses of economic agents to price signals. For global energy and environmental policy analysis, price parameters are deemed important in that they foretell the effectiveness of energy tax and other price-based policy instruments.

With energy data in physical terms and energy price data that comprise the basis for calculating relative prices of energy for different final uses, we can employ the standard model specification developed in the preceding chapter as the logical object. In other words we apply the equation specification object developed for G7 countries to the data objects coming from Asian countries.

2. Energy prices for Asian countries

The models that exist at present do not utilize full information of market signals with which economic agents are confronted, namely changing relative prices of energy. In contrast, the COMPASS approach emphasizes the price-responsiveness of the economic agents. This is deemed important in view of the preference on market-oriented policy measures such as carbon tax in achieving a sustainable development path.

Energy prices for Asian countries are required, to improve the simulation models for selected Asian countries.

The energy balance model for G7 countries developed in Chapter 7 focuses on three major parts of the energy balances, namely: (1) energy intensity (defined as energy consumption per output), (2) conversion efficiency of primary energy into electricity, and (3) energy substitution among different carriers. Of these, parameters concerning (1) and (3) are econometrically estimated, whereas (2) is a technological parameter (explained by time trend in the baseline simulation). In the case of South-East Asian countries we focus on energy intensity in three end-use sectors, i.e. Industry, Transportation, and Residential/Commercial sectors. The shift among primary energy sources (including preference on domestially produced energy) is not a pure market phenomenon but more or less a policy decision in Asian countries.

2.1. *The estimation procedure*

Energy prices

It is noted that Japan, and more recently Korea, are OECD members from Asia. Japan is analysed together with other G7 countries in Chapter 7. Korea is included in Asian countries for this study because of its development stage which, until recently, characterized this country as in a rapid industrialization phase, rather than a mature phase characterized by service orientation, etc. Taiwan, a region of China according to the official status in the international community, is treated as an independent economic entity within the framework of COMPASS. This is justified on the grounds of its relative independence from mainland China in economic terms, at least up until recently, and its significance in terms of economic size. Thus, Asian countries/regions in this study are as follows: Thailand, Malaysia, Singapore, Indonesia, Philippines, Taiwan, and Korea. China is analysed separately in Chapter 9 on the grounds that energy use there can hardly be conceived of as activities of independent economic agents; rather, it is largely determined by central planning.

For Asian countries the energy balance tables are available from IEA *Energy Balances for Non-OECD Countries*. Alternatively, and particularly for the period from 1970 to 1992, the Asian Development Bank has published annual tables for a wide range of countries under its jurisdiction (ADB, 1994). The ADB work is now integrated into the IEA work. This source, which provides time-series price data by energy carrier, is the basis of our energy analysis. The ADB source provides time-series energy prices per physical unit (such as baht/ton, baht/kWh) for various energy categories. What we need in the model is energy price for the final use sectors (i.e. Industry, Transportation, Household, Commerce, etc.). For this, prices by energy carrier are bundled together according to the share in the final use. The energy balance tables are also utilized in calculating the shares of various energy sources for respective final use. The base year is 1990. The shares of energy categories in the base year, obtained above, are multiplied by the time-series price data in order to estimate energy prices in different final use sectors.

The problem is that the ADB price data end in 1992. The availability of data thereafter varies among countries/regions. For a limited number of countries/regions (to be specific, Korea, Taiwan, and Thailand, among other developing economies), data are published in the IEA33 *Energy Prices and Taxes*. Thus, the availability of energy price data is as shown in Table 8-1.

Where data are not available, it is conceivable (1) to 'borrow' the data from neighbouring economies where the development phase and other economic characteristics are more or less similar, and use a dummy variable for 1993 onwards in the regression analysis in order to accommodate any data discontinuity, if necessary; or (2) to use the international price of oil, natural gas, and coal, adjusted by the exchange rates (local currency per US$, from *International Financial Statistics*).

Table 8-1. Availability of energy price data

	1973-1992	1992-1998
Thailand	OK	OK
Indonesia	OK	OK
Philippines	OK	n.a.
Malaysia	OK	n.a.
Taiwan	Partial (1982-1992)	OK
Korea	Partial (1980-1991)	OK
Singapore	n.a.	n.a.

Energy use data correspond to (1) Industry, (2) Transport, (3) Residential, (4) Public/Commerce, (5) Agriculture, (6) Others, and (7) Non-energy use. In the case of energy analysis for G7 countries attempted earlier, items (3) through (7) were combined into one category labelled 'Residential/ Commercial'. However, in order to reflect the response of economic agents to changing price signals it may be desirable to focus on items (3) and (4), without mixing them with other activities.

2.2. *1990 weights from the energy balances*

We use the shares of individual energy sources in total as the weights to aggregate into energy prices. The procedure to have access to the IEA energy balances is shown in Figures 8-1 and 8-2.

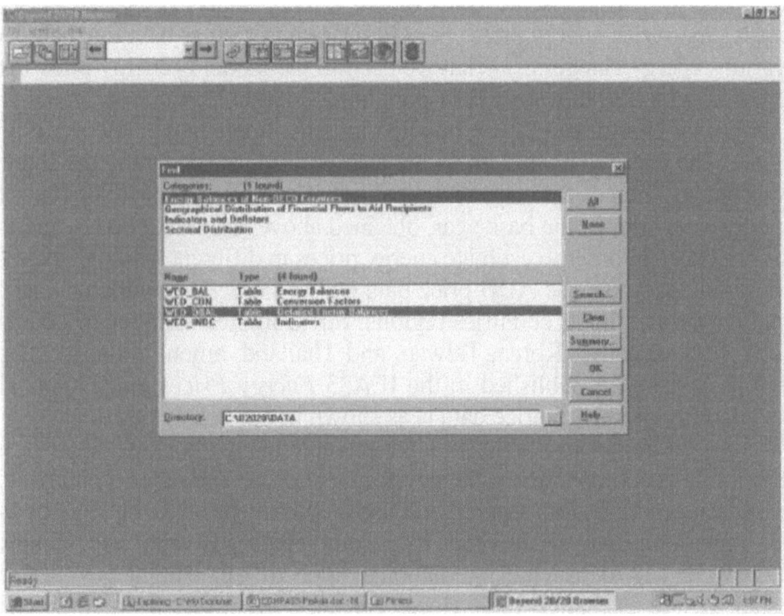

Figure 8-1. Access to the IEA energy balance database.

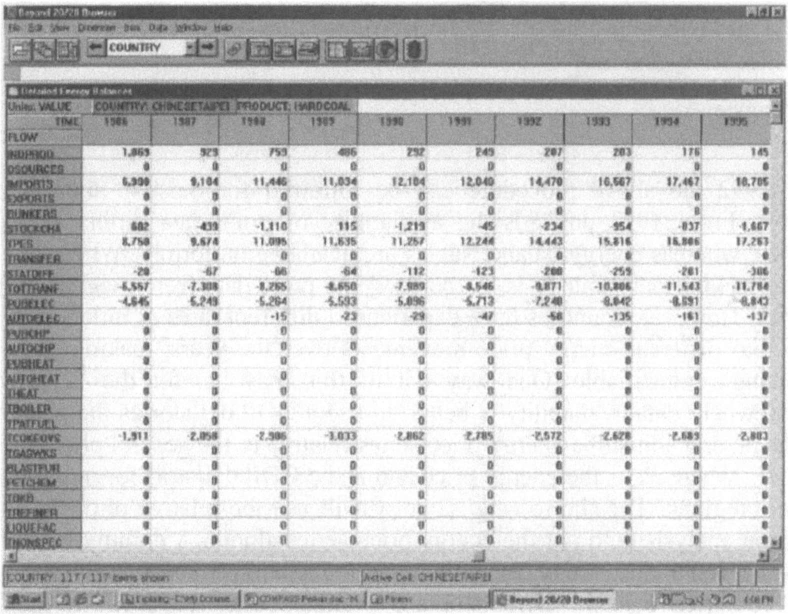

Units: VALUE | COUNTRY: CHINESETAIPEI | PRODUCT: HARDCOAL

TIME / FLOW	1986	1987	1988	1989	1990	1991	1992	1993	1994	1995
INDPROD	1,863	923	753	486	292	249	207	203	176	145
OSOURCES	0	0	0	0	0	0	0	0	0	0
IMPORTS	6,930	9,104	11,446	11,034	12,104	12,040	14,470	16,667	17,467	18,705
EXPORTS	0	0	0	0	0	0	0	0	0	0
BUNKERS	0	0	0	0	0	0	0	0	0	0
STOCKCHA	682	-439	-1,110	115	-1,219	-45	-234	-954	-837	-1,667
TPES	8,750	9,674	11,095	11,635	11,257	12,244	14,443	15,816	16,806	17,263
TRANSFER	0	0	0	0	0	0	0	0	0	0
STATDIFF	-29	-67	-86	-64	-112	-123	-200	-259	-281	-306
TOTTRANF	-6,557	-7,308	-8,265	-8,650	-7,989	-8,546	-9,871	-10,806	-11,543	-11,784
PUBELEC	-4,645	-5,249	-5,264	-5,593	-5,096	-5,713	-7,240	-8,042	-8,691	-8,843
AUTOELEC	0	0	-15	-23	-29	-47	-58	-135	-161	-137
PUBCHP	0	0	0	0	0	0	0	0	0	0
AUTOCHP	0	0	0	0	0	0	0	0	0	0
PUBHEAT	0	0	0	0	0	0	0	0	0	0
AUTOHEAT	0	0	0	0	0	0	0	0	0	0
THEAT	0	0	0	0	0	0	0	0	0	0
TBOILER	0	0	0	0	0	0	0	0	0	0
TPATFUEL	0	0	0	0	0	0	0	0	0	0
TCOKEOVS	-1,911	-2,058	-2,306	-3,033	-2,862	-2,785	-2,572	-2,628	-2,689	-2,803
TGASWKS	0	0	0	0	0	0	0	0	0	0
TBLASTFUR	0	0	0	0	0	0	0	0	0	0
TPETCHEM	0	0	0	0	0	0	0	0	0	0
TOKO	0	0	0	0	0	0	0	0	0	0
TREFINER	0	0	0	0	0	0	0	0	0	0
TLIQUEFAC	0	0	0	0	0	0	0	0	0	0
TNONSPEC	0	0	0	0	0	0	0	0	0	0

COUNTRY: 117 / 117 items shown | Active Cell: CHINESETAIPEI

Figure 8-2. IEA database search results.

In Figure 8-2 the selector window at the top left (now showing COUNTRY in this example) will allow users to select among country, product, flow, time, and units. The left and right arrows allow the users to move within each category (i.e. from country to country, product to product, or year to year).

2.3. *Deriving estimated time-series of energy price data*

The next step is to establish a linkage between price data and the weights.

Note that price data in terms of physical units have been converted to index numbers (1990 as 1.00). The shares of individual energy products have been obtained above. The sum of products of index and weights should generate energy price series. This process is repeated for Industry, Transportation, Household, Commerce and Public, etc. They have to be aggregated into an identical classification of energy sources. In the COMPASS database and model the following categories are used:

1. Coal (coa)
2. Crude oil (cru)
3. Petroleum products (prf)
4. Gas (gas)
5. Nuclear (nuc)

6. Hydro/Other (ren)
7. Electricity and heat (elh)

Difference
Total

Up to 1992 the above procedure can be followed for most of the Asian countries. From 1993 onwards the availability of price data differs among countries, whereas energy balance tables are available uniformly. Where price data are lacking our strategy is to 'borrow' the price data from neighbouring countries. However, countries have experienced different rates of inflation that reflect into different energy price movements over the years, making simple 'borrowing' unjustifiable. One way out of this may be, for the 'lending' country, where data availability is better, to use a GDP deflator as the general price level, and adjust the energy price series relative to the general price level (i.e. *Pe*/*P*, where *Pe* is the energy price and *P* the GDP deflator) for individual countries/regions. The 'borrowing' country will use the relative prices in the regression analysis (and not the energy price series alone). A dummy variable may be needed to absorb any discontinuity in the relation among variables.

For countries/regions where data are available after 1993 (from IEA's *Energy Prices and Taxes*, rather than ADB series for 1973-992), price data have been derived by the following procedure.

Tabulate current prices per physical unit.
Convert into index (1990 as 1.00).
Obtain weights for individual energy sources for the base year, based on IEA's *Energy Balances for Non-OECD Countries* (in the process it was found that inclusion of 'New Energy' leads to wrong results (Singapore, Taiwan, and Korea are the exceptions), total elements being larger than the total, presumably due to misplaced decimals. 'New Energy' data have been omitted. 'Crude Oil' is omitted because only in Thailand does it have a minor share).
Time series data on energy prices are obtained by multiplying weights and price series. (Not all prices for components under energy products are available. Thus, we were obliged to use different items for different countries/regions. The estimated data are included for the period 1992 on.)

The same formula as above is applied for earlier years. This gives better continuity, but at the cost of not including some components where data are available for earlier years but not so for the period 1992 on.

3. Regression results

The preliminary results of regression runs for the Industry sector are reported in Table 8-2a, for the Transportation sector in Table 8-2b, and for Household sector in Table 8-2c. They are preliminary in the sense that energy price data in respective countries are, as we have explained above, not in an ideal situation, although probably there is not much room for improvement given the availability of raw

Table 8-2a. Estimated parameters on energy equations for Asia, Industry sector

	Price Elasticity (*t*-value)	Output Elasticity (*t*-value)	Adjusted RSQ	DW
Thailand	−2.28 (−2.40)	−0.24 (−1.20)	0.44	1.34
Malaysia	−1.38 (−7.41)	*	0.88	1.25
Singapore	−4.66 (−3.87)	−0.35 (−0.86)*	0.61	1.24
Indonesia	*	*	*	*
Philippines	*	*	*	*
Taiwan	−0.21 (−1.24)	−0.33 (−2.74)	0.75	1.59
Korea	*	*	*	*

* indicates that no significant results were observed.

Table 8-2b. Estimated parameters on energy equations for Asia, Transportation sector

	Price Elasticity (*t*-value)	Output Elasticity (*t*-value)	Adjusted RSQ	DW
Thailand	−0.67 (−5.86)	*	0.85	1.78
Malaysia	−1.26 (−6.01)	*	0.89	0.96
Singapore	−0.11 (−0.80)	*	*	*
Indonesia	−0.34 (−0.77)*	*	*	*
Philippines	−0.39 (−1.79)	*	0.74	1.07
Taiwan	−0.35 (−2.38)	*	0.87	1.95
Korea	−0.25 (−1.49)	*	0.88	1.22

* indicates that no significant results were observed.

data for the East-Asian countries. They are preliminary also in the sense that we have employed the functional form developed and tested for G7 countries in Chapters 5 and 7 without searching for the best-fit equations. It is customary in a model-building process to first start from a theoretically accepted standard form and then, after looking at the results, try out alternative functional forms, alternative data, different time lags, and so forth. We chose this approach here in order to experiment with the object-oriented approach in model specification.

The results are mixed. One notable thing is that, in the case of the Transportation sector, although price elasticity is empirically measured for all countries, output elasticity is not. Output elasticity is intended to capture the size effect: it is expected that, as the activities in question expand in scale, there would be

Table 8-2c. Estimated parameters on energy equations for Asia, Residential sector

	Price Elasticity (*t*-value)	Output Elasticity (*t*-value)	Adjusted RSQ	DW
Thailand	*	−0.71 (−3.08)	0.79	0.77*
Malaysia	−0.58 (−12.19)	−0.40 (−3.39)	0.97	0.62*
Singapore	−0.86 (−3.70)	−0.65 (−3.51)	0.56	0.63*
Indonesia	−0.70 (−1.64)	−1.29 (−2.88)	0.84	1.90
Philippines	*	0.67 (−6.22)	0.81	1.05
Taiwan	−0.19 (−1.25)	−0.10 (−0.66)*	0.38	0.65*
Korea	*	−0.28 (−1.89)	0.89	0.65*

* indicates that no significant results were observed.

efficiency gains. If our observation is correct there is no such effect in the case of transportation. This may very well be true in the Asian context where a mass-transit system and large-scale cargo transportation network tend to be lacking. If we look at the Industry sector we realize that there are cases where the scale effect is captured together with responsiveness to relative price change. In the Residential sector price elasticity is observed for four countries out of seven, and output elasticity is uniformly observed, and also significant, in most of the countries under consideration.

4. Simulation experiment

It is necessary to examine the performance of the model before actually conducting policy analysis. We have carried out a simulation experiment by incorporating the energy balance models for South-East Asia into the globally linked COMPASS covering economy–energy–environment spheres. Just as we did in Chapter 7, which dealt with the industrialized countries of G7, we examine whether the simulation results are consistent with other studies in terms of the magnitude and direction of change. The models for South-East Asia focus on changes in energy intensity in response to price signals. As the reference case we chose energy demand and supply outlook for the APEC (APERC, 1998).

This publication is composed of energy balance tables for the APEC, which include all of the countries covered by the energy balance model that is developed in this chapter. The publication includes three cases including baseline, environmentally friendly scenario, and protracted case scenario. We base our comparison on the baseline scenario. The outlook is provided for 1995 (actual), 2000, 2005, and 2010. Our simulation results are available on an annual basis but we only look at year 2010, the terminal year of the experiment.

Let us look at the simulation for Thailand. The total TPES in 2010 is calculated to be 87.4 Mtoe whereas APERC predicts 95.9 Mtoe. In terms of primary energy supply the COMPASS scenario is more optimistic about the introduction of renewables by 2010. We share a similar view regarding the rapid introduction of natural gas, whereas we are more reserved about oil and petroleum products. In the final consumption (TFC) in general, we tend to believe that there would be more energy saving in Industry and Transportation. The Malaysian economy is projected to hover throughout the decade at the level that prevailed during 1998-2000 period, which is lower than in 1995. We predict a lower number than APERC, that predicts 77.2 Mtoe in 2010. Singapore is a small state with market orientation and, perhaps reflecting this, the results from two sources are almost identical. The COMPASS simulation experiment generates 25.7 Mtoe compared to 25.3 from the APERC. Indonesia is a country for which it is extremely difficult to predict the future based on price signals because of the hyperinflation that country has gone through and political turbulence that the country is experiencing. Our macroeconomic indicators predict that the Indonesian economy will be 7% larger in 2010 compared to the 1995 level, and this should be reflected in energy demand. Our problem with Indonesia also lies on the energy supply side. It seems we have an exaggerated view on the availability of gas and renewables in that country if we believe country experts who have produced the APERC projections. As for the Philippines, COMPASS is more optimistic than the APERC, predicting 58.1 Mtoe compared to 48.8 Mtoe in 2010. We believe the Industry sector will be more energy saving while the Residential sector will be using twice as much energy as is predicted by the APERC. We share the similar view concerning the Transportation sector. COMPASS and the APERC have different views on Taiwan. The major difference is that we believe the economy there will become much more energy efficient than is predicted by the APERC. The TFC in Taiwan was 33.4 Mtoe in 1990 and 44.3 Mtoe in 1995 according to COMPASS (or 42.2 Mtoe in 1995 in the APERC). The figure for 2010 is 35.6 Mtoe for COMPASS compared to 84.6 Mtoe for APERC. Industry will be three times as lean in 2010. For Korea we share a common view with the APERC. The TFC in 2010 is predicted to be 179.5 Mtoe by COMPASS whereas the comparable figure from the APERC is 178.0. The number for the Residential sector is identical, and that for the Transportation sector is more or less the same, although we believe the Industry sector will be more energy efficient than is assumed by the APERC. The major difference is found in the energy supply: we believe more gas could be used in Korea and we are more optimistic about the introduction of nuclear energy there. This is again a national policy issue rather than a model assumption.

The comparisons generally prove that the simulation of energy balance tables can actually be done for developing economies. We have limited our modelling attempt to the energy intensity in the final use sectors including Industry, Transportation, and Residential, and excluded the shift among energy sources which was one of the focal points in modelling G7 countries. The reason is that, in the Asian context, the choice among primary energy sources is often a policy decision rather than market phenomena participated by many 'atomic' agents.

5. In lieu of conclusion

The COMPASS approach in database construction and model-building is characterized by an object-oriented approach. Although this concept was established nearly two decades ago among database theoreticians, and some database development tools have become available in the market in the past decade, the content side is still lagging in actually implementing the concept in the empirical context. Statistics the world over are still published in a more or less segmented and self-contained user interface which, once faced with a need to produce an integrated database, proves to be inflexible. What is more lagging is the application of the object-oriented approach to model-building and simulation. It is evident, however, when one wants to develop an empirically based model for the global community with detailed country resolution as well as industrial branch classification encompassing enormous economy–energy–environment spheres, a traditional approach, where a single country model is developed and then international linkage is attempted, cannot be followed. We have tried to fill this gap.

We have demonstrated that the object-oriented approach to database and modelling is indeed possible and effective. The COMPASS approach in model-building is to develop a prototype such as we did in Chapters 5 and 7 for the energy balance model and apply the logic to other cases. This is the focal point of this chapter. It can now be said that the experiment went rather well, although considerable refinement in the regression analysis must follow. We can reasonably hope that this approach can be used more extensively in future developments of the model.

9
The case of China

WANG YINCHU and ZUO LI

1. Progress in the national economy

The Chinese economy has grown at a fast rate since 1978 when the policy of economic reform and openness commenced. The GDP values in nominal, the GDP per capita in nominal, the GDP indexes in real terms and the GDP per-capita index in real terms, from 1978 to 1997, are shown in Table 9-1. It can be seen that, in real terms, the GDP in 1997 was 5.82 times the 1978 GDP. During the period from 1978 to 1997 the average yearly growth rates of GDP and GDP per capita were 9.8% and 8.4%, respectively. The average GDP growth rate is 6.5% higher than the world average growth rate in the same period, 7.3% higher than the developed countries, 6.5% higher than other developing countries, 1.9-3.5% higher than Singapore, Korea, Chinese Taiwan, Malaysia and Indonesia, all of which showed high economic growth rates in the same period.

As a result of the high speed of economic development, Chinese integrative national power increased greatly, the industrial structure improved significantly and Chinese people's quality of life improved considerably.

According to the World Development Report of the World Bank, the Chinese GDP value was 252.23 billion US dollars in 1978 and 1055.4 billion US dollars in 1997. The position of Chinese GDP in the top ten countries in the world was raised from 8th in 1978 to 7th in 1997, as shown in Table 9-2. From the table it can also be seen that the difference between the Chinese GDP and the one of the 6th country (Italy) of the top ten countries has reduced from 36% (1 − 252230/393950) in 1978 to 8.7% (1 − 1055400/1155400) in 1997.

Meanwhile, Chinese foreign trade value increased considerably. The total amount of Chinese foreign trade was 20.64 billion US dollars in 1978 and China held the position of 27th in the world order. Exports were 9.75 billion US dollars and this showed the share of 0.81% in total world export and 28th position. In 1997 Chinese foreign trade had extended to 325.1 billion US dollars and was 15.8 times the 1978 level. The position in world trade rose to 10th. Export values reached 182.7 billion US dollars and this was 18.7 times the 1978 export value. Export share increased to 3.3% of world total; the international reserve also increased greatly from 0.17 billion US dollars in 1978 to 139.9 billion US dollars in 1997. China was then in second position in the world, and just below Japan. The development of Chinese foreign trade 1978-1997 is shown in Table 9-3.

K. Uno (ed.), Economy–Energy–Environment Simulation, 145–161.

Table 9-1. The GDP and per-capita GDP of China in the past 20 years

Year	GDP at Current Price, 100 Million Yuan	Per Capita GDP at Current Price, Yuan	GDP Index at Comparable Price, Preceding Year = 100	Per Capita GDP Index at Comparable Price, Preceding Year = 100
1978	3624.1	172.8	111.7	110.2
1979	4038.2	184.2	107.6	106.1
1980	4517.8	205.0	107.8	106.5
1981	4860.3	211.1	105.2	103.9
1982	5301.8	236.7	109.3	107.5
1983	5957.4	264.9	111.1	109.3
1984	7206.7	327.1	115.3	113.7
1985	8989.1	406.9	113.2	111.9
1986	10,201.4	475.6	108.5	107.2
1987	11,954.5	544.9	111.5	109.8
1988	14,922.3	661.0	111.3	109.5
1989	16,917.8	786.0	104.2	102.5
1990	18,598.4	1147.5	104.2	102.3
1991	21,662.5	1409.7	109.1	107.7
1992	26,651.9	1681.8	114.1	112.8
1993	34,560.5	2123.2	113.1	112.2
1994	46,670.0	2685.9	112.6	111.4
1995	57,494.9	3054.7	109.0	109.3
1996	66,850.5	3494.0	109.8	108.4
1997	73,142.7	3797.2	108.5	107.7

Table 9-2. International comparison of GDP (M. = Million US$, R. = Rank)

	1980		1985		1990		1995		1997	
	M.	R.	M.	R.	M.	R.	M.	R.	M.	R.
US	2,587,100	1	3,946,600	1	5,392,200	1	6,952,020	1	7,690,100	1
Japan	1,039,980	2	1,527,900	2	2,942,890	2	5,108,540	2	4,772,300	2
Germany	819,140	3	624,970	3	1,488,210	3	2,415,764	3	2,319,500	3
France	651,890	4	510,320	4	1,196,780	4	1,536,089	4	1,526,600	4
Britain	522,850	5	454,300	5	975,150	6	1,105,822	5	1,220,200	5
Italy	393,950	6	358,670	6	1,090,750	5	1,086,932	6	1,155,400	6
Canada	253,350	7	346,030	7	570,150	7	568,928	9	583,900	9
China	252,230	8	265,530	8	364,900	10	697,647	7	1,055,400	7
Brazil	237,930	9	188,250	9	414,060	9	688,085	8	773,400	8
Spain	198,320	10	164,250	11	491,240	8	558,617	10	570,100	10
Mexico	166,700	11	177,360	10	237,750	13	250,038	13	348,600	16

Along with the rapid development of the economy, the Chinese industrial structure was adjusted towards the direction of maturity. The ratios among primary industry, secondary industry and tertiary industry were 28.1:48.2:23.7 in 1978. These figures changed to 19.1:50.0:30.9 in 1997. The change was basically consistent with the development of industrial structure throughout the world, i.e. the ratio of primary industry decreasing and the ratios of secondary and tertiary industries increasing. In the sector of primary industry, the situation of farming production was the major part (the ratios in 1978 among farming, forestry, animal

Table 9-3. Total imports and exports (unit: US$ 100 million)

Year	Total Imports & Exports	Total Exports	Total Imports	Balance
1978	206.4	97.5	108.9	−11.4
1980	381.4	181.2	200.2	−19.0
1985	696.0	273.5	422.5	−149.0
1986	738.5	309.4	429.1	−119.7
1987	826.5	394.4	432.1	−37.7
1988	1027.9	475.2	552.7	−77.5
1989	1116.8	525.4	591.4	−66.0
1990	1154.4	620.9	533.5	87.4
1991	1356.3	718.4	637.9	80.5
1992	1655.3	849.4	805.9	43.5
1993	1957.0	917.4	1039.6	−122.2
1994	2366.2	1210.1	1156.1	54.0
1995	2808.6	1487.8	1320.8	167.0
1996	2898.8	1510.5	1388.3	122.2
1997	3251.6	1827.9	1423.7	404.2

husbandry and fishery were 80:3:15:2); these changed to 58:4:28:10 in 1997. As regards secondary industry, China experienced light industry having priority to develop, with the result that light and heavy industries developed in synchronization. The problem of a shortage of daily goods was solved first and the market boomed. The bottleneck problem of the supply capacity of infrastructure sectors such as energy, raw materials production, transportation and communication has been solved to some degree. In the sector of tertiary industry, finance, insurance and real estate developed rapidly. This changed the traditional situation in which tertiary industry mainly consisted of commerce and restaurants. Table 9-4 shows the history of Chinese industrial structure change from 1978 to 1997.

The fast growth of the economy led to a fast increase in residents' income. The per-capita net income of rural households increased from 133.6 yuan in 1978 to 2090 yuan in 1997. As regards price factors, the average yearly growth rate was 8.1%. The per-capita disposable income of urban households rose from 343.4 yuan to 5160 yuan. Adjusting for price factor the average yearly growth rate was 6.2%. The urban and rural household saving deposit was 4628 billion yuan; adding cash in hand and various securities, household financial assets exceeded 6000 billion yuan at the end of 1997. Along with increased income, the life quality of the population has obviously been improved. Household consumption levels increased from 184 yuan in 1978 to 2677 yuan in 1997. The yearly growth rate was 7.7% in real terms. The consumption structure has also showed much change. The Engel coefficient (the ratio between food expenditure and total expenditure) of rural households decreased from 67.7% in 1978 to 55.1% in 1997. The Engel coefficient of urban households decreased from 57.4% to 46.4% during the same period. Table 9-5 shows details of increasing household income.

Table 9-4. Industrial structure changes (%)

Year	GDP	Primary Industry	Secondary Industry	Tertiary Industry
1978	100	28.1	48.2	23.7
1979	100	31.2	47.4	21.4
1980	100	30.1	48.5	21.4
1981	100	31.8	46.4	21.8
1982	100	33.3	45.0	21.7
1983	100	33.0	44.6	22.4
1984	100	32.0	43.3	24.7
1985	100	28.4	43.1	28.5
1986	100	27.1	44.0	28.9
1987	100	26.8	43.9	29.3
1988	100	25.7	44.1	30.2
1989	100	25.0	43.0	32.0
1990	100	27.1	41.6	31.3
1991	100	24.5	42.1	33.4
1992	100	21.8	43.9	34.3
1993	100	19.9	47.4	32.7
1994	100	20.2	47.9	31.9
1995	100	20.5	48.8	30.7
1996	100	20.4	49.5	30.1
1997	100	19.1	50.0	30.9

Table 9-5. Per-capita annual income of households (unit: yuan)

Year	Rural	Index (1978 = 100)	Urban	Index (1978 = 100)
1978	133.6	100.0	343.4	100.0
1980	191.3	139.0	477.6	127.0
1985	397.6	268.9	739.1	160.4
1986	423.8	277.6	899.6	182.5
1987	462.6	292.0	1002.2	186.9
1988	544.9	310.7	1181.4	182.5
1989	601.5	305.7	1375.7	182.8
1990	686.3	311.2	1510.2	198.1
1991	708.6	317.4	1700.6	212.4
1992	784.0	336.2	2026.6	232.9
1993	921.6	346.9	2577.4	255.1
1994	1221.0	364.4	3496.2	276.8
1995	1577.7	383.7	4283.0	290.3
1996	1926.1	418.2	4838.9	301.6
1997	2090.1	437.4	5160.3	311.9

2. The economic problems

Chinese economic development was not plain sailing. For example, China was successful in controlling the fast growth of the population and in transferring the labour force from agriculture to secondary and tertiary industries in the past 20 years. However, China is still under very great pressure from the size of the population, especially in rural areas. This is a long-term problem. There were other problems in the past 20 years, including the following four points that are prominent.

(1) The forming of the unified market was slow and the efficiency of resource allocation was low. China is in the process of changing its economic system. In this process, on the one hand, the traditional system is losing its dominant position. On the other hand, the new market economic system is not fully in place. Local protectionism fragmented the market in reality. This had a bad impact on raising the efficiency of resource configuration. It did not achieve equilibrium in various regional developments in China. There are quite large and obvious differences between the north and the south, between shoreland and backland. There are conflicts and contradictions among developed and undeveloped regions. Local governments involved in the competition process have distorted the competition among enterprises; therefore, the market was segmented and the efficiency of the national economy was lost.
(2) The inequality of wealth resulted in social dissatisfaction. It was pointed out in many research reports that income distribution was becoming increasingly unequal when reforms because deeper. According to the World Bank the Gini coefficient of China was 28.8 in 1981 and reached 38.8 in 1995 (*Sharing the Increasing Income*, World Bank, 1998). The difference in the occupancy of state-owned assets and the difference in the right of using state-owned assets were important factors in expanding the difference of household incomes. They were also important reasons behind social dissatisfaction.
(3) The relative surplus happened under conditions in which the development level was not high. Even though Chinese household income and consumption levels increased during the reform and opening-up process, it is still at a comparatively low level compared to developed countries. According to the World Bank (*World Development Report, 1998/99*), the GDP per capita of China was 860 US dollars in 1997. This means that China has just stepped over the threshold of middle-low-income countries. However, due to the lopsidedness of income distribution, a relative surplus, from consumption materials to production materials, is currently extant. From the point of view of society as a whole the increase in urban household income and saving deposit is insufficient to support purchases of housing and cars; the increase in rural household income and saving deposit is insufficient to support the purchase of modern household goods, further restricting the development of the Chinese economy.
(4) The excess consumption of natural resources and pollution of the environment. The rapid development of the Chinese economy also consumed an ever-increasing amount of natural resources and caused deterioration the living environment. Farming land was reduced significantly. The per-capita area of cultivated land in 1997 was 60% of that in 1957. The water shortage

problem became very serious in many cities and rural areas. The efficiency of resource use, such as energy consumption per unit of output, was much lower than the advanced level of the world. Pollution from waste water, waste gas and solid waste is seriously destroying environmental quality.

3. The economic prospects

From the external point of view the Chinese economy will benefit from the maintained increase in the world economy and the fast process of economic globalization in the next 10 years. At the same time, from the internal point of view the Chinese economy will benefit from the setting up of a market economic mechanism and improvement in government management ability in the macro-economy in next decade. Under these conditions China has her long-term development goals. They are:

(1) To advance the integrative national power.
(2) To increase the ability of international competition.
(3) To improve people's life quality.
(4) To promote sustainable development.

With these goals, and a proper scenario, the COMPASS China model provides the prospect of the Chinese economy from 2000 to 2010.

The Chinese economy will maintain its high speed of development the next ten years. GDP values, in nominal, will reach 16,587 and 30,700 billion Chinese yuan in 2005 and 2010, respectively. The yearly average growth rates, in real terms, will be 7.20% from 2001 to 2005 and 7.23% from 2006 to 2010. The GDP per capita will reach 12,259 Chinese yuan in 2005 and 21,386 Chinese yuan in 2010. According to the exchange rate in year 2000 the GDP value will be 3712 billion US dollars in 2010, which will mean China reaching the 4th position among the top ten countries in the world, behind the US, Japan and Germany, if all the other countries which are in front of China at present have yearly growth rates of about 3.0% from 2001 to 2010. However, due to the huge population, the GDP per capita of China in 2010 will be 2586 US dollars, which will still be lower than the world average level. Table 9-6 shows the details.

Table 9-6. GDP in next 10 years

Year	2000	2005	2010	2000-2005	2005-2010
GDP (Billion Yuan)	9343.9	16,587.4	30,700.2		
GDP index in real term	100	141.60	200.75	7.20%	7.23%
GDP in 2000 USD	1129.9	2005.7	3712.2		
GDP per capita	7306	12,259	21,386		
GDP per capita index in real term	100	133.8	178.84	6.00	5.97%
GDP per capita in 2000 USD	884	1482	2586		

Table 9-7. Industrial structure changes to 2010

Year	Primary Industry	Secondary Industry	Tertiary Industry
2000	16.9	49.6	33.5
2001	16.3	50.0	33.7
2002	15.8	50.0	34.2
2003	15.5	49.8	34.7
2004	15.1	49.8	35.1
2005	14.8	49.9	35.3
2006	14.7	49.7	35.7
2007	14.3	49.8	35.8
2008	14.0	49.4	36.6
2009	13.8	48.9	37.3
2010	13.5	48.6	37.9

The Chinese economy is still in the process of industrialization. The change in industrial structure is one of the main sources to accelerate the growth of the economy. Table 9-7 shows the changes in industrial structure from 2000 to 2010. It can be seen that the share of primary industry in GDP will continue to decline, from 16.9% in 2000 to 13.5% in 2010. The share of secondary industry will show a slight increase and reach a peak value of 50% in the period from 2001 to 2005. This share will show a slight decrease in the subsequent five years. On the other hand the share of tertiary industry in GDP will show a stable increasing trend, from 33.5% in 2000 to 37.9% in 2010.

Adjustment of the internal structure of agriculture will continue. The share of farming will continue to go down, that of forestry will keep the same value as before, while animal husbandry and fishery will go up in the next ten years. The gross output ratios of farming, forestry, animal husbandry and fishery will be about 53:4:31:12 in 2005 and 51:4:33:12 in 2010.

As for the internal picture of manufacturing industry, the structure will show the following changes, as shown in Table 9-8.

(1) In the energy production aspect the share of coal mining in whole industry gross output will go down from 1.45% in 2000 to 1.34% in 2010 while the share of the production of crude petroleum and natural gas will go up from 0.93% in 2000 to 1.0% in 2010. Meanwhile, the share of electricity production and supply, including hot water and steam, in the whole industry gross output will go up from 3.53% in 2000 to 3.80% in 2010; the share of petroleum refining will increase from 2.95% in 2000 to 3.47% in 2010.

(2) The share of food processing will increase from 5.09% in 2000 to 5.77% in 2010, which means the degree of processing food will be greater than before. The increase in the share of beverage shows the Chinese people will have more drinks, such as milk, fruit juice, and so on. At the same time the share of tobacco production will decrease from 1.59% in 2000 to 1.37% in 2000.

(3) The traditional industrial sectors, such as textiles and, wearing apparel production, among others, will tend to decrease their shares, but still keep

Table 9-8. Structure changes of industry to 2010 (gross output, selected sectors)

Sector	2000	2005	2010
Coal mining	1.45	1.39	1.34
Crude petroleum and natural gas	0.93	0.97	1.00
Food process & manufacturing	5.09	5.47	5.77
Beverages	3.28	4.22	5.14
Tobacco manufacture	1.59	1.47	1.37
Textiles	7.67	7.11	6.88
Wearing apparel	3.00	2.78	2.69
Leather, fur and their products	1.95	1.97	2.07
Paper and paper products	1.70	1.65	1.61
Printing industries	0.86	0.93	0.97
Cultural, education, sports articles	0.63	0.71	0.77
Petroleum refineries	2.95	3.30	3.47
Chemical industries	6.35	6.61	6.86
Medicines	1.81	1.86	1.88
Chemical fibers	1.04	1.16	1.28
Rubber products	1.08	1.05	1.03
Plastic products	2.35	2.49	2.62
Primary iron and steel	5.22	4.39	3.71
Primary non-ferrous metals	1.59	1.24	1.01
Metal products	4.17	3.77	3.42
Ordinary machinery	7.35	6.90	6.43
Motor vehicles	5.48	6.27	6.73
Electric machinery and instrument	4.55	4.34	4.17
Electronic and communication equip.	4.55	4.08	3.84
Electricity, steam and hot water	3.53	3.72	3.80
Production and supply of water	0.29	0.32	0.34

large ratios, more than 9.5%, in the whole industry gross output. Meanwhile, the leather, fur and their product will increase the share, reflecting the change of demand for clothes.

(4) The shares of chemical industry, medicine, chemical fibre and plastic production, will increase continuously. Rubber production will maintain almost the same share. Including the situation of petroleum refining, we can conclude that the share of heavy chemical industry will rise in the Chinese economy in next ten years.

(5) The shares of primary iron and steel, primary non-ferrous metals, metal products and ordinary machinery will go down and the share of motor vehicles will go up. This is a reasonable change in manufacturing structure. However, the decrease in shares of electric machinery and electronic products is not consistent with the globally observed development trend. This reflects the fact that there is a shortage in supply capacity of electric machinery and electronic products, suggesting the need to strengthen investment in these sectors.

Within the tertiary industry sector the structure changes towards the direction of raising the tertiary industry level. Table 9-9 shows the changes in shares of selected sectors in the whole tertiary industry gross outputs.

Table 9-9. Structure changes of tertiary industry to 2010 (gross output, selected sectors)

	2000	2005	2010
Railway transportation	2.07	1.66	1.42
Highway transportation	8.77	9.22	9.85
Air transportation	1.07	1.10	1.14
Communications	2.58	2.91	3.18
Commerce	30.50	28.41	27.62
Restaurants	7.65	8.20	8.04
Finance and insurance	13.95	14.61	15.83
Real estate and social services	12.31	13.47	14.47
Health care, sports and social welfare	4.09	3.70	3.24
Education, culture, arts, radio, film, TV	5.14	4.67	4.23
Scientific research and technical services	3.33	2.57	2.04

The share changes in transportation, i.e. the railway decreasing while highway and air increase, reflect the general trend of the development speed of different transportation modes. The traditional commerce service goes down significantly, but still has a high share. Along with the change of lifestyle, restaurant service will raise its share continuously. Communication, finance and insurance, as well as real estate and social services, will increase their shares steadily. It should be noticed that the sectors of health care and social welfare, education and culture service, scientific research and technical services will lose their shares. This points to the fact that the development speed of these sectors is slower than the whole tertiary industry. It also foretells that there is a possibility that some aspects of social development will lag behind in economic development if there are no special policy measures in place.

After successfully resisting the Asia financial crisis and joining the WTO, Chinese foreign trade will maintain its tendency to increase. The average yearly growth rate of exports and imports, in real terms, will be more than 10%, which is larger than the GDP growth rate. However, comparing the positive values of the net export from 1998 to 2003, net export will become negative from 2004. Table 9-10 shows a forecast of total imports and exports in current prices.

The import and export structures will also change. Shares of imported crude petroleum and petroleum refining in total imports will increase, while the share of metal manufacturing (primary iron and steel, primary non-ferrous metals and metal products) will decrease. The import share of machine manufacturing (ordinary machinery, electric machinery and instruments, motor vehicles, electronic and communication equipment) will keep its high value, which is around 50%. The export share of primary products (agriculture, mining) in total export will go down while the share of textile products, including clothes, will go up. The export share of machinery equipment and electronic products will go up.

As a result of the fast economic growth household income and consumption levels are expected to rise quickly. The per-capita net income of rural households will increase from 2700 yuan in 2000 to 7500 yuan in 2010. Adjusting for the

Table 9-10. Exports and imports to 2010 (current prices, 100 million Chinese yuan)

Year	Export	Import	Net Export
2000	17,767	15,898	1869.1
2001	20,642	19,332	1309.4
2002	23,956	22,900	1056.2
2003	27,771	27,043	727.9
2004	32,158	32,159	−1.6
2005	37,196	37,647	−450.7
2006	42,977	43,828	−851.6
2007	49,600	50,457	−856.8
2008	57,180	58,645	−1464.2
2009	65,846	68,358	−2512.0
2010	75,741	79,493	−3752.2

price factor the average yearly growth rate will be 7%. The per-capita disposable income of urban households will rise from 7000 yuan in 2000 to 13,000 yuan in 2010. As regards the price factor, the average yearly growth rate will be 5.5%. Household consumption levels will increase from 3530 yuan in 2000 to 10,500 yuan in 2010. The yearly growth rate will be 6% in real terms. The consumption structure will improve continuously, and the Engel coefficient of rural households will decrease from 52.4% in 2000 to 43% in 2010, while that for urban households will decrease from 44.8% in 2000 to 32% in 2010.

4. Progress in the energy sphere

Energy is the important physical basis of economic and social development. In the procedure of economic opening and reform since 1978 the energy industry has made a great contribution to Chinese economic achievement.

(1) The capacity of energy production extended rapidly and the position in world energy rose continuously. In the process of economic development the Chinese government placed the energy industry as the highest priority sector in the investment list in order to increase the capacity of energy production. From 1978 to 1997 the output of primary energy increased from 628 million tons to 1324 million tons of standard coal equivalent. Of this total the output of coal increased from 618 million tons to 1387 million tons; the output of crude oil increased from 104 million tons to 163 million tons; the output of natural gas increased from 13,730 million cubic metres to 24,110 cubic metres; and the output of hydropower increased from 44,600 million kWh to 196,000 million kWh. Nuclear power came into use in 1995 and its output was 14,420 million kWh. Details are shown in Table 9-11.

Following its development in the past 20 years, China has become one of the biggest countries in the world in energy production. The output of primary energy is number 3 in the world and in excess of 10% of the world total. From

Table 9-11. Production of primary energy

Year	Coal Production, 100 Million Tons	Crude Oil, 100 Million Tons	Natural Gas, 100 Million Cube Meters	Power, 100 Million KWh	
				Hydro-power	Nuclear Power
1978	6.18	1.04	137.30	446	
1980	6.20	1.06	142.70	582	
1985	8.72	1.25	129.30	924	
1990	10.80	1.38	152.98	1267	
1995	13.61	1.50	179.40	1906	128
1996	13.97	1.57	201.14	1880	143
1997	13.87	1.63	241.10	1960	144

Table 9-12. Total production of energy and its composition

Year	Total Energy Production, 10,000 Tons of SCE	As Percentage of Total Energy Production (%)			
		Coal	Crude Oil	Natural Gas	Hydro-power
1978	62,770	70.3	23.7	2.9	3.1
1980	63,735	69.4	23.8	3.0	3.8
1985	85,546	72.8	20.9	2.0	4.3
1986	88,124	72.4	21.2	2.1	4.3
1987	91,266	72.6	21.0	2.0	4.4
1988	95,801	73.1	20.4	2.0	4.5
1989	101,639	74.1	19.3	2.0	4.6
1990	103,922	74.2	19.0	2.0	4.8
1991	104,844	74.1	19.2	2.0	4.7
1992	107,256	74.3	18.9	2.0	4.8
1993	111,059	74.0	18.7	2.0	5.3
1994	118,729	74.6	17.6	1.9	5.9
1995	129,034	75.3	16.6	1.9	6.2
1996	132,616	75.2	17.0	2.0	5.8
1997	132,410	74.1	17.3	2.1	6.5

1978 to 1997 the position of coal production rose from number 3 to number 1 in the world, while the position of crude oil production rose from number 8 to number 5. The position of generated electricity rose from number 7 to number 2. The improvement in energy supply greatly alleviated the bottleneck restriction on economic development.

(2) The structure of energy starts to change and the demand for clean energy increases. Tables 9-12 and 9-13 show the structure changes in energy production and consumption. It can be seen that, while coal still shows a high ratio in energy structure, the demand for clean energy shows an increasing tendency. The structure change of energy consumption showed that the ratio of coal had decreased since 1991, whereas the ratio of crude oil and natural gas had increased since 1996. The continuously increasing demand from the domestic market for crude oil products stimulated the import of crude oil. China has become a net import country of crude oil since 1993. There is a trend towards increasing the import of crude oil.

Table 9-13. Total consumption of energy and its composition

Year	Total Energy Consumption, 10,000 Tons of SCE	As Percentage of Total Energy Production (%)			
		Coal	Crude Oil	Natural Gas	Hydro-power
1978	57 144	70.7	22.7	3.2	3.4
1980	60 275	72.2	20.7	3.1	4.0
1985	76 682	75.8	17.1	2.2	4.9
1986	80 850	75.8	17.2	2.3	4.7
1987	86 632	76.2	17.0	2.1	4.7
1988	92 997	76.2	17.0	2.1	4.7
1989	96 934	76.0	17.1	2.0	4.9
1990	98 703	76.2	16.6	2.1	5.1
1991	103 783	76.1	17.1	2.0	4.8
1992	109 170	75.7	17.5	1.9	4.9
1993	115 993	74.7	18.2	1.9	5.2
1994	122 737	75.0	17.4	1.9	5.7
1995	131 176	74.6	17.5	1.8	6.1
1996	138 948	74.7	18.0	1.8	5.5
1997	138 173	71.5	20.4	1.7	6.2

The increasing demand for clean energy stimulated the development of the industry of electricity production and supply. From 1978 to 1997 the power-generating capacity increased from 57 million kW to 237 million kW. The total electricity generated from 256,600 million kWh increased to 1,135,553 million kWh.

(3) The efficiency of energy use improved considerably and the coefficient of energy elasticity decreased continuously. Simultaneously with developing energy resource, China paid more attention to the popularization of the technology of saving energy and made considerable efforts towards that end. From 1980 to 1995 energy consumption per-capita GDP decreased, with a yearly average of 4.4%. The total energy saved was about 620 million tons of standard coal equivalent. The improvement in efficiency of energy use helped aided economic development under the condition of limited energy resources. Since the commencement of reform and opening, China's coefficient of energy elasticity has decreased continuously. The average level of the coefficient was 1.67 from 1971 to 1975. This decreased to 0.96 in the period from 1976 to 1980. In the period from 1981 to 1990 the coefficient further decreased to 0.56. The average level in the period from 1991 to 1997 was 0.50.

(4) Energy consumption per capita showed an obvious rise. Table 9-14 shows the situation of primary energy consumption per capita in China. This shows that primary energy consumption per capita rose from 594 kilograms of standard coal equivalent in 1978 to 1165 kilograms of standard coal equivalent in 1997. However, it is still at a low level. There is quite a large difference compared with developed countries and some developing countries. In 1996 primary energy consumption per capita in China was half of the world

Table 9-14. Per capita energy production and consumption (unit: kilogram of standard coal equivalent)

Year	Production	Consumption
1978	652	594
1980	646	611
1985	808	724
1986	820	752
1987	835	793
1988	863	838
1989	902	860
1990	909	863
1991	905	896
1992	915	932
1993	937	979
1994	991	1024
1995	1065	1083
1996	1084	1135
1997	1071	1118

average. It was less than 10% of the level of the USA, Canada and Korea, less than 20% of the level of Japan, Germany, France and the UK. Furthermore, the ratio of energy used for general living was low, being about 11%.

5. Problems in the energy sphere

(1) The relationship between energy development and national economy development is not harmonized. From the history of developed countries' economic development and the trend of present world development, there is a pronounced relationship between energy consumption level per capita and economic development level. To date there is no such precedent in the world of reaching high a economic development level with very low energy consumption per capita. In other words, under the technology and consumption mode that prevails at present, a high level of energy consumption per capita is necessary for reaching the economic development level of industrialization. Since energy consumption per capita is still at a very low level, and there is a huge population in China, the demand for energy will show a big increase during the process of modernization. However, the development speed of the energy industry is not adequate for the development speed of the national economy. In the period from 1978 to 1997 the average yearly growth rate of GDP was 9.8% and the average yearly growth rate of energy production was 4.1%. The coefficient in energy elasticity was 0.42. The improvement in energy efficiency had positive promotional effects in decreasing the coefficient of energy elasticity. However, from the long-term point of view, the prolonged shortage of energy had restrained the demand for energy. To realize the elevation of economic and social development levels it is necessary that the energy industry shows faster development.

(2) The structure of energy production is not fit for the demand from economic development. The shortage of high-quality energy is the main concern. The energy structure in which coal plays the main part has not changed fundamentally in China in recent years. However, the demand for clean energy rises steadily. That high-quality energy plays a major part in the energy structure is a consequence of social and economic development. In a relatively closed energy system such as China the production structure and consumption structure of energy are basically the same. Compared with the increasing demand for high-quality energy, the production structure in which coal plays a major part will become worse in satisfying the demand. This is one of the main difficulties in developing the energy industry in China.
(3) The efficiency of energy use is low. The energy consumption per unit of output in China is higher not only than that of developed countries, but also than the average level of other developing countries. There are several reasons for this, e.g.: (a) the share of manufacturing industry in GDP is higher than in other developing countries; (b) solid fuel comprises the main energy source, which lowers energy efficiency; (c) a low technical and management level in energy use. Improvement in the efficiency of energy use depends not only upon progress in technology, improvement in management, and adjustment in the industrial structure, but also on elevation of national policies.
(4) A contradiction between demand and supply of energy in rural areas emerges. There are 850 million people living in rural areas in China; 70% of the energy for the livelihood of the rural people is biology-nature energy. This destroys the forest vegetation in vast areas. Mass of straw cannot be returned to the field, and this lowers the amount of organic material in soil. Furthermore, the electrification level is low in the countryside and there are about 100 million people who have no chance to use electricity. Third, the energy consumption of agriculture reaches a high level along with the increase in output per unit of land area. To meet these demands is a big task in energy supply in China.
(5) The environment problem resulting from energy production is serious. Development of the energy industry with the energy structure where coal plays the main part places huge pressure on the environment. Coal has a share of about 75%, and 80% of this is burned directly. The SO_2, CO_2 and soot created from burning coal result in air pollution of the fuliginous type. The extension of SO_2, hydronitrogen and volatile hydrocarbon harmed the environment and caused some regional acidification, which was due to emissions that were not fully treated. The acid rain area accounts for 29% of the country's land area. As the biggest country in coal consumption in the world, the volume of greenhouse gas emission was at the position of number 2 in the world according to the report of IEA in 1997. If there is no big change in energy structure, along with the increasing demand for energy, the volume of greenhouse effect gas emission will double and become

number 1 in the world in 2020. The environmental issue will be one of the important factors restricting energy development in China.

6. The energy prospect

(1) Energy consumption will increase continuously in China. Even considering the factors of technical progress in energy saving, changes in industrial structure and improvement in production management, total energy consumption in China will increase continuously along with the rapid economic development and the considerable improvement in people's quality of life. According to our analysis the total energy consumption of China in 2010 will be 2050 million tons of standard coal equivalent and this is 1.49 times the 1997 energy consumption. The average yearly growth rate is 3.12%. Coal consumption will be 1800 million tons in 2010, i.e. 1.29 times the 1997 coal consumption, or an average yearly growth rate of 1.98%. Crude oil consumption will be 370 million tons in 2010, i.e. 1.88 times the 1997 crude oil consumption. The average yearly growth rate is 4.96%. Natural gas consumption will be 60,000 million cubic metres in 2010, i.e. 3.07 times the 1997 natural gas consumption with an average yearly growth rate of 9.02%. Electricity consumption will be 2560 billion kWh in 2010, i.e. 2.27 times the 1997 electricity consumption, with an average yearly growth rate of 6.50%.

(2) Coal will still from the main part of the structure of energy consumption. Since the limitation in energy resource, the main part of the structure of energy consumption in China will still be coal. This type of energy consumption framework was formed for long time because China thought it was a great threat to the national security to import energy in large quantities, especially crude oil and natural gas, and therefore took the strictly autarkic energy guarantee strategy at all times. The large part played by coal will see no essential change in next 10 or 20 years. To reduce the pressure on environment and transportation, new large investments in energy transportation system, such as pipelines for coal slurry, will be carried out. The carefully washed coal will be widely used and its ratio in total coal consumption will rise from 20% at present to more than 50% in 2010.

(3) It will be imperative to greatly extend the imports of crude oil and natural gas. The demand for crude oil and natural gas, which exceeds its supply, will be another basic character of China's energy market in the future. China is one of the countries in which there is a lack of crude oil resources. Expansion of the imports of crude oil will continue and the ratio between net imports and total consumption of crude oil will rise from 20% in 1997 to 33% in 2010. China becoming a main importer of crude oil will have an important influence on the international oil market. This is probably good news and a new opportunity for international oil companies and for many oil-producing countries.

Table 9-15. Forecast of power supply capacity to 2010

		2000		2010	
			%		%
Thermal Power	Capacity (million kW)	235	76.87	413.8	77.24
	Output (billion kWh)	1175	83.17	2069.0	82.33
Hydropower	Capacity (million kW)	67.5	22.08	100.0	18.66
	Output (billion kWh)	220.0	15.57	318.0	12.66
Nuclear Power	Capacity (million kW)	2.7	0.88	20.0	3.73
	Output (billion kWh)	16.2	1.15	120.0	4.78
New Energy	Capacity (million kW)	0.5	0.17	2.0	0.37
	Output (billion kWh)	1.6	0.01	5.8	0.23
Total	Capacity (million kW)	305.7	100	535.8	100
	Output (billion kWh)	1412.8	100	2512.8	100

(4) The supply capacity of electricity will increase to meet the demand for it. The forecast for power supply capacity is shown in Table 9-15. It can be seen that the main part of the power supply will still be generated from burning fuel (coal, oil and gas), which is more than 80% of the total. The capacity of power-generating units of thermal power will increase from 235 million kW in 2000 to 413.8 million kW in 2010. The average yearly growth rate is 5.85%. The capacity of power-generating units of hydropower will increase from 67.5 million kW in 2000 to 100 million kW in 2010. The average yearly growth rate is 4.1%. Nuclear power capacity will increase from 2.7 million kW in 2000 to 20 million kW in 2010. Its share in the total capacity of power-generating units will rise from 0.9% in 2000 to 3.7% in 2010. The capacity of power-generating units using other new energy, such as wind and solar, will increase from 0.5 million kW in 2000 to 2.0 million kW in 2010. The total capacity of power-generation units will increase from 305.7 million kW in 2000 to 535.8 million kW in 2010. The total output of power could reach 2512.8 billion kWh in 2010, which is close to the demand for electricity consumption (2560 billion kWh).

(5) Air pollution is serious in some regions. As mentioned earlier, the main source of energy in China is coal. Due to a lack of strong consciousness in protecting the blue sky, people did not pay attention to the elimination of soot in many factories. SO_2 emissions increased from 1550 million tons in 1990 to 1970 million tons in 1997. Air pollution is serious in some regions. Acid rainfall sometimes occurred. Meanwhile, CO_2 emissions rose from 670 million tons in 1990 to 750 million tons in 1997.

Air pollution will be controlled and the air quality will be improved in the next ten years based on the factors listed below:

(1) The energy structure will change continuously in the direction of reducing coal usage. This can be seen from section 4 of this chapter.

(2) Investment in equipments to eliminate smoke and particles will increase, since stricter air-protection measures will be implemented. SO_2 emissions will be reduced from 21.5 million tons in 2000 to 18.0 million tons in 2010.

(3) CO_2 emissions will still increase from 780 million tons in 2000 to 850 million tons in 2010. However, comparing the growth rate of the CO_2 emission from 1990 (670 million tons) to 2000 and the expanded scale and maturity of the economy, the growth rate of CO_2 emissions will slow down and CO_2 emissions per unit of GDP will decrease greatly. According to a forecast report from the Research Team on Sustainable Development, China Academy of Sciences, CO_2 emissions will reach a peak value (900 million tons) in 2020 and go down from that point on.

10
The case of Russia

ALEXEY KOLTSOV and VLADIMIR VOLKOV

1. Results of eight years of the Russian economy reforming

During eight years of reforms, in relatively short terms, the critical mass of changes carried out makes the reforms irreversible. The Russian economy has passed from a centrally operated economic system to a different one, though this one is very far from the systems established in developed market economies. This time large-scale privatization was carried out in Russia. In 1999, according to one evaluation, more than 75% of GDP was produced in the non-state sector of the economy. The share of production for which prices and tariffs are adjusted at the federal level did not exceed 15-16% of GDP. Securities and currency markets were formed, despite some problems, and are functioning, as well as financial institutions including the banking sector. The state monopoly in foreign trade was abolished, and Russian participation in world markets of commodities, capitals and services has become more active.

In 1997 it seemed changes in operation of the Russian economy had begun to produce positive results. Rates of inflation were reduced in 1997 to less than 1% per month. In the middle of the year interest rates have begun to decline values which allowed enterprises to make money borrowings for implementation of efficient projects in the real sector of the economy. After the deep decline which continued during some seven years, growth of real volumes of GDP and industrial production was observed. This happened mainly because of money borrowings by enterprises to increase their circulating capital.

However, the financial and economic crisis, with its sharpest phase coming in August 1998, has shown up the fragility and instability of positive changes in the Russian economy which were obtained beginning in 1997. The ruble was devalued sharply, and as a result the rate of inflation increased significantly, production declined sharply, the real incomes and standard of living of the population decreased perceptibly, the crisis in bank liquidity sharpened, and the confidence of investors and population in state and financial institutions was undermined. As regards the main macroeconomic parameters Russia was thrown back for some years (see Table 10-1).

At the same time the crisis had some positive consequences (admittedly through high prices), because it 'on appearance' has removed some accumulated deformations and disproportions which could not be removed by the government or the Bank of Russia.

K. Uno (ed.), Economy–Energy–Environment Simulation, 163–184.

Table 10-1. Macroeconomic parameters of the Russian economy in the 1990s (1992-1999, percentage of the previous year)

	1992	1993	1994	1995	1996	1997	1998	1999	1999 to 1991, %
Gross Domestic Product	85.5	91.3	87.3	95.9	96.6	100.9	95.1	103.2	62.5
Inflation (consumer's prices index, %):									
by the end of year	2609	940	315	231	122	111	184	136.5	6080 times
In average per month	31.2	20.5	10	7.2	1.7	0.9	5.2	2.6	
Industrial production	82	86	79.1	96.7	96	102	94.8	108.1	54.1
Price index of industrial production producers:									
by the end of year	3378.0	995	333	275	126	107.5	123	167	8570 times
In average per month	34.0	21.1	10.6	8.8	1.9	0.6	1.75	4.4	
Agriculture production	90.6	96	88	92	94.9	101.5	86.8	102.4	60.3
Investment in basic capital	60	88	76	90	82	95	93.3	104.5	27.4
Retail trade	100.3	101.6	100.2	93.6	99.5	103.8	96.7	92.3	88.1
Value of payable services to population	82	70	62	82	94	105.6	99.5	107.4	30.9
Real disposable income per capita	52	116	113	85	100.1	105.3	83.7	85.2	43.5
Unemployment (% to economically active population, by the end of year):									
by the ILO methodology	–	6.0	7.8	9	10	11.2	13.3	12.2	
officially registered	0.8	1.1	2.2	3.2	3.4	2.8	2.6	1.7	
Exports	–	111	113.2	120	109.3	99.6	84.4	100.7	139.5*
Imports	–	103.1	114	120.6	112.9	107	81.1	67.2	93.3*

* 1999 to 1992, %

Source: Here and below (if this is not stipulated specially) the actual data are from the Russian Statistic Committee.

Analysis of the August 1988 crisis usually begins for external reasons – change for the worse of the world markets, conjuncture of fuels, and the negative impact of the 'Asian' financial crisis. Certainly these factors have complicated the financial-economic situation. However, the reasons for the crisis are to a marked degree associated not with external reasons but with fundamental problems of the Russian economy which were not overcome during reforms.

Generalized comprehension of the Russian economy leads to the following observations that may be called lessons of reforms:

(1) Persistent problems of budget balancing, delays with structural and tax-budget reforms that have caused an avalanche-type growth of state debt, which has determined the long duration of difficulties experienced by Russia in this most important area. At the same time the good performance of financial institutions, which was supported not as a result of their operations with real sectors of economy but as a result of the high profitability of operations with state securities (really – at the expense of state budget) turned out to be an illusion and has disappeared for many banks, along with removal of the short-term state obligations pyramid. A result of the crisis was removal of disproportion by high prices.
(2) Errors in the exchange rate policy of the Bank of Russia, especially immediately before the crisis. As a result of perceptible strengthening of the ruble to dollar real exchange rate (when internal inflation is taken into account) the ruble exchange rate was evidently abnormally high. However, the Bank of Russia has continued to support the ruble exchange rate persistently, even under significant worsening of Russian trade and international payment balances with a strong negative impact of the abnormally high ruble exchange rate on domestic production. In this context overcoming the crisis by ruble devaluation (though it would be better to 'release' the ruble gradually) also had positive consequences.
(3) Inefficiency of foreign trade and tariff policy in combination with exchange rate policy has made the situation worse in a number of very important directions. In particular, the inconsistency between final demand and production was increased sharply, which was characterized by a deeper decline in production compared with consumption. At the same time the moderate dynamics of household consumption was supported, first of all, through increased imports and state loans. The volume of imports depended on exports of fuels and raw materials, causing production for domestic market to 'shrink'.
(4) Hypertrophied concentration on macroeconomic policy and insufficient attention to the reforms at micro level (in terms of real actions, if not openly declared) was one of the most important reasons which hindered the removal of structural disproportion, through the growth of production and improved efficiency.

(5) Weakness of state social policy, long-standing social problems at the periphery of reforms, and perceived intangibility of reform results for most of the population.
(6) Keen regional problems are not solved; this is manifested, in particular, in significant differentiation of Russian regional social and economic development, in insufficient sharing of authorities between federal and regional government bodies and interaction among them.
(7) Inconsistency of reforms, in particular, 'fluctuations' in carrying out macroeconomic and stabilizing policies when periods of rigid financial and monetary-credit policy are followed by periods of significant slackening, and vice-versa. This lowered the level of confidence in government from potential investors and population, a barrier not so easy to break.
(8) Serious problems with formation of the market economy norms based on laws and rules (mainly because of keen disparities between presidential and government structures from one side, and State Parliament (Duma) from the other side), as well as the inefficiency of laws and rules enforcing mechanisms. This, in particular, relates to the disability of the legal system protecting the rights for owners and investors, insufficient anti-monopoly regulation and measures aimed at formation of competitive environment, non-completion of laws and rules for accomplishment of private ownership rights for land and immovable fixed properties.
(9) The simplified notion on the quick transition from administrative-command system to market economy in Russia as well as on complexity of associated problems); heightened expectations of rapid, positive, and almost automatic results from inclusion into the market mechanism. Unfulfilled expectations led to 'fatigue from reforms'.

The reasons for the real sector economic revival, and for rapid financial budget improvement, can be traced to the impact of ruble devaluation and favourable world markets for fuels. Physical values of exports were raised essentially (more than 10% compared to previous year), and imports reduced. As a result the GDP growth rate for 1999 was 5.4% compared to the previous year, and the growth rate of industrial production was 11.0%.

For a period since the beginning of ruble devaluation to the end of 1999, the dollar on the Russian market strengthened two times in real terms. Along with that, the tendency of some (reasonable from our point of view) ruble strengthening was observed: as a whole the ruble/dollar exchange rate has grown at 30.8% with inflation rate equal to 36.5%.

2. The Russian economic outlook: 2000 and the first half of 2001

During the year 2000 and the first half of the current year the trends in improvement of the economic and financial situation that appeared during the post-crisis period of 1998-1999, on the whole, continued. This reflected in positive dynamics

Table 10-2. Macroeconomic parameters of the Russian economy, 2000-2001 (percentage compared to the appropriate period of the previous year)

	2000	2001 1-st half	2001 (estimate)
Gross Domestic Product	108.3	105.0	105.5
Inflation (consumer's prices index,%):			
by the end of year	20.2	12.7	18
In average per month	1.5	2.0	1.4
Justrial production	111.9	105.5	105.5
Price index of industrial production producers:			
by the end of year	31.6	8.8	16.5
In average per month	2.3	1.4	1.3
Agriculture production	107.0	101.0	106
Investment in basic capital	117.4	107.2	106.0
Retail trade	108.7	109.5	110.0
Value of payable services to population	106.0	103.0	103.0
Real disposable income per capita	109.3	104.7	105.5
Exports	139.5	106.3	101.0
Imports	113.5	120.5	112.5

of the most macroeconomic figures in 2000-2001 (see Table 10-2). GDP and industrial production growth continued, and the growth rate in 2000 was the highest during the past 30 years. Investments into fixed capital increased with a high rate of growth, first at the expense of the enterprises' own resources, and because of the inflow of foreign direct investments into the Russian economy.

A rise of budget incomes was significant. The year 2000 was the first time from the commencement of reforms when the surplus of the federal budget had been reached, and this tendency has continued in the current year. Usage of barter and other non-monetary accounts in 2000 and the first half of the current year has diminished substantially.

Dynamics of real income and consumption of goods and services by the population rose; population savings increased. Under the conditions of overall economic revival the unemployment rate reduced.

In spite of continued capital outflow from the country high volumes of external trade balance enabled the government of Russia to serve external debt almost without additional loans from international finance organizations.

At the same time positive tendencies in social-economic development in 2000 and the first half of 2001 were to a great extent based on favourable international trade situation and the consequences of ruble devaluation, although some activation of governmental measures took place in this period (particularly in budget, tax and custom spheres, de-bureaucratization of the economy, etc.).

Under monetary growth stimulated by the Bank of Russia making intensive purchases of foreign currency from the exporters to prevent excessive strengthening of the ruble, inflation has been rather high. For all that, in spite of the current growth in real supply of money in the economy, the period of its active bank investments into production has not come materialized. Financial results of enterprise activity (in the real sector of the economy) have gradually worsened.

Under some positive shifts in population living conditions, differentiation of the population by incomes and consumption remains significant, with a very high poverty rate. The situation in the agricultural sector remains unstable, especially in livestock breeding. Reduction in cattle numbers continued, mainly because of low provisions of feed.

GDP volume in the first half of 2001 increased by 5.0% compared to the corresponding period in 2000 and, according to estimations of the Ministry of Economic Development and Trade of Russia, will increase for the whole of 2001 by 5.5% (in 2000 this figure was 8.3%). The part of economic growth based on ruble devaluation and import substitution is gradually shrinking.

Despite growth of real income of the population, and especially investments into fixed capital, in the structure of final demand characterized by used GDP the share of exports remains high. As a whole more than a third of GDP growth in 2000 was based on external factors. At the same time a component of internal demand and its share in the structure of used GDP has gradually extended.

Volume of **industrial production** in the first half of 2001 increased by 5.5% compared to the corresponding period of the previous year (after a growth rate of 11.9% in 2000). High growth of industrial production in 2000 was also stimulated by improvements in financial conditions of enterprises (increase of balanced profit and monetary resources in current assets, reduction of quantity of unprofitable enterprises). For the whole of 2000 the inflation rate was 20.2% and in the first half of 2001 it was 12.7%.

Federal budget incomes during January-December 2000, as estimated by the Russian Ministry of Finance, amounted to 16.0% of GDP and was much higher compared to 1999 (comparable figure for last year amounted to 13.4% of GDP). In the current year the gradual increase of volume of federal budget earnings with respect to GDP has been taking place (it comprised 17.9% of GDP in the first half of 2001). **Expenses of federal budget** during January-December 2000 were financed at the volume of 14.6% of GDP and 14.5% of GDP in the first half of the current year.

During January-December 2000 18% of all expenses of the federal budget were spent to serve government debt (in the first half of the current year this figure was 22.2%). Disbursements for the external debt service were 67.2% of volume of percentage expenses in 2000 (82.2% in the first half of 2001). **Federal budget profit** was 1.4% of GDP in 2000 and 3.4% of GDP in the first half of the current year. So the budget policy has been rather austere since the crisis.

The state of **balance of payments** during 2000 and the first half of the current year remained stable. According to data (2000 year) and estimations (the first half of 2001 year) from the Bank of Russia, in 2000 and the first half of the current year the surplus of current account operations was US$ 46.3 billion and US$ 21.2 billion, respectively. The global market environment was favourable for the development of the external economic sphere, significant export and trade balance growth. New foreign capital investments have not improved, and this resulted in a reduction of external liabilities of economy. Negative balance of financial account became US$ 37.5 billion in 2000 and US$ 11.0 billion in the first half of 2001, respectively. Official gold and hard-currency reserves in Russia by 1 July 2001 amounted to US$ 35.1 billion approximately, which is the maximum since the beginning of reforms. This considerably exceeds internationally accepted standard of 3-month reserve sufficiency (in comparison with import of goods and non-factor services).

3. Features of Russian industry and international comparisons

Some positive tendencies as a result of reform measures are observed in the dynamics and structure of Russian industrial production during 1998-2000.

To bring the Russian economy out of crisis the Government of the Russian Federation and the Bank of Russia during the last months of 1998 realized some measures on stabilization of the social-economic situation, money circulation and state finances. The most important measures were as follows:

(1) elimination of many of the payment defaults in the bank system by inter-bank clearing,
(2) normalization of the situation on the currency market,
(3) reduction of import customs duties for many commodities of everyday consumer demand,
(4) some emergency measures, ensuring steady food supplies to the population, having a special purpose financing and state support of the investment process.

Negotiations with foreign investors were begun on conditions of adjustment and restructuring of the Russian internal debt, resident obligations and others.

These measures have allowed enterprises to stabilize and even to increase volumes of production, including export-oriented products, along with reducing barter operations. As a result rates of production decline are significantly slowed compared with the corresponding period of the previous year, but taking into account the seasonal component since the October 1998 declining trend in production was reversed.

In 1999 the positive changes of main production-economic activity indexes of industry enterprises which have been marked in the last months of 1998 continued. The index of physical volume of total industrial production as a whole was 108.1%, including: July 112.8%, August 116%, September 120.2%, October 110.3%, and December 111.1% compared with the corresponding period of the previous year. The economic performance of the last months of 1998 was associated basically with the compensation effect – in August 1998 production was at a rate of 88.4%, in September 1998 it was 85% and in October 88.3% compared to the corresponding months of 1997.

The increase in total industrial production was caused mainly by increasing exports and the development of imports-substitution processes that promoted growth of the domestic products share on the home market. An increase in volume of production in machine building has brought an increase in demand on the product of adjacent sectors – metallurgy, chemistry and oil chemistry, and an improvement in the situation in the investment complex that has caused growth of construction materials production, forestry and woodworking industry production. The tendency of the fuel and raw materials sectors share to decrease continued.

In 2000, basically, because of a demand increase on the home market on domestic products and activations of export, the growth rate of production was 109.2% compared with the previous year, or 111.9% compared with the 1997 (in 1999 it was 108.1% and 102.5%, respectively).

The growth of industrial production during 1999-2000 was caused by the following factors:

(1) Expansion of internal demand because of budget incomes increase, improvement of enterprises' financial conditions, growth of population incomes and expenses, trade turnover and capital investment.
(2) Increase in the competitive advantages of domestic commodity producers, defined, first of all, by home product price and quality.
(3) Import-substitution processes, which have caused a growth in demand for the domestic product and, accordingly, growth of production in many industrial sectors.

However, rates of production growth are slowed at the end 2000-beginning of 2001. In the first quarter of 2001 the growth rate was 103.3% compared with the corresponding period of the previous year. Together with that, in April of the current year the growth rate was 5.2%. As a whole in 2001 the expected growth rate of production is 4.5%. Decline in growth rates is stipulated basically by cessation of impact of the growth-promoting in 1999-2000 associated with the devaluation of the ruble (export-promoting) and import-substitution processes.

Growth of industrial production was not accompanied by qualitative changes in the structure of industry, goods or services. Russian-produced goods have won in the competitive struggle in the home market, not due to superior quality, but simply because they are cheaper compared with imported ones. At present, when economic environment becomes less favourable, we observe a trend of imports increasing because of the low competitiveness of domestic goods.

As before, the raw materials orientation of the economy prevails, high energy and other resources consumption per unit of production remains, and efficiency of 'human capital' use is also low.

Lack of financial resources continues, stipulated by low profitability and insufficient amortization. The absence of ownership protection particularly for minor shareholders and foreign investors remains. as well as negative impact on production has lack of separate types of raw materials, half-finished products and others.

3.1. Production trends

As mentioned above, some positive tendencies as a result of progress reform measures are observed in the dynamics and structure of the Russian industrial production during 1998-2001. Prolongation of these historical changes taking into account the main tendencies of the perspective development of the Russian economy allows to evaluate the medium-term perspectives of industrial production to 2004 (Table 10-3 and Figures 10-1(a-d), where indexes of industrial production are shown).

Figure 10-1a shows that in 1999-2001 recovery was begun, with growth of electricity and fuels production being slower than growth of total industrial production. Extrapolation of these tendencies leads to faster growth of total industrial production compared with growth of electricity and fuels production.

The growth of chemistry and oil chemistry production, which is mainly the manufacturing sector (Figure 10-1b), is faster than growth of total industrial production, and growth of iron and non-iron metallurgy production is slower than growth of total industrial production.

The production of the basic manufacturing sector – machinery and metal-working sector (Figure 10-1c) – grows faster compared with total industrial

Table 10-3. Growth rates of the industrial production by sectors (percentage of previous year)

	1997	1998	1999	**2000**	2001	2002	2003	2004
Industry, total	1.020	0.95	1.11	**1.12**	1.05	1.05	1.06	1.05
Electricity	0.980	0.98	0.99	**1.02**	1.03	1.03	1.03	1.03
Fuels	1.003	0.98	1.03	**1.05**	1.02	1.01	1.01	1.01
Iron metallurgy	1.100	0.92	1.17	**1.16**	1.02	1.02	1.02	1.02
Non-iron metallurgy	1.060	0.95	1.10	**1.15**	1.02	1.02	1.02	1.02
Chemistry and oil chemistry	1.040	0.93	1.24	**1.13**	1.06	1.07	1.08	1.08
Machinery and metal-working industry	1.040	0.93	1.17	**1.20**	1.07	1.09	1.10	1.09
Forestry, wood-working and cellulose-paper industry	1.009	1.00	1.18	**1.13**	1.05	1.05	1.06	1.06
Construction materials industry	0.960	0.94	1.10	**1.13**	1.08	1.08	1.08	1.08
Light industry	0.980	0.89	1.12	**1.21**	1.07	1.06	1.05	1.05
Food industry	0.992	0.98	1.04	**1.14**	1.09	1.09	1.09	1.09

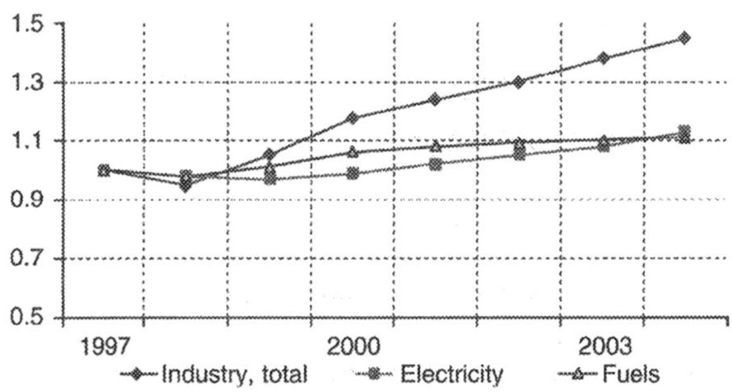

Figure 10-1a. Indexes of production by sectors; 1997 = 1.

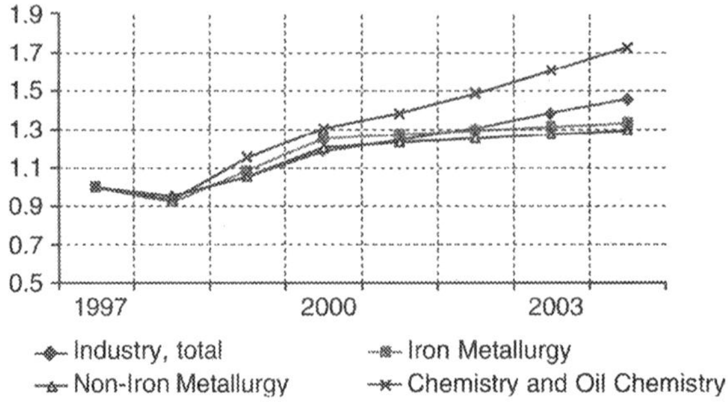

Figure 10-1b. Indexes of production by sectors; 1997 = 1.

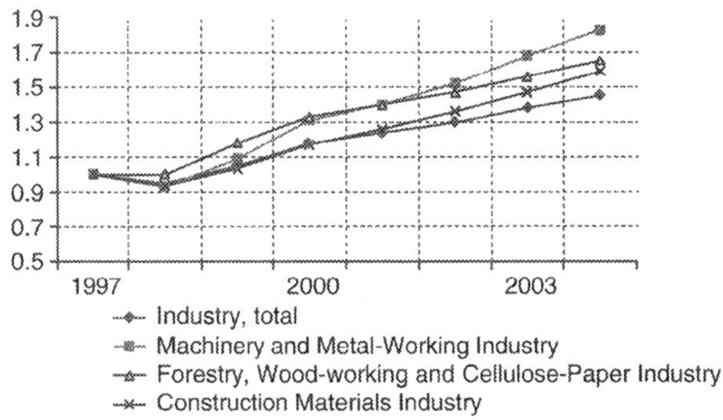

Figure 10-1c. Indexes of production by sectors; 1997 = 1.

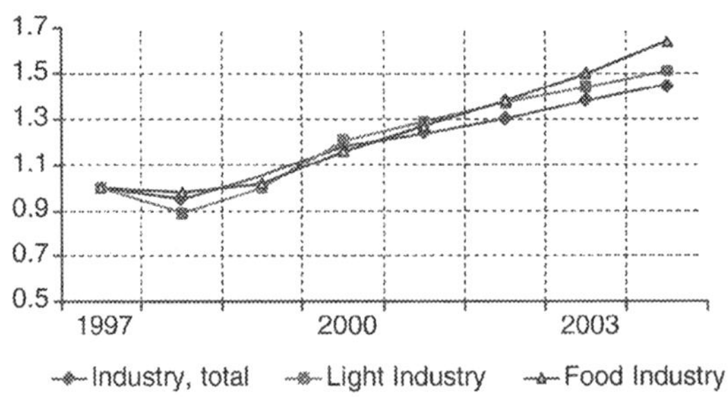

Figure 10-1d. Indexes of production by sectors; 1997 = 1.

production and all sectors. At the same time growth of forestry, woodworking and cellulose-paper industry production, which is partially manufacturing and partially produces raw materials (raw wood), is approximately as fast as growth of total industrial production. The growth of the construction materials industry, which produces the raw materials partially, is also approximately equal to total industrial production.

Table 10-4. Shares of production by sectors in total industrial production (%)

	1997	1998	1999	**2000**	2001	2002	2003	2004
Electricity	17.1	17.0	10.1	**9.1**	9.0	8.8	8.6	8.4
Fuels	17.4	15.5	16.9	**19.9**	19.3	18.6	17.9	17.2
Iron metallurgy	7.9	7.7	8.3	**9.0**	8.7	8.4	8.1	7.8
Non-iron metallurgy	5.5	7.6	10.1	**10.1**	9.8	9.6	9.2	8.9
Chemistry and oil chemistry	7.2	7.3	7.3	**6.4**	6.5	6.6	6.7	6.9
Machinery and metal-working industry	18.8	18.0	19.2	**18.6**	19.0	19.7	20.5	21.2
Forestry, wood-working and cellulose-paper industry	3.7	3.9	4.8	**4.5**	4.5	4.5	4.5	4.6
Construction materials industry	4.1	3.5	3.0	**3.1**	3.2	3.3	3.3	3.4
Light industry	1.8	1.8	1.7	**1.7**	1.7	1.7	1.7	1.7
Food industry	12.4	13.3	14.6	**12.6**	13.1	13.6	14.0	14.5

Light industry and the food industry (Figure 10-1d) are the main producers of consumer commodities. The growth of the food industry is the same as the growth of total industrial production and growth of light industry production, where the biggest decline during the reform years was observed, is slower than the growth of food industry production.

Such dynamics of production, when growth of manufacturing sectors is faster than growth of non-manufacturing and raw materials-producing sectors, resulted in some positive **structural shifts** in industrial production (Table 10-4).

This analysis allows us to draw the following main conclusions concerning the dynamics and structure of industrial production in Russia during 1997-2004.

First, the growth of total industrial production and industrial production by sectors will be continued but with decreasing rates after 2000, that is associated with delay of the import-substitution by home-made products processes and aggravation of conditions for exporters at external markets caused by rapprochement of internal and world prices as well as delay (or even falldown) in world oil prices, which is one of the most important Russian export commodities.

Second, most manufacturing sectors are growing faster compared with fuels and raw materials-producing sectors, with coincidence of growth tendencies (but with different rates) for total industry and its sectors.

Third, because of such production dynamics, some positive structural shifts are observed associated with decrease or stabilization of raw materials-producing and non-manufacturing sectors production shares and increase in manufacturing sectors production shares.

3.2. International comparisons: production

More detailed characteristics of the Russian economy give a comparative analysis of production, exports and imports structures (shares in GDP) for G7 countries, China and Russia, which is based on the input-output tables of these countries for 1995 using the OECD 36-sector classification.

The structure of production for selected G7 countries plus China and Russia by sectors in 1995 is shown Tables 10-5a, b, and c.

The Russian economy has large shares (8-9%) compared with G7 countries for food products, mining, and agriculture, and only China has larger share of agriculture (around 12%). This reflects the fact that Russia is a large producer of oil and gas, forestry, food including beverages and tobacco, fishery and livestock, and is evidence of the important role that these sectors play in the Russian economy on the one hand, and raw materials orientation of the economy as a whole on the other.

Table 10-5a. International comparison of economic structure in 1995, production (unit: %)

Industrial Branches	Germany	Japan	USA	Canada	China	Russia
1 Agriculture	1.72	2.17	2.66	3.67	12.38	9.18
2 Mining	0.56	0.18	1.87	4.17	3.46	7.53
3 Food	5.45	4.81	4.10	4.45	6.26	8.97
4 Textiles	1.64	1.82	1.46	1.40	8.12	3.23
5 Wood and wood products	1.33	0.87	1.07	1.96	1.43	1.93
6 Paper	1.86	2.60	3.02	4.14	1.52	0.67
7 Chemicals	3.59	2.49	2.60	2.22	3.93	3.64
8 Medicine	0.00	0.70	0.49	0.36	0.76	0.42
9 Petroleum and coal products	1.35	1.23	1.72	1.78	1.63	0.20
10 Rubber plastics	1.45	1.64	1.08	1.00	1.85	0.13
11 Nonmetallic mineral products	1.25	1.32	0.67	0.82	4.29	2.33
12 Iron and steel	1.45	3.30	1.11	1.24	4.22	4.15
13 Nonferrous metals	0.55	0.68	0.74	1.68	1.50	3.08
14 Metal products	2.33	2.03	1.56	1.83	2.54	1.12
15 Machinery	4.25	3.19	2.16	1.31	4.21	2.69
16 Office and computing machinery	0.56	1.24	0.67	0.39	0.97	0.67
17 Electrical machinery	3.90	2.26	0.92	0.74	2.51	1.01
18 Radio TV communication equipment	0.00	2.89	1.28	1.07	0.98	0.78
19 Ship building and repairs	0.15	0.24	0.16	0.14	0.34	1.23
20 Other transportation equipment	0.00	0.32	0.17	0.21	0.16	0.62
21 Motor vehicles	5.16	4.69	2.55	5.39	3.66	2.75
22 Aircraft	0.34	0.06	1.31	0.70	0.34	0.95
23 Personal goods	0.58	0.53	1.28	0.25	0.72	0.45
24 Other manufacturing	1.52	0.51	0.40	0.41	0.83	1.06
25 Electricty	2.73	2.12	2.51	2.23	3.13	8.78
26 Construction	6.67	9.12	7.14	8.90	8.32	9.59
27 Whole and retail trade	9.01	9.77	11.11	9.41	6.48	1.14
28 Restaurants and hotels	1.86	2.88	2.97	2.64	1.26	0.49
29 Transportation	3.13	5.96	3.74	4.56	2.63	4.12
30 Communication	1.53	1.02	1.77	1.67	0.45	1.76
31 Finance and insurance	4.35	4.10	5.18	8.36	1.57	3.30
32 Real estate business	17.17	10.40	14.50	3.76	4.32	0.95
33 Social and personal services	3.55	6.93	10.93	4.82	0.92	7.50
34 Government	6.11	3.51	5.18	0.38	2.29	3.30
35 Other activivites	2.11	1.27	−0.07	11.93	0.00	0.30
36 Discrepancy	0.80	1.16	0.00	0.00	0.00	0.00

Table 10-5b. International comparison of economic structure in 1995, exports (unit: %)

Industrial Branches	Germany	Japan	USA	Canada	China	Russia
1 Agriculture	0.83	0.09	3.87	3.62	5.47	0.89
2 Mining	0.36	0.03	1.24	10.15	1.52	24.04
3 Food	4.70	0.46	4.81	4.22	4.35	3.81
4 Textiles	4.24	1.60	1.68	1.20	27.53	1.38
5 Wood and wood products	1.17	0.09	1.22	4.06	3.10	5.07
6 Paper	2.18	0.65	4.52	8.68	0.17	1.35
7 Chemicals	11.69	4.93	6.16	3.44	1.18	9.39
8 Medicine	0.00	0.22	0.85	0.17	0.95	1.08
9 Petroleum and coal products	0.67	0.47	2.26	1.57	0.43	0.16
10 Rubber plastics	3.05	1.56	1.57	1.49	2.30	0.32
11 Nonmetallic mineral products	1.54	1.02	0.72	0.63	2.81	0.33
12 Iron and steel	2.70	3.35	1.35	1.25	0.50	13.04
13 Nonferrous metals	1.81	0.66	1.52	5.63	0.47	15.14
14 Metal products	2.19	1.33	3.94	2.29	2.78	1.29
15 Machinery	14.34	10.68	6.35	3.13	18.73	3.10
16 Office and computing machinery	1.91	6.82	4.37	1.73	4.01	0.77
17 Electrical machinery	10.28	5.84	3.03	1.15	3.93	1.16
18 Radio TV communication equipment	0.00	13.75	4.78	3.14	4.10	0.90
19 Ship building and repairs	0.52	1.95	0.35	0.06	0.82	1.42
20 Other transportation equipment	0.00	1.24	0.32	0.64	0.38	0.71
21 Motor vehicles	16.06	20.34	7.91	22.55	0.33	3.16
22 Aircraft	1.10	0.20	7.25	2.62	0.10	1.10
23 Personal goods	1.61	2.85	3.15	0.62	8.86	0.01
24 Other manufacturing	3.31	0.70	0.84	0.75	0.59	0.49
25 Electricty	0.20	0.05	0.11	0.40	0.01	0.72
26 Construction	0.38	0.00	0.04	0.02	0.00	0.16
27 Whole and retail trade	3.74	4.61	7.54	3.01	0.00	0.71
28 Restaurants and hotels	0.47	0.81	0.08	1.85	0.28	0.30
29 Transportation	5.89	8.65	8.37	2.14	2.62	4.93
30 Communication	0.41	0.09	0.63	0.40	0.48	2.11
31 Finance and insurance	0.07	0.96	3.14	1.02	0.06	0.11
32 Real estate business	1.50	1.05	4.06	1.65	1.12	0.01
33 Social and personal services	0.96	0.10	1.99	1.02	0.01	0.69
34 Government	0.12	0.00	0.00	0.00	0.00	0.11
35 Other activities	0.00	0.09	0.00	3.70	0.00	0.01
36 Discrepancy	0.00	2.75	0.00	0.00	0.00	0.00

The share of petroleum and coal products in Russia is one of the lowest between all countries in spite of the large oil mining constituent (Figure 6). This is explained by the large amount of oil exports and the relatively low internal prices of petroleum products compared with world prices.

Russia and China have the largest share (around 4%) of iron and steel production compared with G7 countries and this is further confirmation of the raw materials and half-finished orientation of the Russian economy as a whole. This is also confirmed by the relatively high share in Russia (more than 2%) of non-metallic mineral products; China has the largest share of this sector (around 4%).

The share of machinery for Russia, which is one of the basic manufacturing sectors and determines the technological level of all sectors of the economy and the possibility of producing high-quality and competitive commodities, is

Table 10-5c. International comparison of economic structure in 1995, imports (unit: %)

Industrial Branches	Germany	Japan	USA	Canada	China	Russia
1 Agriculture	5.81	6.44	1.75	1.59	4.34	3.68
2 Mining	5.14	18.00	7.65	4.22	2.18	2.49
3 Food	6.60	9.72	3.99	4.01	0.76	19.13
4 Textiles	9.11	6.00	8.77	5.43	3.50	16.74
5 Wood and wood products	1.95	2.22	1.65	1.43	2.04	1.81
6 Paper	2.98	1.23	2.54	3.16	1.49	0.64
7 Chemicals	9.39	4.02	3.44	4.41	7.91	6.13
8 Medicine	0.00	0.96	0.93	0.67	0.35	0.70
9 Petroleum and coal products	4.59	4.60	3.47	1.50	0.63	0.07
10 Rubber plastics	2.72	0.91	1.94	2.48	0.79	0.21
11 nonmetallic mineral products	1.84	0.73	1.16	1.23	0.41	1.47
12 Iron and steel	2.72	1.53	2.04	1.56	2.19	3.87
13 Nonferrous metals	3.11	4.82	1.27	2.18	1.65	1.88
14 Metal products	1.45	0.62	2.03	4.36	0.51	2.76
15 Machinery	5.97	4.40	5.93	7.63	27.48	6.62
16 Office and computing machinery	3.67	1.67	3.72	4.03	4.65	1.66
17 Electrical machinery	9.48	1.48	3.36	3.04	5.13	2.48
18 Radio TV communication equipment	0.00	2.61	8.69	5.25	5.52	1.93
19 Ship building and repairs	0.17	0.12	0.16	0.16	0.82	3.03
20 Other transportation equipment	0.00	0.10	0.37	0.54	0.08	1.52
21 Motor vehicles	8.14	2.54	14.76	20.11	11.17	6.76
22 Aircraft	1.72	1.32	1.58	1.45	2.55	2.35
23 Personal goods	1.59	1.24	3.21	1.63	5.73	0.56
24 Other manufacturing	3.05	1.05	3.23	2.24	1.17	1.05
25 Electricty	0.24	0.01	0.37	0.36	0.32	0.07
26 Construction	0.19	0.00	0.00	0.05	0.00	2.34
27 Whole and retail trade	0.92	0.76	2.40	1.07	3.38	1.48
28 Restaurants and hotels	0.50	5.43	0.00	3.51	0.49	0.64
29 Transportation	2.90	5.84	0.65	2.57	1.83	3.08
30 Communication	0.65	0.13	0.00	0.46	0.47	1.32
31 Finance and insurance	0.12	1.77	0.65	2.74	0.01	0.19
32 Real estate business	2.51	2.44	0.31	3.10	0.45	0.02
33 Social and personal services	0.74	0.69	0.14	1.82	0.00	0.75
34 Government	0.00	0.00	0.00	0.00	0.00	0.19
35 Other activities	0.00	0.07	12.57	0.00	0.00	0.37
36 Discrepancy	0.00	4.54	0.00	0.00	0.00	0.00

comparable with this one for all countries, and equal to around 2.7%. Shares of other countries vary from around 1.3% for Canada to more than 4% in Germany. This sector should be developed to around 4% as is the case in Germany and China. The share of metal products for Russia (i.e. the manufacturing sector) (around 1%) is less compared with shares for all other countries, which vary from around less than 1.5% for the UK to around 2.3% for Germany. This sector, as well as machinery, should be rapidly developed to improve the structure of the Russian economy in the direction of a manufacturing sectors share increase. On the other hand, in Russia production of the non-ferrous metals sector, which is mostly non-manufacturing, amounts to 3% of GDP compared with a range from 0.5% for Germany to around 1.5% for Canada and China.

Russia is a small producer of high-technological commodities such as production of the radio, TV and communications sector. The share of this sector (less than 1%) is the lowest of all countries being compared. The situation with electrical devices is a little better and the share of this sector production for Russia (1%) is more than for the USA and Canada, and less than for other countries. The largest producers in this area are Japan (more than 2%), China (2.5%) and Germany (around 4%). A rather better situation concerns the production of office and computing machinery, of which the share for Russia is around 0.5%, i.e. more than for Italy, France and Canada, and approximately equal to the share for the USA and Germany and less than for Japan, China and the UK (around 1%).

The share of motor vehicles production in Russia, which is around 3%, is comparable with the USA, more than in Italy and the UK and less than in Canada, Japan, Germany, France and China.

The electricity, gas and water supply sector production plays a large role in the economies of all countries. In Russia the share of this production is around 9%, that is three or four times more compared with G7 countries and China, and is indirect evidence of excessive inputs of these resources. Some measures towards their preservation are urgent.

A very low share of the restaurants and hotels sector in Russia, which is less than 0.5% and much lower compared with all countries, is evidence of its underdevelopment, and the low attention paid to this sector, which may be a large source of revenue. Poor development of market services in Russia is indicated by the very low share of the wholesale and retail trade sector (around 1%) which is much less compared with other countries (around 11% for the USA, around 10% for Italy, the UK and Japan, etc.) Much construction activity in Russia shows the biggest share of the construction sector (around 10%), that is more than for all countries being compared (from 6-7% for USA, Germany and Italy to 8-9% for the UK, China, Canada and Japan.

A sufficiently high level of financial and insurance activity in Russia is confirmed by the relatively high share of this sector, which is over 3%, i.e. more than in China and a little less than for the USA, Japan, Germany and Italy (4-5%). The biggest share is characteristic for France, Canada and the UK (7-8%). The share of the communication sector in Russia (around 2%) is comparable with those for the USA, Canada, Germany and the UK, a little greater than for France, Italy and Japan (around 1%) and significantly less than for China (around 0.5%). The development of a transportation system, judging by the shares, is sufficiently high in Russia and comparable with all other countries (3-4%) except Japan, where this share is more, and accounts for around 6%, and China where this share is the least (less than 3%).

The government services sector in Russia is sufficiently large (around 3%) and is more compared with Canada, China and Italy and less compared with other countries (from around 4% for Japan and to around 9% for France). The social and personal services sector in Russia is more than twice as big compared with the government service sector and its share accounts for around 7%, i.e. greater than the shares for Canada, Japan, Germany, France and China, and lower than those for the USA, UK and Italy, where it is largest (around 14%).

3.3. International comparisons: exports

The structure of Russian exports, which depends strongly on production, is very similar to the industrial structure, as shown in Table 10-5b. Russia is the largest exporter of mining products (first of all oil and gas) which accounts for almost 25% of total exports. The share of mining products is noticeable for Canada (10%) and the UK (around 5%) and for other countries is significantly less. As for food exports Russia is a very small exporter of agricultural products (around 1% of total exports) because of the poor level of agricultural development and shortages of these products.

Iron and steel exports in Russia take a large share, which is around 13% of total exports, and much less than for all other countries. For Canada and the USA this share is around 1% and around 3% for Japan, Germany, Italy, France and the UK. The lowest exporter of iron and steel products is China (less than 1%). Russia is the lowest exporter of non-metallic mineral products and rubber and plastics products (shares are less than 1%). Shares of these products for other countries are not significant (in the limits 1-3%).

Russia is a relatively small exporter of machinery and metal products, and these shares are around 3% and 1% of total exports, correspondingly. Canada is a similar (by the share) exporter of machinery, and other countries have much bigger shares of these exports (from around 6% for the USA and France to 14% for Germany and over 18% for Italy). As regards metal products exports, all countries have small shares in total exports, which range from around 1% for Japan to 4% for Italy and the USA. On the other hand, Russia is a large exporter of non-ferrous metals, with the share of total exports being around 15%, the second position is taken by Canada, around 5%, and all other countries have shares which range from less than 1% for China, Italy and Japan to around 2% for France, the UK and Germany.

Russia is also the lowest exporter of such high-tech products of manufacturing sectors as Radio, TV and communications, electrical machines and office computers. The exports share of these products for Russia is around 1% while shares of these range from around 3% of radio, TV and communications products exports for Canada to around 14% for Japan. The shares of electrical devices exports range from around 1% for Canada to around 10% for Germany, and of office computers shares of exports which range from around 2% for Canada, Germany, Italy and France to over 6% for Japan.

All G7 countries are large exporters of motor vehicles. The shares of these products exports are range from around 8% for the USA to 20% for Japan, and around 22% for Canada, while such shares for Russia are around 3%, and for China less than 1%.

3.4. International comparisons: imports

Imports are important to supplement the national economies with insufficient internal demand for production of goods and services. The structure of imports for selected G7 countries, Russia and China by OECD sector classification is shown in Table 10-5c.

Russia is a largest importer of food because of low agriculture and food industry efficiency, and the share of food imports is around 19%, that is much more compared with all other countries. The lowest food imports share is for China (less than 1%). These shares for other countries range from 8% for Italy to around 10% for Japan. Being a large exporter and producer of mining products Russia, along with China, are small importers of these products i.e. around 2%. The largest importers of mining products are Italy (around 19%) and Japan (around 18% of total imports). Russia is also a relatively small importer of agricultural products (around 4% of total imports). The lowest importers of these products are Canada and the USA (less than 2%) and the UK (2%), and the largest ones are Italy (7%) and Japan (6%).

Russia is the lowest (almost zero) importer of petroleum and coal products. For other countries the share of these products varies from 0.5% for China to 4.5% for Japan, Germany and Italy. The shares of drugs and medicine exports are small for all countries and do not exceed 1%. On the other side, all countries, including Russia (whose share of these products is 6%), are large importers of industrial chemicals, that is more compared with Canada, the USA and Japan (3-4%) and less compared with the UK, China, France, Italy and Germany (around 7-9%).

Russia, Canada, France and China are the largest importers of machinery products (with shares around 6-7% for the first three countries and over 25% for China). These shares are less for other countries (around 6% for the USA, Germany and France) and are the least for Italy and Japan (around 4%). All countries are small importers of metal products and suitable shares do not exceed 3% (for Russia around 3%), except Canada, for which these shares account for around 4%). The same situation occurs with non-ferrous metals – shares of these products imports do not exceed 3% (except Japan, around 5%) and for Russia this share accounts for around 2%.

Russia is a moderate importer of radio, TV and communications equipment and its share accounts for 2%, while other countries are bigger importers (over 2% for Japan, around 5% for Italy, Canada and France, 6% for the UK and around 9% for the USA). Imports of electrical devices are over 2% for Russia, which is more compared with Italy and Japan (around 1%, respectively), less than for other countries (around 3% for the USA, Canada, France and the UK) and China and Germany (5% and over 9%, respectively). Russia is a small importer of office computers and the share of this import is around 1.5%, that is similar to that for Japan, and for other countries this share is more (around 3% for Germany and the USA, and around 4% for Canada, France, the UK and China).

Motor vehicle imports into Russia and Italy are around 7%; that is more than into Japan (around 2.5%) and less than into Germany (around 8%). Much larger importers are the UK and France (8-9%), China (around 11%) and the largest ones are the USA and Canada (15% and 20%, respectively). Russia is the largest importer of other transportation means (less than 2%) and for other countries these shares are insignificant (less than 1%). The same situation is characteristic for shipbuilding and repairing: Russia and China are the largest importers of this sector products (around 3% and 1%, respectively) and for all other countries these shares are insignificant and less than 1%.

4. Long-term social and economic programme of the government of the Russian Federation

In June of 2000 the government generally approved the Long-Term (to 2010) Social and Economic Policy Guidelines. In July the government adopted a social policy and economy modernization action plan for 2000-2001 designed to secure the completion of strategic developmental milestones laid down in the Guidelines.

4.1. Current state assessment and selection of a development path

Despite a number of major negative trends, Russia currently finds itself in a completely new environment opening up a 'window of opportunity' for solving the fundamental problems of this country. The political situation has become stable, and the economy is showing signs of recovery.

Russia still boasts significant potential for accelerated growth. Its tremendous intellectual capital, the basic market institutions framework in place, vast natural resources, core components of production and communication infrastructure and the accumulated financial savings all add up to favourable preconditions for national renewal.

In the medium-term perspective we need to prevent a further gap between Russia and industrialized nations, and in the longer-term context to restore and strengthen Russia's status as one of the leaders in global development. In order to capitalize on this 'window of opportunity' we must identify a path that will lead to the realization of Russia's potential and balanced growth. There are three realistic development options.

The first scenario includes deregulation of virtually all areas of economic activity, opening up the country to the outside world and privatization of most social functions. This scenario can lead to significant economic growth. However, it is the least probable one, given the impossibility of the quick creation of an institutional market framework which is critical for a liberal economy, and the inability of most Russian enterprises to compete on an equal footing with foreign market players.

The second scenario presumes more direct involvement of the state in the regulation of economic and social relations, which will necessitate a major increase in tax collection and investment activity of the state. This will hardly create a favourable environment which would encourage entrepreneurial activities. This option also implies the preservation of a closed economy, which will make it impossible to participate in the international division of labour and may lead to further technological lagging behind industrialized nations. Any attempts to increase the effective tax burden will result in additional capital flight and further criminalization of the economy and society. Broader functions of the government will make the state machine even less manageable than it is now.

The third, modernization, scenario is based on the idea of freeing up private initiative and strengthening the role of the state in promoting a favourable business environment, including financial and social stability, and is a balanced combination of the elements of the other two strategic options. Instead of a social state (paternalism) and privatization of social functions (radical liberalism), a 'subsidiary' state is created to provide social guarantees to the extent that society is unable to do so on its own. Instead of complete isolation or openness, the focus is on Russia's active integration into the world community and economy while protecting the interests of Russian producers from unfair competition on the part of foreign market players.

4.2. Goals of long-term policy

The primary goal of the government's long-term policy is to ensure sustained improvement in the general standard of living, reduce social inequality, safeguard and promote Russia's cultural values, and restore the economic and political role of Russia within the world community.

The Russian Government and general public face the following key challenges:

(1) implement social reform, institute a new social contract (based on trust between citizens and the state, employees and employers, business entities and individuals);
(2) reform the state system;
(3) modernize the economy.

By now Russia has a market framework in place, but in order to be able to function efficiently it also requires a seamless operation of market support infrastructure, including the judicial system, compliance enforcement, law-enforcement institutions and federal relations. The strengthening of the state or the *reform of the state system* is a prerequisite condition for modernization of the economy. However, the critical necessity of reforming the state system is stipulated not only by economic growth requirements, but first and foremost by the need to safeguard the rights, freedoms and safety of citizens and to secure public trust in authorities.

The programme's goals are achievable only provided there is *economic modernization*. The only way to bridge the gap between Russia and the most developed nations, and to create the basis for enhancing the standards of living, is economic growth at a rate higher than the growth of the world economy. Such growth can be secured by combining the accumulation of intellectual and capital resources with their more efficient utilization and freeing up entrepreneurial initiative. A reduction of the administrative and tax burden on businesses will ensure (as early as at the first stage) a return on the short-term resources for growth achieved through increasing labour productivity and utilization of idle but viable production capacity. At the second stage economic freedom will result in increased efficiency of investments, the promotion of which will become the focus of economic policy. The major components of the economy modernization programme include, first of all, the creation of a favourable investment climate, sensible budgetary policy and structural reform.

The *social policy* goal is to complete transition from the paternalist to subsidiary state model. This implies accessible and free-of-charge basic social services for all citizens, first and foremost in the area of education and health care; re-allocation of social expenses of the state in favour of the most vulnerable members of the public while reducing support for reasonably prosperous households; gradual elimination of social inequality; and providing individuals with opportunities for a higher level of social consumption from the sources of personal income.

The critical areas of social policy during economic modernization are: first, the support of vulnerable social groups and secondly, investments in human development, primarily in education which is a prerequisite condition for Russia to become competitive within the globalized, information-based and dynamic world economy.

4.3. Expected main quantitative results of the long-term programme

The implementation of the programme is expected to ensure GDP growth at a minimum annual average rate of 5% during a 10-year period. By 2010 the GDP should increase by approximately 70% (compared to 1999). In dollar terms GDP will grow by a factor of 2.5 by 2010, calculated at the current exchange rate.

The funding of economic growth will greatly depend on reduced capital flight. If capital flight declines two-fold, investments can be increased by approximately one-third provided that capital remains in the country and is used for investment purposes.

Gross savings in 2000-2010 will amount to approximately 30% of GDP, enabling Russia to retain high growth potential throughout the planned period. This implies that economic growth, at least in the medium-term perspective, should be accompanied by increased investments in capital assets outpacing the pattern of consumption. Thus, this strategy is focused on the maximum utilization of growth potential to make the national economy attractive for long-term investors. Foreign investments will constitute another source of funding.

The period of investment growth outpacing consumption may continue for at least 6-8 years. By the end of the decade the growth of investments and economic growth in general may slow down to a certain extent.

The implementation of the programme will make it possible to gradually overcome distortions in the distribution of *personal income* and, in addition to general growth in disposable income, increase the percentage of medium-level income social groups in total personal income. This will create a basis for sustained mass domestic demand and transition to a qualitatively new growth pattern, with consumption outpacing investment.

As a result, by the end of the decade the pattern of private consumption will steadily outpace investments so that in 2010 consumption will grow at a somewhat higher rate than GDP. Overall, GDP growth will lag behind both consumption and gross savings. This lag will be offset by reduced net exports and, consequently, reduced capital flight.

A decline in *net exports* and the resulting reduction of the current account balance may start as early as in 2001, triggered by both decreasing exports due to the anticipated unfavourable situation on the world commodity markets comprising traditional Russian exports, and increasing imports, including the import of investment resources. However, this trend will have no destabilizing effect on the economy as it will be accompanied by reduced capital flight. By 2004 the balance of trade and the current account balance will decline to less than $20 billion and $10 billion, and by 2010 to under $10 billion and $5 billion, respectively.

Notably, in the nearest decade the ruble will be becoming stronger in real terms and market rate will approximate purchasing power parity, thus encouraging imports which may grow two-fold over 10 years. By 2010 we may expect growth in the real ruble rate by a factor of at least one-third.

In parallel, there are expectations of a stable accumulation of foreign currency reserves by monetary authorities (Bank of Russia).

Judging from the present-day *sectoral structure of the Russian economy*, the GDP development pattern during the next few years will be largely shaped by the development of the manufacturing sector. Information and communication sectors, which are expected to grow at a presumably higher rate, account for some 2% of GDP. Although an increase in the scope of those and other market services in the medium-term perspective is likely to accelerate GDP growth, the above sectors will not yet become dominant. A noticeable increase in the export of services can be expected only by the end of this decade, so the traditionally negative value of the balance of services may approximate zero.

Net exports may fall to 5-7% of GDP by 2004 and may total some 2-3% of GDP by the end of this decade. Investments in capital assets will account for some 26-27% of GDP by 2010 and will tend to fall gradually thereafter. Consumption as a percentage of GDP will remain almost unchanged until 2004 – an increased percentage of gross savings will be offset by a proportionate decrease in net exports. Following 2004 there will be a steady growth of consumption which may exceed 70% as a percentage of GDP. This ratio is well over the percentages of 1991-1995 and approximates the proportions of 1996-1998 (however, in those years Russia's high percentage of consumption was maintained at the expense of external borrowings).

In parallel, state consumption as a percentage of GDP will gradually decline while growing in real terms, albeit at a pace which is expected to be significantly slower than the GDP growth rate. These developments will be accompanied by a reduction of the nominal tax burden and state debt which is not being served at present. As a result private consumption will grow at least 80% provided that overall consumption grows at a slightly faster rate than GDP.

A reduction of expanded production as a percentage of GDP will help to redistribute resources available to the private sector more efficiently. The state will be able to perform its obligations regarding *the repayment and servicing of its foreign debt.* Those expenses as a percentage of GDP will gradually decline as a consequence of GDP growth in real terms and a stronger national currency. Interest expenses will account for less than 3% of GDP by 2010. Non-interest expenses as a percentage of GDP will account for some 30% of GDP by the end of the period.

Beginning from 2004, *inflation* will stay at less than 10%, despite the higher rate of increase in money supply which will be offset by increased demand for money as a result of economic growth and the substitution of non-monetary forms of settlements with monetary ones.

The development of financial markets within the deficit-free budget policy framework will also help contain the pressure of money supply on price levels, so a potential 'excessive' inflow of foreign capital may be counterbalanced by issuing financial instruments with relatively low yields. Increased monetization of the economy will contribute to higher capitalization of the banking system, and the share of debt in the structure of investments will grow gradually.

11
Carbon tax and labour compensation – a simulation for G7

BERND MEYER and CHRISTIAN LUTZ

1. The scenario

A worldwide harmonized climate policy needs target lines for the greenhouse gas emissions of all countries. One perspective is to establish a world market for pollution rights, on which only governments are suppliers and demanders. To reach the national targets every country might introduce energy taxes. World climate policy will only be successful if the burden to reach the targets is acceptable for the different countries. In this chapter we use the system to calculate the maintenance costs for the group of G7 countries, which are the most important energy users.

We assume that, in each of the G7 countries, a carbon tax is introduced, which has the following characteristics. Every producer and importer of fossil energy carriers has to pay taxes in relation to the CO_2 emissions which are set free when the specific energy carrier is burned. Producers and importers of fossil energy carriers raise their prices due to the tax payment. The governments start with a tax rate of 1 US dollar in 2001 per ton of CO_2 and raise the tax rate linearly to 10 dollars in 2010. The governments use the tax revenue to lower social security contributions. The idea – implemented in the environmental fiscal reform in Germany starting in 1999 – is to raise energy costs and to lower labour costs in the hope of obtaining rising employment as a second dividend of climate policy.

The taxes will raise the prices of the different energy carriers depending on their carbon concentration. Coal is levied more than petroleum and gas. The relation is about 4:3:2. Nuclear and renewable energy is not taxed at all. Higher energy prices will reduce energy demand and carbon emissions. Further energy prices will spread all over the economy and raise the prices of all goods depending on their direct and indirect energy inputs. On the other hand wages will be reduced, because the government uses the tax revenue to lower social security contributions. This effect lowers costs, and thus prices, especially in the labour-intensive sectors. Structural change will be induced, which favours labour-intensive industries and depresses energy-intensive industries.

The aggregate impact on the economy cannot be predicted without a model simulation, the reactions on GDP and employment depend on contrary effects.

K. Uno (ed.), Economy–Energy–Environment Simulation, 185–189.

Of course, the economies will develop in different ways, because of their different structures of production. This will induce further effects on other contries via international trade.

2. Specification of the labour market in COMPASS

Since the labour market plays a specific role in the analysis, we want to discuss the model structure here in more detail. We assume that in every sector and every country the nominal wage of the sector depends with elasticity 1 on the actual price of the sector and its productivity trend. The real labour input coefficient is dependent on the real wage rate with a negative elasticity. Because of data restrictions we were not able to estimate these elasticities econometrically. For Germany the elasticities of labour demand on the real wage for most industries are close to −0.7 (Meyer *et al.*, 1999). We assume this value for all industries in all countries.

For every country and every sector we specify the following labour market model. The model is sector- and country-specific, but we omit sector and country indices.

Real wage rate is growing exogenously:

$$\log(w/p) = \log a + g * t \quad a > 0, g > 0 \tag{1}$$

w = nominal wage rate, index 1990 = 100
p = price index of output, 1990 = 100
t = time.

Real labour demand is dependent on output, the real wage rate and a time trend:

$$\log L = \log b + \log x + \beta * \log(w/p) + v * t \quad b>0,\ \beta<0,\ v<0. \tag{2}$$

L = real labour input measured in 1990 wage rates
x = real gross production.

(1) in (2):

$$\log(w*L) = \log b + \log x + \beta \log(w/p) + v*t + \log a + g*t + \log p \tag{3}$$

$$\log(w*L/p*x) = \log b + (1+\beta)*\log a + [(1+\beta)*g + v]*t \tag{4}$$

Assuming $\beta = 0.7$, an experience from estimations with German data, $g = 1.5$ or 2.0, a plausible long-run growth rate for real wages, can be calibrated in equation (1) by data of 1990, when w and p are 1 and g and t are given. Running the regression

$$\log(w*L/p*x) = a_0 + a_1*t \quad a_0 = b*a*\exp(1+\beta),\ a_1 = (1+\beta)*g + v \tag{5}$$

v and b can be identified by the regression results.

3. Simulation results

A uniform tax rate of up to 10 US$/t CO_2 in 2010 induces different tax revenues in G7 countries. In the US carbon taxes reach in 2010 about 60 billion US$, about 1% of GDP. In the other countries taxes are much lower. In Japan the carbon tax amounts to 12 billion US$ or 0.25% of GDP, because of the already high energy productivity of the Japanese economy. Energy efficiency in European countries is much higher than in the USA, but much lower than in Japan. In 2010 carbon taxes reach 9 billion US$ in Germany and 0.5% of GDP, in France (4/0.4), Italy (3/0.5) and Britain (5/0.7) the carbon tax is similiar. Canada (6/1.0) with its low energy productivity is comparable to the USA. Over all G7 countries the carbon tax reaches about 100 billion US$ in 2010. So the USA would bear about 60% of the tax.

Impacts on the different G7 countries are shown in Table 11-1. Carbon emissions are reduced by around 5% against the business as usual in six countries. In the USA and Canada energy consumption is obviously less price-dependent than in European countries. Higher taxes as a percentage of GDP have more pressure on overall prices in northern America than in Europe. GDP is at the same time affected much more. On the labour market two different effects can be seen. Lower labour costs induce more employment, whereas lower GDP reduces employment. For the four European countries the labour cost effect is more important. Especially Germany, France and Italy, that still suffer from high unemployment, can improve the situation on the labour market. For the USA and Canada both effects more or less equalize. Japan, compared to the other countries, shows different effects. Energy productivity has shown almost no price dependency in the past. Higher energy prices reduce carbon emissions only slightly. As the taxes are low compared to GDP, impacts on GDP and labour are relatively small.

A comparison of the effects on the industry level shows some general features of the carbon taxes. The production decline of petroleum and coal products is the strongest. Construction, on the other hand, is reduced only slightly. Except for Italy and Britain the reduction in all other countries is below average.

4. Environmental tax reform: the concept and its realization

We have just discussed the impact of a carbon tax on economy and environment as we find it in textbooks. Of course from the ecological point of view it

Table 11-1. Effects of carbon taxes on GDP, labour, prices and emissions in G7: percentage differences against the business as usual in 2010

	USA	Japan	Germany	France	Italy	Great Britain	Canada
GDP	−1.72	−0.23	−0.35	−0.31	−0.34	−0.75	−1.61
Labour	0.08	0.27	0.89	0.90	0.93	0.56	0.19
Output prices	1.82	0.60	0.52	0.38	0.66	0.77	1.92
Carbon	−5.38	−1.14	−4.5	−4.18	−6.11	−4.31	−5.90

Table 11-2. Output effects of carbon taxes on different sectors in G7: percentage differences against the business as usual in 2010

	USA	Japan	Germany	France	Italy	Great Britain	Canada
Food processing	−2.02	−0.27	−0.32	−0.36	−0.29	−0.69	−1.83
Petroleum and coal products	−2.87	−0.33	−0.82	−0.50	−0.47	−2.42	−3.67
Iron and steel	−1.35	−0.28	−0.33	−0.45	−0.48	−0.82	−1.60
Machinery	−1.06	−0.22	−0.26	−0.29	−0.48	−0.72	−1.11
Motor vehicles	−1.41	−0.42	−0.33	−0.47	−0.40	−0.74	−1.92
Construction	−1.01	−0.02	−0.13	−0.21	−0.39	−0.78	−1.06
All industries	−1.74	−0.18	−0.32	−0.33	−0.35	−0.75	−1.71

is the most effective way to tax primary energy consumption, because it induces improvements in energy efficency on all stages of production and in the energy transformation processes. Since we have assumed that all G7 countries introduce the carbon tax, the following problems could be avoided.

If a single country introduces such an ecotax, the problem might arise that energy imports could rise. To avoid this substitution, imports also have to be taxed. This will work for primary energy but not for electricity, because in this concept electricity is not taxed. Nevertheless, the price for electricity in the domestic country will rise, because the costs for primary energy increase. So there will be growing imports of electricity, because in the foreign countries we have stable prices for electricity. If our country in question belongs to the European Union the possible reaction – taxation only on energy imports – is forbidden, because it would be a discrimination.

Further restrictions derive from conflicts between the environmental and general economic targets. Very often several regions or branches of the economy have to be sheltered. So we should not be surprised, that the political process creates environmental taxes with a confusing variety of tax rates and exemptions, that are differentiated for energy carriers and user groups.

This is the case for Denmark, Finland, the Netherlands, Norway, Sweden and Germany. For the latter country a model simulation study (Bach *et al.*, 2001) is available, which calculates the impact of the German Environmental Tax Reform on economy and environment. It might be interesting to compare our results of the textbook carbon tax for Germany with that of a real-world tax with all its exemptions and differentiations. This comparison makes sense, since the tax revenue is of a comparable order and the tax is also used to reduce payments for social security.

The tax revenue that will be reached in Germany with the 'real-world ecotax' in the year 2003 is about 15 billion US$. In our simulation of a textbook taxation tax revenues for the last simulation period for Germany amount to 10 billion US$. The effect on GDP is nearly the same with −0.35% in our simulation and −0.43% in the 'real-world case'.

Since the German Environmental Tax Reform is aimed to the end-use of energy, and hits primarily households, we have relatively strong negative production effects for labour-intensive consumer industries and service sectors. So in spite of the stronger wage reductions the employment effects are, at 0.5%, a little bit weaker than in the textbook simulation (0.89%).

The rise of production prices is correlated with the level of taxation. In the 'real-world case' we have as an average of all industries 0.64% higher prices and in the textbook case 0.52%.

The emissions of CO_2 are affected not only by the level of taxation, but also by its design. In the textbook case the tax rate per ton CO_2 is the same for all energy carriers. This means that the 'dirtiest' carrier coal has the highest taxation; thus its consumption will be reduced most strongly. In the German Environmental Tax Reform coal is not taxed at all. Further industries with a high energy intensity are excluded from tax payment. This difference in the design of the tax explains why the CO_2 emissions induced by the German Environmental Tax Reform reduce only at 2.3%, whereas in the textbook case this figure is 4.6%.

Summarizing, we see that the results of both simulations for Germany seem to be consistent, which also gives strong confirmation of the results of our analysis for the other countries.

Part IV

Methodologies

12
Energy projections: comparison of methodologies

KIMIO UNO

1. Introduction

The global-warming issue has passed through several phases, starting from scientific fact-finding to the worldwide recognition of the nature of the issue as constituting a sustainability of human society and the ecosystem, then to the debate on options open to the world community and presently to the phase of policy formulation and implementation. It has been established that CO_2 constitutes the largest contributor among major greenhouse gases and its emission is largely traced to energy use. Of the greenhouse gases (GHGs) with direct effect on radiative forcing, CO_2 contributes 64% of 1990 emissions to global warming. Of this, 80% is attributed to energy use, with other factors such as deforestation and changing land use and cement production contributing 17% and 3%, respectively. Combustion of fossil fuels is also blamed for emissions of CH_4 and N_2O with direct effects on radiative forcing and CO and NO_x with indirect effects on radiative forcing. This explains strong interest in evaluating CO_2 emission potentials and in limiting greenhouse gas emissions in general, and CO_2 emissions in particular. This review provides an overview of energy projections conducted by various research groups from the methodological point of view.

The purpose of this review is two-fold. First, it is intended to highlight the main features of COMPASS compared to the existing energy projection framework. Second, methodological comparison in itself is useful in interpreting the projection outcomes. The capabilities, as well as the limits, of models are already embedded in the model framework itself. It is imperative to make explicit (1) scope of the model, (2) the theoretical standpoint, and (3) assumptions that went into the simulation experiment. In addition, the main thrust of the individual modelling exercise is represented in accompanying tables reproduced from the original literature.

We may summarize some of the characteristics of COMPASS as follows, after cross-checking with other models.

(1) COMPASS has feedback channels running through economy–energy–environment spheres.
(2) Its framework is disaggregated almost to the maximum extent that is feasible from the data point of view.

K. Uno (ed.), Economy–Energy–Environment Simulation, 193–298.

(3) COMPASS incorporates, and is capable of endogenously generating, a relevant accounting scheme, including national income accounting, balance of payments and saving-investment balance, input-output tables, trade matrices, and energy balance tables. It can also generate prices for individual industrial branches, final demand, and energy. Exchange rates and interest rates are endogenously determined.
(4) Technological factors are represented as changes in input coefficients for each industrial branch and changes in energy intensity.
(5) An object-oriented approach is maintained throughout the project, ranging from naming convention, statistical database, model-building software, and simulation engine, making the construction and maintenance of large-scale models considerably systematic.

The main exogenous variables needed for the current version of COMPASS are as follows:

(1) Population.
(2) Exchange rate (this can be endogenized).
(3) Service trade (commodity trade is fully explained by the model) .
(4) Unit labour cost ((wage rate * employment) / output, or wage rate * labour productivity) at branch level (this can be endogenized, but this variable is often determined by wage policy).
(5) US interest rate.
(6) World oil price (model also calculates world oil prices to monitor the market force)

The scope of the COMPASS extends over economy, energy, and environmental spheres:

Economy: Input structure, output structure, final demand, value added, prices. Technological change (input coefficients, conversion efficiency, shares among primary energy sources), technology transfer, lifestyle shift. Balance of payments, trade shares matrices, exporting country, importing country for each branch. Saving-investment balance, flow of funds. Exchange rates, interest rates.

Energy: Total final consumption (TFC), energy conversion, total primary energy supply (TPES), required energy supply, energy trade, energy prices (global as well as domestic).

Environment: CO_2 emissions, emission trade. (Price change through emission trade, CO_2 taxes, etc.)

It is planned to present the raw data and simulation results in various formats, among them internationally linked environmental accounting.

COMPASS also provides information at detailed country levels, an advantage over various other models that provide information at aggregated regional groups. Analytical needs exist concerning developing countries and oil-producing countries as well as for developed industrial countries, to evaluate the impact of policy measures that are in most cases implemented by individual governments.

2. Comparison of methodologies

The methodology used can be broadly categorized as follows. In general the word *projection* is used to mean future outcome based on econometric or other quantitative analysis which relies on observed behaviour in the past, adjusted by evolutionary perspectives in order to take account of discontinuities such as introduction of new technology, new lifestyle, or new policy instruments. *Scenarios* refer to examination of potential alternative development over time, or a series of sensitivity analysis, contingent upon various assumptions pertaining to variables in question, presented in objective terms. The word *outlook* can be more judgemental observation pertaining to the future course of events, not necessarily based on formal analytical framework. In contrast, *forecast* and *prediction* are used interchangeably, and are inclined to deterministic description of future outcomes presented as a single, point estimate. *Prescription* emphasizes suggestions for policy response to an anticipated future situation, with or without quantitative description. *Goal* or *target* refers to policy objectives, with the implication that its realization is desirable. *Model solution* or *simulation result* refers to quantitative outcomes of a mathematically or econometrically formulated system of causal relationship, both during the observation period and in the future, without judgemental observation. It is noted, however, that underlying assumptions of models could be highly judgemental and subjective, without empirical foundation.

2.1. Scenarios

Projection based on various sources, without formal modelling. Studies by international organizations, with periodical updates, are often of this type (e.g. IEA, EIA, APERC. For a systematic examination of scenarios, see Kram *et al.*, 2000). IPCC report (IPCC, 2000b) as well as IIASA's work (IIASA, 1995) also belong to this category. There is a case where technological scenarios are incorporated into an IO model (World Model by Duchin and Lange, 1994). The work based on various elasticities can also be classified in this group, or it can be viewed as a hybrid of scenario writing and econometric modelling (IMAGE 2.0).

Scenarios can be more structured by decomposition of indicators. In other words, definitional identity can be broken down into relations among various indicators, whereby projection of individual components can be synthesized into a single, consistent projection (Unander and Schipper, 1998, 1999).

2.2. Computable general equilibrium (CGE) models

This approach, which assumes market-clearing equilibrium for individual economic agents as well as at macro level, dominates the academic scene. The term 'general equilibrium', however, is used to refer to a variety of approaches by different researchers. One can properly distinguish two schools in this camp: one using parameters from one specific point in time and seeking optimization over time under certain constraints, and the other using econometric models based on time-series observations employing market-clearing assumptions (i.e. disequilibrium is not dealt with by assumption). This approach can solve for optimality for a very long period of time, with a result that all long-term projections (typically 2100 as the terminal year) are of this variety.

2.3. Input-output plus econometric models

This approach employs the input-output framework, sometimes from one point in time and sometimes in time-series, to describe the structure of the economy, which is combined with a model embodying econometrically estimated parameters (COMPASS, GREEN, and G-cubed). It is thus capable of detailed intersectoral description as well as macroeconomic analysis. West (1995) referred to this type as IOE, as opposed to static IO and CGE. The availability of empirical data for econometric estimation limits its ability to project into the future, and models based on this methodology usually look at 10 to 20 years into the future. In return, it can deliver a time path of various structural adjustments as well as responses to disequilibrium in the system.

We have examined the models with focus on economy–energy–environment interaction. It is noteworthy that many models with energy-environment orientation treat the linkage with economy as exogenous. In other words they take the economic projection/scenario from an independent source and build energy and environmental analyses based on it. Naturally, the feedback from energy-environment spheres to the economy is typically lacking. The framework that closes this feedback channel employs an econometric modelling approach (COMPASS, GREEN, G-cubed, and Oxford Model; GTAP employs IO; the PRIMES is a group of country models).

The number of regions that are explicitly treated in the analysis range from several to about 60. The relation among them can best be described by trade matrices (COMPASS, GREEN, Duchin and Lange, WorldScan, GTAP, G-cubed). Some models use a pooled approach rather than describing the trade structure (International Futures). Other works do not specify the global linkage through trade. There are some cases where international linkage through financial flow is incorporated in the model framework. Most models, however, ignore this aspect despite the fact that growth potentials of various regions are constrained by global availability of financial resources. It is important to recognize that any investment plan and public policy has to be financed to be viable, including those in energy and environmental spheres.

The environmental impact of trade liberalization is actively discussed, with the discussion at the WTO as the backdrop. We have examined various studies as to their applicability to fields other than energy. Agriculture, forest products, and various service activities are the cases in question. The frameworks which demonstrate considerable flexibility include COMPASS, E3ME (Barker, 1998, 1999; Barker *et al.*, 1999), World Model (Duchin and Lange, 1994), GemWTraP (Bernard and Vielle, 1999, 2000), GTAP (Hertel and Tsigas, 1997), G-cubed (Bagnoli *et al.*, 1996), and IMAGE 2.0 (Alcamo, 1994a, b), followed by GREEN and International Futures. Other works employ an analytical framework strictly limited to energy-environment spheres.

The Kyoto Protocol calls for global effort aimed at curtailing CO_2 emissions to the 1990 levels among Annex I countries, and there are several approaches that trace the effects of international emission permit trading, such as MS-MRT (Bernstein *et al.*, 1999), GemWTraP (*ibid.*), Rose-Stevens Model (Rose and Stevens, 1999), and WorldScan (Bollen *et al.*, 1999), with COMPASS trailing. It is interesting to note that they were all published in 1999, except for pioneering work RICE by Nordhaus which came out in 1996.

3. Summary of models and projections

The field of economy-energy modelling and forecasting is developing rapidly, and the list is not intended to be comprehensive. This review is based on published materials that were available to the author at the time of writing, and may not reflect the current development of individual projects. The description is solely the responsibility of the author of this chapter.

In the following, particular emphasis is placed on the examination of economy-environmental linkage. Quantitative results are not emphasized; rather, the focal points of various projects are presented so that the whole array of methodologies will be exhibited. The review will also reveal that the end-results are contingent upon a set of assumptions that went into the analysis, as well as the scope of the conceptual framework being employed.

In order to facilitate comparisons, individual models are described in terms of the following aspects:

(1) Time span: target year of the analysis.
(2) Regions: regional details provided.
(3) Branch classification: economic sectors (industry, household, etc.) and disaggregation of industry sector into branches.
(4) Energy sources: primary energy, secondary energy.
(5) Scenarios: purpose of the simulation runs and a set of major assumptions.
(6) Technology: treatment of technology in the model framework in view of the importance of the technological factor.
(7) Energy intensity improvements: related to technology, but more precisely on energy.
(8) Lifestyle: consumption pattern, modal shift of transportation, recycling, among other things, related to economy–energy–environment.

(9) Policy tools: introduction of new policy measures and how this is incorporated in the model.
(10) Environment linkage: impact of economic activities and energy consumption on environment, with particular emphasis on CO_2 emissions.
(11) Economic linkage: feedback from energy and environment to economy.
(12) Modelling logic: theoretical background of the model, causal flow among economy–energy–environment, the method of achieving global linkage.

The specific references in the original work pertaining to the above points are referred to with page numbers. The summary is intended to highlight methodological characteristics as well as main findings of the original work. The results of methodological comparisons are listed in Table 1-4 in Chapter 1. The quantitative projections into the future are summarized in Table 12-25 below.

COMPASS (Comprehensive model for policy assessment)

Source

(A) Uno, Kimio (ed.). *Economy–Energy–Environment Simulation: Beyond the Kyoto Protocol*. Dordrecht: Kluwer Academic Publishers, 2002.

Time span: Annual to 2010-(2020).
Regions: 61 (54 countries plus 7 aggregate regions).
Branch classification: 36 for IO countries, 26 world trade for OECD (20 for APEC).
Energy sources: 6 (coal, crude oil, petroleum products, gas, nuclear, new and renewables) plus secondary (petroleum products, gas, and electricity and heat).
Scenarios: Alternative development in technology described through changes in input coefficients, changes in energy intensity (which is endogenous but can be exogenized for experiment), world oil price development (equilibrium world oil prices endogenously calculated but without direct feedback to the model). Policy simulation has not been attempted.
Technology: Scenario pertaining to input coefficients. Technology transfer through input coefficients.
Energy: Final consumption of energy endogenous. Substitution among energy carriers (shares) endogenous. Energy intensity improvements endogenous.
Lifestyle: (Endogenous household consumption).
Policy tools: Carbon tax, emissions trading.
Economic linkage: GDP for individual regions endogenous. International trade endogenous (endogenous trade matrices also tested), energy export and import endogenous. Prices, exchange rates, domestic energy prices endogenous. Saving-investment balance for the region and the world as a whole endogenous.
Environment linkage: Carbon emissions endogenous.

Modelling logic: Econometric models based on time-series IO tables, trade matrices, energy balances, financial flows, linked internationally through trade, price changes, and exchange rates.

Summary

COmprehensive Model for Policy ASSessment (COMPASS) is an econometric simulation model designed to capture the interaction among 3E spheres encompassing economy, energy, and environment. COMPASS is designed for simulation experiments up to 2020, but current experiment focuses on the period up to 2010. The COMPASS is characterized by regional and structural details based on econometrically obtained parameters, which enables the model to trace the actual time-path of the global economy. The model achieves international linkage for about 60 regions, covering 99.5% of the world GDP and energy consumption. The country models are based on time-series of a 36-sector input-output framework, the SNA, and financial accounts. The linkage is achieved through time-series of 25-commodity trade. The model also generates energy balances and CO_2 emissions for individual countries. The energy-related parameters include final consumption demand (industry sector, transport sector, and other sectors including residential, etc.) and substitution among energy carriers. The COMPASS closes the world in terms of goods and services (production and trade), financial flows (saving and investment), and energy supply and demand.

The COMPASS focuses on medium-term transition, not short-term (cyclical) analysis or long-term (general equilibrium) analysis, but traces how the paths would evolve in different regions and in different industrial branches responding to introduction of policy measures (such as tradable permit or carbon tax), technological change (such as new input structure and better conversion efficiency), and lifestyle shift (such as service orientation or energy conservation effort).

The COMPASS framework, because of its branch disaggregation and regional breakdown, and because of its simultaneous equation system, allows bottom-up and top-down interaction. The inclusion of branch prices (36 branches), final demand prices (consumption, investment, government spending, exports, and imports), exchange rate, interest rate, and various energy prices as endogenous variables, the adjustment mechanism by economic agents based on relative prices are described explicitly. The model treats world petroleum prices as an exogenous variable, but equilibrium world petroleum prices are calculated within the model in order to monitor the market forces behind the assumed price levels. Technology factors such as production technology (described by input coefficients), energy substitution (among different energy carriers), and changes in energy intensity are fully explained endogenously. The model can project input-output tables and energy balances in the future, providing structural information of the world community.

The energy block will be linked to environmental block via carbon emission factors (Asia Pacific Energy Research Centre, 1998, p.22)

Coal: 1.08 tC/toe
Gas: 0.64 tC/toe
Oil: 0.84 tC/toe

The policy experiments using the energy model for environmental analysis (CO_2 emissions, tradable permits, and AIJ (activities implemented jointly)) are being formulated.

Adam Rose-Brand Stevens model

Sources

(A) Adam Rose and Brandt Stevens. 'A Dynamic Analysis of the Efficiency and Equity of Tradable Greenhouse Gas Emission Permits.' Paper presented at the Joint Meeting organized by the International Energy Agency, Energy Modelling Forum, and the international Energy Workshop (IIASA), 16-18 June 1999, Paris.
(B) Adam Rose and Brandt Stevens. 'A Dynamic Analysis of Fairness in Global Warming Policy: Kyoto, Buenos Aires, and Beyond.' *Journal of Applied Economics*, 1998.
(C) Adam Rose, B. Stevens, J. Edmonds, and M. Wise. 'International Equity and Differentiation in Global Warming Policy: An Application to Tradable Emission Permits.' *Environmental and Resource Economics*, 12 (1), 1998.

Time span: 25-year time horizon (A, p.2). 1990 as the base year (A, p.27).
Regions: 14 (Africa, Australia and New Zealand, Canada, China, Eastern Europe, Former Soviet Union, India, Japan, Latin America, Middle East, Asian Tigers, South Asia, United States, Western Europe) (A, p.2, p.4).
Branch classification: No.
Energy sources: No disaggregation.
Scenarios: Stepwise, from the most basic (emission quotas for Annex B countries as set forth in the Kyoto Protocol to a globally comprehensive and fully dynamic CO_2 emissions trading) (A, p.10).
Technology: Positive and increasing marginal cost as percentage of abatement increases (A, p.8).
Energy intensity improvement: Not explicit.
Lifestyle: No.
Policy tools: Banking, inter-period trading, interregional trading.
Economic linkage: The number of constraints, which is 364 under business as usual case, is reduced to one as trading extends temporally and spatially (A, p.2, p.9).

Environment linkage: Benefit functions are tied to CO_2 concentration levels in climate equations (A, p.2).

Modelling logic: Nonlinear programming model, incorporating 364 constraints (A, p.2). The objective function is to minimize total global abatement cost which is endogenous.

Summary

Scenarios consist of the following (A, pp.10-11).

(1) No trading (Kyoto based emission quotas for Annex B countries).
(2) Trading by Annex B countries only (Kyoto-based tradable quotas):
   (a) trading among countries or groups (interregional trading)
   (b) banking
   (c) banking and borrowing (complete inter-period trading)
   (d) interregional and inter-period trading
   (e) interregional and inter-period trading and intergenerational equity.
(3) Trading among Annex B countries and developing countries (equity-based quotas for developing countries):
   (a) interregional trading (no harm to developing countries)
   (b) banking
   (c) banking and borrowing
   (d) interregional (no harm) and inter-period trading
   (e) timing of developing country emission reduction commitments
   (f) interregional and inter-period trading and intergenerational equity.

It is concluded that 'the greatest efficiency gains in the design of a GHG tradable permit policy stem from utilizing the low-cost mitigation options of developing countries (even if no additional mitigation is forthcoming from this group itself) and then from requiring emission reductions from developing countries at some future date' (A, p.25).

Inter-period permit trading (banking or borrowing) yields relatively small incremental gains. This is partly due to assuming constancy of mitigation costs for each country/region over time. Even if mitigation costs decreased by 1% per year over time, relative gains from inter-period trading improve only a factor of one- to two-fold (A, p.25).

In the case with interregional permit trading, compared to banking and inter-period trading cases, mitigation costs are lowered for each Annex B country whereas gross benefits for each country stay the same (A, p.13). The inclusion of developing countries into permit trading enlarges the global gains. In this case, however, Eastern Europe and the FSU are permit buyers rather than permit sellers (A, p.16).

AIM: Asian-Pacific integrated model

Sources

(A) National Institute for Environmental Studies, Japan. 'Asian-Pacific Integrated Model AIM' (mimeo). No publication date given.
(B) Intergovernmental Panel on Climate Change. *Special Report on Emissions Scenarios*. Cambridge: Cambridge University Press, 2000.
(C) Tsuneyuki Morita *et al. Global Carbon Dioxide Emission Scenarios and Their Basic Assumptions: 1994 Survey*. CGER-1101-94. Tsukuba: Center for Global Environmental Research, National Institute for Environmental Studies, 1994.

Time span: 2000-2100, with 5-year steps before 2030, and 2050, 2075, and 2100 subsequently (B, p.336).
Regions: 19 regions (A, p.21). 17 regions for the bottom-up and land equilibrium models (Economies in transition, OECD-West, USA, Oceania, Japan, Korea, Indonesia, Thailand, Malaysia, Other East Asia, India, Other South Asia, China, Other Centrally Planned Asia , Middle East, Africa, Latin America), 9 regions for the energy-economy model (B, pp.336, 338).
Branch classification: Varies by country (typically, industrial sector, residential sector, commercial sector, transportation sector, power plant).
Energy sources: Detailed.
Scenarios: Economic growth ranging from 1.1% to 3.0% per annum to 2100 (A, p.13). As defined in the Special Report on Emissions Scenarios (SRES) (B, p.5).
Technology: Technology selection from 100 alternatives, based on purchase prices and fuel costs, and the timing of replacement (A, p.19).
Energy intensity improvements: Varied but not explicit.
Lifestyle: No.
Policy tools: Not explicit. The case of carbon tax (30,000 yen per tC) is discussed for Japan (A, p.23).
Environment linkage: Modules on SO_2 emissions and deposition (A, pp.39-42), desulfurization (A, pp.43-44), climate change (A, pp.51-53), global carbon cycle (A, pp.55-57), water resources (A, pp.61-66), natural ecosystems (A, pp.67-70), malaria (A, p.71). CO_2 and other GHGs (B, p.336).
Economic linkage: Integrated bottom-up national models and top-down global model (A, p.3).
Modelling logic: Economic spheres based on country macro models and global linkage. Feedback from economy to energy is through technological selection which is endogenized. Various independent modules covering details on energy-environmental spheres, including spatial description.

Summary

AIM is a computer simulation model for scenario analyses of GHG emissions an the impacts of global warming, with focus on the Asia-Pacific region. The project was started in 1991 by the National Institute for Environmental Studies, Tsukuba, Japan, Dr Tsuneyuki Morita as the leader, with funding from the Environmental Agency of the Japanese government. The AIM model was used to provide global and regional emission scenarios to the IPCC. The new version links bottom-up models, traditionally the strong point of the AIM, with top-down models (an energy economy model based on Edmonds-Reilly-Barns model, and a land equilibrium model. The bottom-up components are linked with top-down models through a linkage module (B, p.29, pp.336-338). See Figure 12-1.

The entire project consists of three major models, focusing on emission processes (AIM/emission), climate change process (AIM/climate), and the impact analysis (AIM/impact) (B, p.336).

The emission model consists of several national models and provides predictions on the emission of several global warming gases. This in turn is linked to a rest-of-the-world model. Assumptions on population, economic trends, and government policies are entered into the model, providing estimates of energy consumption, land use, etc. The country model contains a module that generates end-use energy demand by multiplying the energy service by an energy efficiency factor. This factor is derived by a series of assumptions about the introduction of new technologies which is induced by energy prices. The energy selection sub-module decides which particular technologies will be introduced from the arsenal of more than 100 technologies. Energy demand estimates are then linked to a top-down economic model.

Economic scenarios entered into the emission model assume annual growth rates of 1.1% to 3.0% by the end of the 21st century, or an expansion of 3 to 25 times. A population growth rate is assumed to be 0.7%, although the estimates for the year 2100 range from 3.6 billion (due to higher death rates attributable to environmental pollution) to 109.4 billion (based on the current fertility rate).

It is predicted that China will become the largest emitter of CO_2 before 2025. India will become the fourth largest emitter, following the United States and former Soviet Union. Japan is placed fifth. The Asian-Pacific share of global CO_2 emissions, which currently stands at 25%, will reach 36% in 2025 and 50% by the end of the next century.

The emission model has a land use change model to take account of the effects of deforestation on CO_2 levels and CH_4 emissions. The climate change model deals with carbon cycles and climatic phenomena. The mean global value of temperature changes is calculated, which then becomes the input into the regional models.

The impact model examines the interaction between the emission and climate change models. It calculates the impact on primary industries (water supply, agriculture, forest products, etc.) and human health.

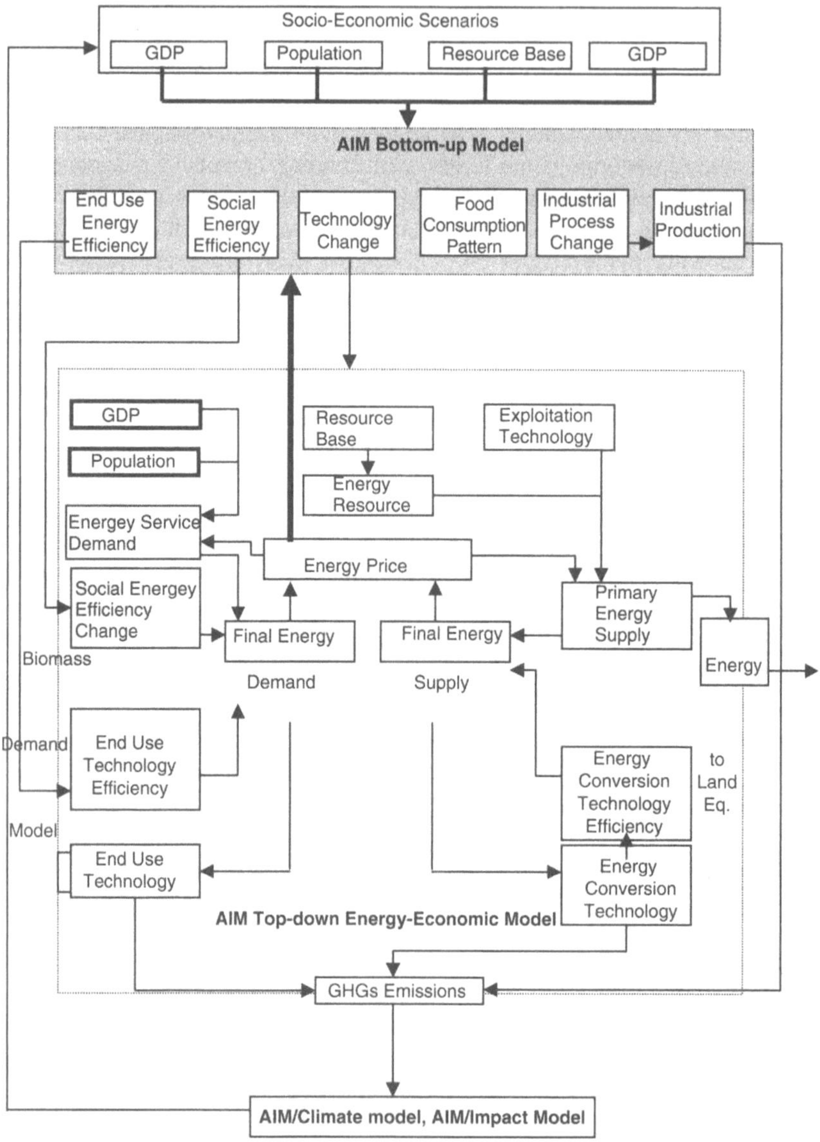

Figure 12-1. Bottom-up and top-down linkage in AIM.

Source: Intergovernmental Panel on Climate Change (2000). Modified by the present author to focus on bottom-up top-down linkage.

Asia-Pacific GIS provides a spatial dimension to the AIM.

Regarding policy tools, the report says as follows referring to the CO_2 emission model for Japan. 'In order to reduce emissions to the 1990 level with a single carbon tax, a high tax rate of 30,000 yen/tC is required. … An alternative would be to use the revenues from the tax to subsidize investment in energy-efficient technologies. In this case, the tax rate would only need to be 3,000 yen/tC. This combination of tax and subsidy policies can be evaluated with a two-dimensional optimization technique. The AIM model can solve for these sophisticated algorithms internally' (A, p.23).

APEC energy outlook by APERC

Sources

(A) Asia Pacific Energy Research Centre. *APEC Energy Demand and Supply Outlook*. Tokyo, 1998.
(B) Australian Bureau of Agricultural and Resource Economics. *The MEGA-RARE Model: Interim Documentation*. Canberra, 1996.

Time span: 2010 (A, p.iii).
Regions: 6 (United States, Other Americas (Canada, Chile, Mexico), China, Other East Asia (Hong Kong (China), Japan, Korea, Chinese Taipei, Oceania (Australia, New Zealand, and Papua new Guinea), and Southeast Asia (Brunei Darussalam, Indonesia, Malaysia, the Philippines, Singapore, Thailand) (A. p.1). 30 in the model (B).
Branch classification: 3 (industrial sector, transportation sector, and residential and commercial sector (A, pp.4-7). 37 in the model (B).
Energy sources: 6 (coal, oil, natural gas, hydro, nuclear, new and renewable energy (NRE)) and electricity (A, pp.9-14).
Scenarios: 3 scenarios: baseline based on MEGABARE model; protracted crisis scenario (PCS); and environmentally friendly scenario (EFS) (A, p.iii).
Technology: Not directly included in the framework; assumed in energy intensity improvements below.
Energy intensity improvements: Assumed. See Table 12-1.
Lifestyle: No.
Policy tools: Not directly included in the framework. Infrastructure requirements, regulatory reform as background information. (A, pp.17-18).
Environment linkage: CO_2 from coal, gas, and oil are derived based on emission factors (A, p.22).
Economic linkage: No feedback. Economic scenarios are from a separate global model MEGABARE.
Modelling logic: Scenarios in energy sphere. Economic performance as exogenous.

Summary

The APEC Energy Action Programme adopted at the APEC Economic Leaders Meeting in November 1995 assigned the task of compiling an Asia-Pacific

Table 12-1. Energy intensity assumptions and energy requirement in APEC

	Actual 1995	B98			EFS		
		2000	2005	2010	2000	2005	2010
Energy Intensity (TPEC / GDP)(toe per million 1987 US$)							
APEC	0.39	0.39	0.38	0.36	0.38	0.35	0.32
United States	0.39	0.37	0.35	0.33	0.36	0.33	0.30
Other Americas	0.57	0.55	0.52	0.51	0.53	0.49	0.45
China	1.20	1.02	0.88	0.77	0.94	0.78	0.67
Other East Asia	0.20	0.21	0.20	0.19	0.20	0.19	0.18
Oceania	0.36	0.39	0.37	0.37	0.37	0.34	0.32
Southeast Asia	0.53	0.56	0.53	0.52	0.54	0.50	0.47
Total Final Energy Consumption (Mtoe)							
APEC	3130.2	3486.8	3901.2	4368.5	3412.8	3699.2	4039.4
United States	1507.3	1651.0	1766.3	1880.9	1632.8	1700.9	1767.3
Other Americas	284.5	316.3	368.6	435.4	307.9	345.9	392.2
China	623.4	750.9	882.1	1014.3	716.6	808.9	922.3
Other East Asia	505	538.3	601.3	675.1	530.1	577.6	629.6
Oceania	83.9	95.4	110.8	130.1	93.5	104.3	116.8
Southeast Asia	126.1	134.8	172	232.8	131.8	161.7	211.3
Net Energy Imports by APEC Region							
APEC	691.1	829.8	1064.6	1327.8	656.5	654.2	647.0
United States	439.4	531.3	629	708.4	490.4	498.7	482.5
Other Americas	−168.6	−185.3	−158.1	−108.6	−198.0	−193.5	−179.9
China	−12.0	49.4	99.5	121.4	−31.3	−46.5	−68.0
Other East Asia	580.4	634.3	708.4	788.3	609.4	645.3	676.0
Oceania	−94.7	−120.2	−139.8	−148.4	−126.4	−154.2	−181.7
Southeast Asia	−53.7	−79.9	−74.4	−33.2	−87.6	−95.5	−81.7

Source: APERC (1998).

Economic Cooperation (APEC) regional energy outlook to the Asia-Pacific Energy Research Centre (APERC) which was established in 1996. Energy experts from the APEC countries (United States, Canada, Chile, Mexico, China, Hong Kong (China), Japan, Korea, Taiwan (Chinese Taipei), Australia, New Zealand, Papua New Guinea, Indonesia, Malaysia, Philippines, Singapore, Thailand, and Brunei Darussalam are stationed at the APERC.

The projection consists of three scenarios. First, 1998 baseline scenario (B98) is built around the GDP growth projections from the MEGABARE model provided by the Australian Bureau of Agricultural and Resource Economics (ABARE). The MEGABARE model has been developed at ABARE to provide a global perspective on major policy issues facing Australia, such as enhanced greenhouse effect and trade liberalization under the Uruguay Round and APEC initiatives. It looks especially into international flows of commodities and capital from a global perspective. Second, a protracted crisis scenario provides a more pessimistic view assuming continued economic turmoil in Asia, with more severe ramifications for APEC economies. Third, the environmentally friendly scenario (EFC) is based on the same GDP as the baseline scenario but assumes improved energy efficiency throughout the region and switching of fuels in order to minimize the use of carbon-intensive fuels.

It is indicated that the APEC region's demand for energy is expected to outpace its energy production by a wide margin, pointing to the region's growing energy import. The policy conclusions drawn are four-fold: (1) securing adequate energy supplies to meet growing energy demand when the economies in the region recover from the current downturn; (2) minimizing environmental degradation resulting from energy production and use; (3) using energy resources efficiently; and (4) promoting infrastructure investment and regulatory reform in the energy sector (A, pp.15-19).

'APERC's outlook projects CO_2 emissions to increase 42 per cent in the period to 2010 and increase by 22 per cent under the more energy efficient EFS case' (A, p.15). 'The APEC forum is ideally suited for development cooperative policies that encourage investment in energy conservation practices and facilitate the introduction of new and more efficient technologies.' 'The EFS case highlights the gains from energy efficiency improvements across the APEC region with reduction in TPEC in 2010 of 10 per cent, net energy imports of 51 per cent and CO_2 emissions of 14 per cent' (A, pp.16-17).

ASF: the atmospheric stabilization framework

Sources

(A) Intergovernmental Panel on Climate Change. *Special Report on Emissions Scenarios*. Cambridge: Cambridge University Press, 2000.
(B) D. Lashof and D. A. Tirpak. *Policy Options for Stabilizing Global Climate*. Washington, DC: US Environmental Agency, 1990.
(C) W.J. Pepper *et al.* 'No-policy Greenhouse Gas Emission Scenarios: Revising IPCC 1992.' *Environmental Science and Policy*, 1998.

Time span: 2100.
Regions: 9 (Africa, Centrally Planned Asia, Eastern Europe and Newly Independent States, Latin America, Middle East, OECD-East, OECD-West, South East Asia and Oceania, USA) (A, p.339).
Branch classification: Energy plus agriculture and forestry.
Energy sources: Various.
Scenarios: As defined in the Special Report on Emissions Scenarios (SRES) (A, p.5).
Technology: Conversion efficiency.
Energy intensity improvement: Implicit in conversion efficiency.
Lifestyle: No.
Policy tools: Tax policies by region.

Economic linkage: Supply and demand of energy, energy supply prices, secondary energy prices, agricultural products (A, p.339).
Environment linkage: Energy, agricultural, and deforestation models leading to estimation of GHGs, which then feeds into atmospheric model.
Modelling logic: Energy supply and demand adjusted by price iteration.

Summary

The ASF study is conducted by the ICF Consulting in the USA.

The ASF treats GHG emissions from energy, agricultural, and deforestation origins, feeding into an atmospheric model. The supply and demand for energy is achieved by energy price adjustments. Energy prices differ by region and by type of energy, reflecting market conditions, supply constraints, conversion costs, etc. An iterative process is employed to determine supply prices. The producer prices, in turn, determine the secondary energy prices in each region. The secondary prices reflect interregional transportation cost, refining and distribution costs, and regional tax policies. The electricity price reflects relative proportions of each fuel, secondary prices of those fuels, and conversion efficiency (A, p.339).

The ASF agricultural model is driven by population and GNP growth, and estimates the production of major agricultural products. It is linked to the ASF deforestation model which estimates the land deforested annually as a function of population growth and agricultural demand.

The ASF GHG emission model uses results from energy, agricultural, and deforestation models to estimate GHG emissions in each region. The ASF atmospheric model then estimates the GHG concentrations and corresponding radiative forcing and temperature effects (A, p.339).

Decomposition of energy scenarios by Unandar *et al.*

Sources

(A) Fridtjof Unander and Lee Schipper. 'Which Road from Kyoto? Decomposition of Emission Scenarios.' Paper presented at the Joint Meeting organized by the International Energy Agency, Energy Modeling Forum, and the international Energy Workshop (IIASA), 16-18 June 1999. Paris 1999.
(B) Fridtjof Unander and Lee Schipper. 'Past and Future Trends in CO_2 Emissions from Energy Use: The Indicator Approach.' *ENER Bulletin* No.22. European Network for Energy Economic Research, 1998.
(C) International Energy Agency. *Energy Efficiency Initiative. Energy Policy Analysis*, Volume 1. Paris: OECD, 1998.

Time span: 1972, 1990, 1994, 2000, 2010 (A, p.2).
Regions: 7 (Australia, Denmark, Germany, Japan, Norway, United Kingdom, United States) (A, p.8); 6 additional countries in (B).

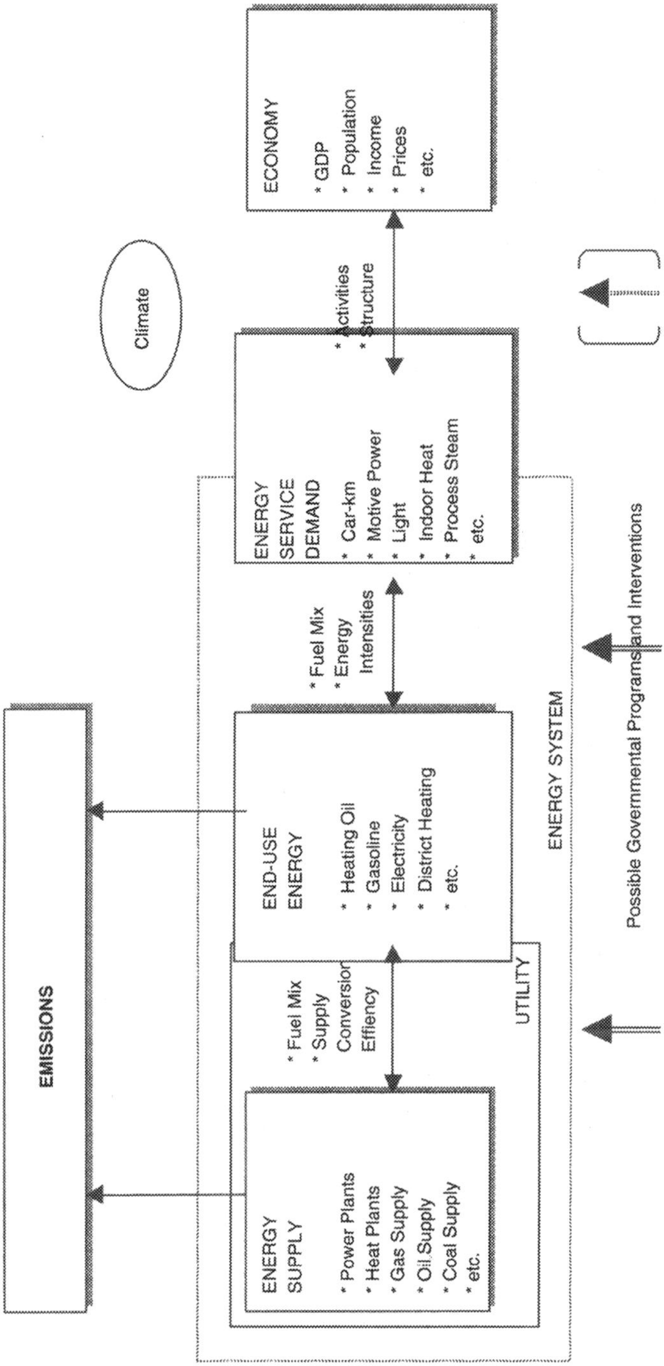

Figure 12-2. Decomposition model of energy-emission indicators, Unander and Schipper.
Source: Unander and Schipper (1999).

Table 12-2. Decomposition results (unit: annual average rate of change)

	Actual Emissions	GDP	Emissions/ GDP	Activity	Structure	Utility Fuel Mix	Energy Intensity	Final Fuel Mix	Carbon Intensity	Energy Services
1973-1990										
Denmark	−0.1	1.8	−1.9	1.1	0.3	−0.1	−2.3	0.5	−1.9	1.50
Norway	−0.9	3.3	−4.2	1.4	−0.1	−0.1	−0.3	−1.7	−2.1	1.30
W. Germany	−0.6	2.2	−2.9	1.4	0.6	−1.1	−2.2	0.3	−2.8	2.00
UK	−0.7	2.0	−2.8	1.0	0.2	−0.5	−1.8	0.4	−1.8	1.20
USA	0.2	2.7	−2.5	1.9	0.1	−0.4	−1.8	0.1	−2.0	2.00
Japan	0.9	3.7	−2.9	2.7	0.0	−0.8	−1.8	0.3	−2.1	2.70
Australia	2.5	3.0	−0.4	2.0	1.0	−0.4	−0.7	0.5	−0.5	3.00
1990-1994										
Denmark	0.0	1.6	−1.6	0.6	0.4	−1.2	0.3	0.0	−0.8	1.00
Norway	−1.1	3.0	−4.0	0.7	−0.8	0.0	−0.5	−0.5	−0.6	−0.10
W. Germany	0.6	1.8	−1.2	0.3	1.0	−0.3	−0.3	0.0	−0.6	1.30
UK	−2.4	0.9	−3.2	0.3	0.2	−2.3	−0.6	0.2	−2.7	0.50
USA	0.9	2.3	−1.3	1.8	0.7	−0.6	−0.8	0.1	−1.3	2.50
Japan	1.6	1.4	−0.2	−0.4	0.8	−0.3	1.5	0.1	1.2	0.40
Australia	1.5	3.3	−1.7	1.4	0.5	0.0	−0.4	0.1	−0.3	2.00

Source: Unander and Schipper (1999).

Branch classification: Nearly three dozen activities and economic branches (A, p.1). Manufacturing, other industries, households, services, travel, freight are specifically discussed (A, pp.9-10, p.12).

Energy sources: Final energy mix, utility fuel mix described as carbon intensity of fuel *j* used in sub-sector *i* (A, p.1, p.5).

Scenarios: Decomposition method used for constructing scenarios in terms of economic/energy/emissions futures originating in various agreements reached at COP-3 or from national sources (A, pp.10-12).

Technology: Energy intensity exogenous.

Energy intensity improvement: End-use energy intensities specified from various sources (A, p.1, p.5, p.10).

Lifestyle: Can be partially reflected in the term representing overall activity levels and sectoral structure.

Policy tools: Policies can be directed towards:

(1) more efficient end-use energy,
(2) changes to end-use fuels with lower carbon contents,
(3) supply side such as improvement of conversion efficiency, moves to renewable energy and low-carbon fuels,
(4) reduced activity levels, and
(5) changes in the structure of energy use, such as improving public transportation (A, p.6).

Economic linkage: Overall activity and sectoral structure (A, p.5).

Environment linkage: Energy intensity and the carbon contents of each fuel (A, p.5).

Modelling logic: Decomposition. Emissions are related to four multiplicative terms (sectoral activity levels, shares of sub-sectors, energy intensity in each sub-sector, and carbon contents of each fuel) (A, p.5). See below.

Summary

'We calculate the relative importance [of] changes in energy intensities, final fuel mix, utility fuel mix, sectoral activity level and structure had on energy use and carbon emissions over time' (A, p.1). The decomposition analysis is not applicable to the past experience. 'The approach will identify what kinds of changes must take place in both fuel mix and energy intensities to achieve a given level of energy use and emissions in a future year. Some important information can be based on results from calculations with energy and macro-economic models.' At the same time, 'this approach can help bring together results from top-down and bottom-up analysis in a very transparent manner' (A, p.10). The decomposition of changes in CO_2 emission can be summarized by the relation as follows:

$$G = A * Si * Ii * Fij$$

where

G = missions in a sector,
A = overall sectoral activity (as value added in manufacturing),
S = share of output by sub-sector i,
I = energy intensity of each sub-sector i (e.g. energy use/real value added)
F = carbon content of each fuel j used in sub-sector i.

'Carbon intensity' refers to the product of the terms I and F, and 'energy services' can be defined as the product of A and S (A, p.5).

'It is important to separate the components of the carbon intensity effect (which is related to energy efficiency and fuel choices) from those related to people and enterprises' demand for energy service, since they change for different reasons and in response to different stimuli, as energy prices' (A, pp.5-6).

It is observed that, between 1970 and the mid-1990s, falling energy intensities, shifts away from coal and oil, and (in some countries) a reduction of energy-intensive manufacturing contributed to lowering emissions per unit of output. The trend has been slowed or reversed due to renewed economic growth, slowdown in the decline in energy intensities, and growth in the size and power of personal vehicles (A, p.13). If Kyoto targets are to be met, 'most IEA countries would need to decarbonize faster than any previous time period, at a time when oil prices are at historically low levels' (A, p.13).

DEMETER: the DEcarbonisation model with endogenous technologies for emission reductions

Source

(A) B.C.C. van der Zwaan, R. Gerlagh, G. Klaassen, and L. Schattenholzer. 'Endogenous Technological Change in Climate Change Modelling.' Paper presented at the Joint EMF/IEA/IIASA Meeting, Stanford University, June 2000.

Time span: 2100.
Regions: No.
Branch classification: No.
Energy sources: Carbon energy.
Scenarios:

	Temperature Constraint	Endogenous Technology	Endogenous Energy Demand
BAU	no	yes	yes
METH1	2C	yes	no
METH2	2C	yes	no
METH3	2C	no	yes
METH4	2C	yes	no

Technology: Technological change as a function of cumulative capacity (A, p.3). Two different energy technologies (carbon and carbon-free) (A, p.5).

Energy intensity improvement: Not assumed, but explained by learning-by-doing. The cost of the carbon-free technology depends on the cumulative capacity installed (A, p.5).

Lifestyle: No.

Policy tools: Indirectly, through subsidies for investments in non-carbon renewables, such as solar, biomass and wind (A, p.18). Subsidies are not incorporated in the model itself.

Economic linkage: Production function incorporates *A*(t), representing the level of technological progress for the capital/labour composite, and *B*(t), representing the specific level of energy technological progress or energy technology stock. *B*(t) is chosen so as to reproduce an exogenous path for the AEEI (A, p.6).

Environment linkage: Impact on optimal CO_2 abatement and carbon tax levels of introducing endogenous technological change in a macroeconomic model.

Modelling logic: An optimization programme is solved in which the objective function of the total discounted sum of consumption is to be maximized (A, p.8).

Summary

'We employ a macroeconomic model that distinguishes between two different energy technologies (carbon and carbon-free), which, as in the bottom-up case, are subject to learning-by-doing. The costs of the carbon-free technology depend on the cumulative capacity installed' (A, p.5). 'The bottom-up and top-down approaches of endogenizing technological change are fundamentally different, both having their advantages. The purpose of this paper is to utilize the advantages of both approaches by combining them in a single model' (A, p.5).

'Effects of emission reduction measures and endogenous technological progress on energy demand are now explicitly reflected' (A, p.5).

'Our results are stronger than suggested by existing bottom-up and macroeconomic models. Earlier reductions are warranted since in contrast to bottom-up models total energy demand reductions are included as an additional carbon abatement option.' 'The model results show that the development of carbon-free technologies turns out to be the most important emission reduction option. The inclusion of endogenous technological progress implies that earlier investments in the non-fossil carbon-free technology are warranted, than traditional models suggest' (A, p.18).

Simulation results show that large differences start to occur between taxes among four scenarios. The carbon tax in METH1 and METH3 increases rapidly during the second half of the 21st century and reaches values of around 450 \$/tC and 380 \$/tC, respectively, in the year 2100. In METH2 and METH4, which incorporate learning-by-doing, the cost of energy produced by non-fossil fuel is lower, the carbon taxes increase much more moderately, carbon taxes reach approximately 230 $/tC and 175 $/tC, respectively, in 2100.

ECN (The Netherlands energy research foundation) study on the Kyoto mechanism

Source

(A) J.P.M. Sijm *et al.* 'Kyoto Mechanism: The Role of Joint Implementation, the Clean Development Mechanism and Emissions Trading in Reducing Greenhouse Gas Emissions.' Patten, the Netherlands: Netherlands Energy Research Foundation, 2000.

Time span: 2010.
Regions: 30 (Western Annex I (23: Netherlands, Belgium, Germany, Italy, Sweden, Finland, Denmark, France, Greece, Portugal, Spain, United Kingdom, Austria, Ireland, Luxembourg, Switzerland, Norway, Japan, USA, Australia, Canada, New Zealand, Iceland), CEE+FSU Annex I, Latin America, Africa, Asia, FSU non-Annex I, Middle East, Oceania) (A, p.38).

Table 12-3. Emission trends, Kyoto targets, and reduction commitments of western Annex I countries (unit: Mt)

	Emissions			Reduction Requirements	
	1990 (Mt)	2010 (Mt)	Kyoto Target (Mt)	(Mt)	(% of 2010 Emissions)
Netherlands	225	258	212	47	18
Belgium	130	144	120	24	16
Germany	1203	976	951	25	3
Italy	511	592	478	113	19
Sweden	67	76	69	7	9
Finland	64	83	64	19	23
Denmark	73	79	58	21	27
France	501	515	501	14	3
Greece	101	147	126	22	15
Portugal	68	86	84	2	3
Spain	294	358	338	19	5
United Kingdom	752	699	658	42	6
Austria	79	86	69	17	20
Ireland	53	69	60	9	13
Luxembourg	16	12	12	0	0
Switzerland	53	66	49	17	26
Norway	43	52	44	8	16
Japan	1333	1587	1253	334	21
USA	6187	7751	5754	1997	26
Australia	423	496	456	40	8
Canada	340	402	320	82	20
New Zealand	69	105	69	36	34
Iceland	3	4	3	1	29

Source: J.P.M. Sijm *et al.* (2000).

Table 12-4. Domestic reductions and emission trade effects

	Reduction Requirements (Mt)	Domestic Reductions (Mt) Case A	Domestic Reductions (Mt) Case B	Trade in Emission Credits (Mt) Case A	Trade in Emission Credits (Mt) Case B	Trade (% of Requirements) Case A	Trade (% of Requirements) Case B
Netherlands	47	19	10	−27	−36	59	78
Belgium	24	8	4	−16	−20	66	84
Germany	25	59	29	34	4	−135	−14
Italy	113	25	11	−88	−102	78	90
Sweden	7	2	1	−5	−6	76	87
Finland	19	7	3	−13	−17	65	86
Denmark	21	4	2	−17	−19	83	91
France	14	36	17	22	3	−158	−22
Greece	22	11	6	−10	−16	48	74
Portugal	2	7	4	4	1	−183	−49
Spain	19	28	14	9	−5	−48	26
United Kingdom	42	20	10	20	10	−48	−23
Austria	17	5	3	−13	−14	73	80
Ireland	9	6	3	−3	−6	29	64
Luxembourg	0	0	0	0	0	n.a.	n.a.
Switzerland	17	3	3	−14	−15	80	84
Norway	8	3	2	−6	−7	70	78
Japan	334	57	48	−278	−286	83	86
USA	1997	734	341	−1263	−1656	63	83
Australia	40	23	13	−17	−27	43	69
Canada	82	31	18	−51	−64	63	78
New Zealand	36	21	9	−15	−27	43	76
Iceland	1	1	0	−1	−1	81	108
CEE + FSU Annex I	0	254	900	254	900	n.a.	n.a.
Latin America	0	151	177	151	177	n.a.	n.a.
Africa	0	113	68	113	68	n.a.	n.a.
Asia	0	1116	1001	1116	1001	n.a.	n.a.
FSU non-Annex I	0	96	41	96	41	n.a.	n.a.
Middle East	0	57	159	57	159	n.a.	n.a.
Oceania	0	1	0	1	0	n.a.	n.a.

Note: In Case A, reduction options at negative marginal costs in non-Annex I and CEE/FSU Annex I countries are excluded from the analysis, whereas this potential is included in Case B.

Source: J.P.M. Sijm *et al.* (2000).

Branch classification: Not explicit.
Energy sources: Not explicit, but the ones leading to CO_2 emissions.
Energy intensity improvement: Not explicit.
Lifestyle: No.
Policy tools: Flexibility mechanisms as stipulated in the Kyoto Protocol (A, p.8).
Economic linkage: Baseline emissions in 2010 assumed, and contrasted to the Kyoto Targets (A, p.33).
Environment linkage: CO_2 emissions, equilibrium price for emission credits, average cost before and after international trade (A, p.37).
Modelling logic: No formal model. Based on cost curves and demand curves.

Table 12-5. Costs effects of using emissions trading

	Reduction Cost before Trade (mUS$95)	Reduction Cost after Trade (mUS$95)		Net Gains of Trade (mUS$95)		Net Gains (% of GDP 2010)	
		Case A	Case B	Case A	Case B	Case A	Case B
Netherlands	506	266	117	240	389	0.04	0.07
Belgium	298	147	64	151	234	0.04	0.06
Germany	20	−69	19	88	1	0.00	0.00
Italy	7551	764	320	6788	7232	0.46	0.49
Sweden	128	46	19	82	109	0.03	0.04
Finland	266	122	54	144	212	0.09	0.13
Denmark	859	143	58	715	800	0.31	0.35
France	11	−50	9	60	1	0.00	0.00
Greece	162	115	53	47	109	0.04	0.09
Portugal	0	−13	0	13	0	0.01	0.00
Spain	33	19	29	14	4	0.00	0.00
United Kingdom	595	−88	−19	683	614	0.05	0.04
Austria	1167	104	43	1063	1125	0.34	0.36
Ireland	46	40	20	6	26	0.01	0.03
Luxembourg	0	−1	0	1	0	0.00	0.00
Switzerland	1061	111	44	950	1017	0.23	0.25
Norway	242	49	20	193	222	0.10	0.11
Japan	32222	2179	864	30043	31357	0.45	0.47
USA	28830	12154	5341	16676	23490	0.18	0.25
Australia	371	196	92	175	278	0.04	0.06
Canada	1090	474	205	616	886	0.08	0.11
New Zealand	282	192	93	90	190	0.13	0.28
Iceland	14	10	5	4	9	0.04	0.10
Total Western Annex I	75753	16911	7449	58842	68304	0.20	0.23
CEE + FSU Annex I	0	−1029	−2557	1029	2557	0.09	0.23
Total Annex I	75753	15882	4892	59871	70861	0.20	0.23
Latin America	0	−580	−442	580	442	0.02	0.02
Africa	0	−432	−138	432	138	0.06	0.02
Asia	0	−4277	−2347	4277	2347	0.08	0.05
FSU non-Annex I	0	−370	−68	370	68	0.17	0.03
Middle East	0	−217	−444	217	444	0.02	0.03
Oceania	0	−5	−1	5	1	0.05	0.01
Total Non-Annex I	0	−5881	−3439	5881	3439	0.06	0.03
World	75753	10002	1453	65751	74299	0.16	0.18

Source: J.P.M. Sijm *et al.* (2000).

Summary

This is a work by ECN (the Netherlands Energy Research Foundation) at Petten, the Netherlands, which is an independent market-oriented organization for research, development, and consultancy in energy and related fields.

The study analyses the 'potential role of the Kyoto Mechanism in meeting the commitments of Annex I countries to limit their greenhouse gas (GHG) emissions in the period 2008-2012' (A, p.2). The Kyoto Mechanism or 'flexible instruments' include Joint Implementation, the Clean Development Mechanism, and Emissions Trading.

Table 12-6. GHG emissions 1990-2010 by region

	GHG			CO_2		
	1990 (Mt)	2010 (Mt)	Growth (%)	1990 (Mt)	2010 (Mt)	Growth (%)
Western Annex I	12,588	14,645	16	9920	11,702	18
CEE/FSU Annex I	4885	4813	−1	4124	3962	−4
Total Annex I	17,473	19,458	11	14,044	15,664	12
Latin America	2141	2572	20	1021	1284	26
Africa	1589	2877	81	767	1876	145
Asia	7290	10,535	45	4051	6337	56
FSU non Annex I	1232	1330	8	341	442	30
Middle East	1113	1702	53	838	1317	57
Oceania	9	13	40	4	4	5
Total non-Annex I	13,374	19,028	42	7022	11,260	60
World	30,848	38,486	25	21,065	26,924	28

Source: J.P.M. Sijm *et al.* (2000).

Following Skea (1999), the authors point to the distinction between the 'what' flexibility, the 'when' flexibility, and the 'where' flexibility (A, p.8).

'What' flexibility: basket of six greenhouse gases (GHGs)consideration of sinks
'When' flexibility: base year choice banking/early crediting
'Where' flexibility: joint implementation (JI) clean development mechanism (CDM) emissions trading (ET)

Those flexibility mechanisms have been introduced in order

(1) to reduce the abatement costs of the Annex I countries as well as
(2) to encourage the sustainable development of non-Annex countries by means of technology transfers and institutional capacity building, and
(3) to facilitate an agreement among all countries involved.

At first sight these commodities seem to be traded on separate markets, which are distinguished institutionally by different systems of governance, monitoring, validation, etc. In fact, however, these markets are closely interrelated as they deal in the same basic commodity, namely additions to – or subtractions from – the amount of GHG emissions assigned to Annex I countries for the period 2008-2012 (p.16). The term 'emission credits' is used as the collective concept for credits generated and transferred by JI, CDM and/or ET. It is assumed that this basic commodity is traded on an integrated market (p.21).

A spreadsheet model has been developed to simulate a market for trading *emission credits* in order to indicate the potential role and cost impact of each Kyoto Mechanism separately and, subsequently, for all mechanisms together in a global setting.

As a first step, data have been gathered or estimated with regard to:

(1) National or regional GHG emissions in a reference year (i.e. 1990/1995) and a future year (i.e. 2010, as representative of 2008-2012). By means of these data and certain reduction targets – such as the Kyoto commitments for 2010 – national or regional reduction requirements have been calculated in terms of physical quantities of GHG emissions.
(2) The potential and costs to reduce GHG emissions in a certain country or region. These data have been used to determine individual *cost curves* for the reduction of GHG emissions in a particular country or region. Subsequently, these individual curves have been added up and combined into aggregated cost curves covering several or all GHGs, countries and/or regions. Marginal cost curves for reducing CO_2 emissions are available for most western Annex I countries. These curves have been derived from ETSAP and COHERENCE studies, based on detailed energy and technology bottom-up models such as MARKAL and EFOM (p.24).
(3) Finally, this process of adding up cost curves has resulted in the construction of a world-wide cost curve for the reduction of all GHG emissions (p.21).

The *demand curve* of emission credits depends on the amount of emissions that a country is obliged to reduce according to the Kyoto Protocol. This curve is represented by a vertical line.

By aggregating the supply and demand curve of all countries concerned, the same procedure has been used to determine the equilibrium price on the international market of emission credits.

EDGE model

Source

(A) Jesper Jensen *et al.* 'The Economic Effects of the European Ceilings Proposal.' Paper presented at the Joint EMF/IEA/IIASA Meeting, Stanford University. 20-22 June 2000.

Time span: 2030.
Regions: 8 (European Union; United States; New Zealand, Australia, and Canada; Japan; Eastern Europe and Former Soviet Union; China; Major Oil Exporters; Rest of the World).
Branch classification: 7 (coal, petroleum, crude oil, natural gas, electricity, energy intensive sectors, other sectors).
Energy sources: 5 (coal, petroleum, crude oil, natural gas, electricity).
Scenarios:
Annex B trade:
No trade: No trade. Only domestic abatement.

Proposal B: The European proposal excluding trade with non-Annex B countries.
Hot air: The European proposal excluding trade with non-Annex B countries and no restriction on hot air.
Annex B: Free trade between Annex B countries.

Global trade:
No trade: No trade. Only domestic abatement.
Proposal G: The European proposal including trade with non-Annex B countries.
Global: Free global trade (A, p.12).

Technology: Production processes are represented by nested constant elasticity of substitution (CES) functions to characterize the choices between: (1) the region-specific imports and (2) the composite import goods and the domestically produced good (A, p.9).
Energy intensity improvement: Autonomous energy efficiency improvement to match growth in GDP and emissions is assigned (A, p.11).
Lifestyle: No.
Policy tools: International trade in emission reductions, and a set of concrete ceilings for selling and buying emission reductions for all Annex B countries.
Economic linkage: Interregional flows of goods and services.
Environment linkage: Climate policy analysis (A, p.9). The model reproduces a set of paths for emissions growth, GDP growth, energy production, and energy prices (A, p.11).
Modelling logic: Applied general equilibrium model (A, p.4).

Summary

EU members have proposed a set of guidelines for emissions reductions trade within the Kyoto Protocol. The European proposal limits imports of emission reductions by Annex B countries to 50% of the total abatement requirements and eliminates exports of hot air. The economic effects of the European proposal are analysed using the EDGE model.

'In the short run, the proposal drives up global prices for emission reductions, because the elimination of hot air reduces the supply of emission reductions. In the longer run, the proposal drives down global prices for emission reductions because binding import ceiling on emission reductions reduce the demand for emission reductions' (A, abstract).

The benchmark steady-state is based on 1995 data from the GTAP database.

The only greenhouse gas in the current version is carbon dioxide. Incorporation of other greenhouse gases and sinks is planned (A, p.10).

The model is under construction and all results are preliminary (A, p.4). The work is part of the project 'Quantitative Instruments for the Analysis of Global

Environmental Issues', an environmental modelling project financed by the Environmental Directorate-General of the European Commission.

G-cubed

Source

(A) Philip Bagnoli, Warwick J. McKibbin, and Peter J. Wilcoxen. 'Future Projections and Structural Change.' In N. Nakicenovic, W.D. Nordhaus, R. Richels, and F.L. Toth, eds. *Climate Change: Integrating Science, Economics, and Policy.* CP096-1. Laxenburg, Austria: International Institute for Applied Systems Analysis, 1996.

Time span: 1990-2020 (A, p.185).
Regions: 8 (USA, Japan, Australia, Other OECD, China, LDCs, Eastern Europe and the Former USSR (EEB), Oil Exporting Developing Countries (OPEC)) (A, p.188).
Branch classification: 12 (Electric utilities; Gas utilities; Petroleum refining; Coal mining; Crude oil and gas extraction; Other mining; Agriculture, fishing, and hunting; Forestry and wood products; Durable manufacturing; Non-durable manufacturing; Transportation; Services) (A, p.188).

Table 12-7. Emission stabilization taxes by the year 2000 (unit: US$1990 per ton of carbon)

	1995	2000	2010
USA			
Scenario 1	2.80	16.80	44.80
Scenario 2	1.40	8.40	22.40
Japan			
Scenario 1	10.50	63.00	168.00
Scenario 2	5.50	33.00	88.00
Australia			
Scenario 1	3.80	22.80	60.80
Scenario 2	2.60	15.60	41.60
Other OECD			
Scenario 1	6.80	40.80	108.80
Scenario 2	3.70	22.20	59.20
China			
Scenario 1	1.15	6.90	18.40
Scenario 2	0.24	1.44	3.84
Developing Countries			
Scenario 1	2.60	15.60	41.60
Scenario 2	1.15	6.90	18.40
Eastern Europe and former USSR			
Scenario 1	1.15	6.90	18.40
Scenario 2	0.35	2.10	5.60

Source: Bagnoli, McKibbin, and Wilcoxen (1996).

Energy sources: Electricity, gas, petroleum, coal (A, p.188).
Scenarios: Scenario1: sectoral productivity growth the same across sectors; Scenario 2: Differential sectoral productivity growth (A, pp.196-203).
Technology: Assumed labour productivity growth (A, p.194).
Energy intensity improvement: Falls in industrial economies, gradually rising beyond 2000; in China and other developing countries, rises initially and then falls over time; in Eastern Europe and Former Soviet Union, more movement due to structural changes. This is assumption, or 'speculation' (A, p.196).
Lifestyle: No.
Policy tools: Carbon taxes, recycling of revenue (A, p.202).
Economic linkage: Carbon taxes (A. p.202), changes in prices (A, p.182).
Environment linkage: The effect of differential sectoral growth on carbon emissions.
Modelling logic: Econometric multisector general equilibrium model (A, p.181, p.184, p.187). Neoclassical growth model (A, p.184).

Summary

The authors point out that projecting energy use based on simple extrapolations of GDP is misleading, and set out to test the sensitivity of the aggregate outcomes to the assumptions about sectoral productivity performance (A, p.181). Of the sources of growth including (1) increases in labour, capital, and other inputs; (2) increases in the quality of these inputs; (3) technical change; (4) improvements in the way that inputs are allocated across industries; and (5) reallocation of inputs among countries, the authors put particular emphasis on (4) and (5) (A, p.185).

The eight regions are linked by flow of goods and assets. Flow of goods are determined by bilateral trade matrices. There is one 8*8 trade matrix for each of the 12 sectors in the model (A, p.190). Capital flows respond to interest arbitrage relations and exchange rates which float freely (A, p.190). The model includes foreign direct investment as well as short-term financial capital (A, p.191).

Producing sectors choose inputs and level of investment so as to maximize stock market values, subject to a multiple-input production function (capital, labour, energy, and materials as inputs) and a cost of adjustment model for the capital stock and prices (A, p.188). The parameters of the production technology are estimated from the input-output tables and price data for the USA (A, p.188). Each industry is assumed to have the same energy, materials, and substitution elasticities regardless of its location in the world. However, the share parameters for regions outside the USA are derived from input-output tables for respective regions and correspond to individual countries.

In Scenario 1 the global emissions rise from 5388 million tons of carbon in 1990 to 15,378 million tons in 2020 (A, p.196). In Scenario 2 the comparable figures are 9818 million tons in 2020, or a decrease of almost 5500 million tons compared to Scenario 1. This is attributable to the fact that 'without rising labor productivity in the energy sectors, energy becomes relatively scarce, which reduces the growth of downstream industries' (A, p.201).

The paper notes that 'different assumptions about the sectoral composition of growth have dramatic effect on the size of the taxes necessary to stabilize carbon emissions in each region' (A, p.203). 'By 2010 the stabilizing tax in the USA is $44.80 (1990 US$) per ton of carbon in Scenario 1 and $22.40 per ton of carbon under Scenario 2' (A, p.202).

GemWTraP: General equilibrium model for assessment of world tradable permits, a new version of GEMINI-E3 (General equilibrium model of international-national-interaction for economy–energy–environment)

Sources

(A) Alain L. Bernard and Marc Vielle. 'Efficient Allocation of a Global Environment Cost between Countries: Tradable Permits VERSUS Taxes or Tradable Permits AND Taxes? An Appraisal with a World General Equilibrium Model.' Paper presented at the Joint Meeting organized by the International Energy Agency, Energy Modelling Forum, and the International Energy Workshop (IIASA), 16-18 June 1999, Paris.
(B) Alain L. Bernard and Marc Vielle. 'Preliminary Results of Sensitivity Analysis on Leakage with GEMINI-E3/GemWTraP.' Paper presented at the Joint EMF/IEA/IIASA Meeting, Stanford, 20-22 June 2000.

Time span: 1990-2020, with yearly steps (A); 2000-2040 (B).
Regions: 7 countries/regions (France, Other European Countries (EU11), USA, Japan, Former Soviet Union (FSU), Energy Exporting Countries (EEC), Rest of the World (ROW).
Branch classification: 12 for France and EU11, 8 for USA, Japan, FSU, EEC, and ROW. 3 institutional sectors (households, firms, and government).
Energy sources: 5, including coal, gas, electricity, crude oil, refined oil products.
Scenarios: Pure domestic taxation: without permit trading;
Iso-Tax OECD: equalization of carbon taxes in OECD countries, on the basis of total abatement of OECD in Kyoto Protocol;
Iso-Tax OECD with hot air: equalization of carbon taxes in OECD countries, on the basis of total abatement of Annex I countries;
Iso-CMA OECD: equalization of marginal abatement cost in OECD;
Iso-CMA with hot air: equalization of marginal abatement cost in OECD countries, on the basis of total abatement of Annex I countries; and
Iso-Tax Annex I: equalization of carbon taxes in Annex I countries, on the basis of total abatement of Annex I countries (A, pp. 7-8).

Sensitivity analysis on leakage for energy-intensive sectors (EIS), based on different values of the elasticity of substitution between imports and domestic production (Armington elasticity, or AE):

Sct0: Domestic taxes and AE in EIS equal to 0;
Sct3 (central scenarios): " 3;
Sct6: " " 6;
Sct9: " " 9;
Scp0: Tradable permits and AE in EIS 0;
Scp3: " " 3;
Scp6:" " 6;
Scp9: " " 9;

ScPworld: Like Scp3, and commitment by non-Annex I countries to BaU CO_2 emissions;

Scpworldm10: like Scp3, and commitment by non-Annex I countries to 10% below BaU CO_2 emissions.

Technology: Not explicit. Costs of adjustment are not taken into account (A, p.16).

Energy intensity improvement: Decreases at about 0.8 to 1.0% in various regions/countries (A, p.21).

Lifestyle: Linear expenditure system.

Policy tools: Indirect taxation and social contributions, 13 categories with rates differentiated: by commodity (taxes on production, on imports); by sector (social contributions, subsidies); by sector × commodity (intermediate consumption); by commodity × institutional sector (final demand); by commodity × sector × IS (investment).

Economic linkage: Annual growth of energy consumption and GDP for 1995-2020 assumed constant (A, p.21). Endogenous real rates of interest (determined by equilibrium between saving and investment) and endogenous exchange rates (determined by constraints on foreign trade deficits and surpluses).

Environment linkage: Elasticities of substitution between imports and domestic production; between capital, labour, and energy; between individual fuels; and between commodities.

Modelling logic: Dynamic general equilibrium model.

Summary

'Internalizing externalities such as pollution can be equivalently performed by taxes or tradable permits. This is not any more the case when there are "distortions", i.e. in an economy which cannot be considered as first-best Paretian.' The paper compares trade of pollution permits based on costs (marginal abatement costs) and on market prices (carbon taxes) (A, p.1). 'As any country would not pay for a pollution right more than it costs to him, and should not sell for less, marginal abatement cost is more likely to be the base for exchanges of permits. A market operating on this base has then for result that marginal abatement costs are equalized across countries, but this is not the case for carbon taxes. Implementing such a market supposes that only governments are traders, or if private agents are the operators, that governments set corresponding taxes (or subsidies) such as the total price paid by firms (eventually households) reaches the desired level' (A, p.3).

Important parameters in the model are the elasticities of substitution. Values of elasticities of substitution were determined according to various sources and econometric estimations (A, p.20). Production function is nested CES with fixed factors for fossil fuel sectors. Import function is nested with domestic production.

The model outputs, by country and annually, include:

(1) carbon taxes, marginal abatement costs, and price of tradable permits;
(2) effective abatement of CO_2 emissions, net sales of tradable permits;
(3) total net welfare loss (net loss from terms of trade, pure deadweight loss of taxation, net purchases of tradable permits);
(4) macroeconomic aggregates including real exchange rates and real interest rates; and
(5) sectoral data such as production and factors of production (A, p.19).

The results of Iso-Tax and Iso-MAC scenarios show that 'abatement in OECD countries is maximum when they cannot benefit from the "contribution" of FSU through the hot air resource. When they benefit from both Hot Air and the additional abatement of FSU induced by a carbon tax (at the same level as OECD countries), abatement is minimum and marginal abatement costs are then very low (from 50 to 100 ECUs in 2010, instead of 140 to 300 in the scenarios of domestic taxes in OECD.' 'Relatively to Iso-Tax scenarios, Iso-MAC equilibria yield a higher abatement for the USA and lower for other OECD countries, but the difference is negligible' (A, p.11). The paper then examines the scenarios of tradable permits where transfers related to sales or purchases of permits are taken into consideration (A, pp.12-16).

Sensitivity analysis on leakage is carried out corresponding to different values of the elasticity of substitution between imports and domestic production (Armington assumption, or Armington elasticity). The analysis is applied to a specific sector or group of sectors, namely energy-intensive sectors (B, p.2).

'Sensitivity analysis on the Armington elasticity confirms that leakage increases with substitutability in international markets of energy intensive goods, but remains of a limited extent when implementation of the Kyoto Protocol incorporates trade of permits between Annex I countries. Total world emissions of CO_2 also increase, mainly through the shifting of production in energy intensive industries from countries induced to reduce carbon intensity of their products toward countries which do not have the same incentive.' In simulation of 'Enlarged Kyoto' scenarios, 'Annex I countries would benefit – and not suffer – from a commitment by non-Annex I countries to stabilize their emissions. Losers would then be energy exporting countries, but this already appears in simulation of Kyoto scenarios' (B, p.16).

Table 12-8. Iso-Tax and Iso-MAC scenarios (abatement in million tons of carbon, costs and taxes in ECUs, 1990)

	2010			2020		
	Abat.	MAC	Tax	Abat.	MAC	Tax
Domestic Taxes						
France	20	292	169	33	514	358
EU 11	220	322	256	297	459	395
USA	522	140	107	663	187	148
Japan	75	228	149	111	336	240
Total or average	838	199	0	1104	285	0
Iso-Tax OECD without HA						
France	18	257	140	24	348	208
EU 11	150	196	140	202	270	208
USA	597	177	140	776	255	208
Japan	73	217	140	103	298	208
Total or average	838	186	0	1104	264	0
Iso-Tax OECD with HA						
France	11	154	65	17	253	145
EU 11	88	107	65	148	186	203
USA	399	92	65	624	167	219
Japan	44	120	65	78	203	181
Total or average	542	98	0	868	175	0
Iso-Tax ANNEX 1						
France	7	102	33	10	161	64
EU 11	51	65	33	87	110	64
USA	262	53	33	430	91	64
Japan	26	76	33	49	118	64
FSU	191	61	33	280	92	64
Total or average	537	59	0	857	96	0

Source: Bernard and Veille (1999).

Table 12-9. Additional emissions of CO_2 by non-Annex I countries and FSU (unit: million metric tons of carbon)

	2010	2020	2030	2040
Sct0	87	131	185	250
Sct3 (reference)	105	157	217	287
Sct6	116	173	238	312
Sct9	124	184	255	332
Scp0	35	66	105	151
Scp3	37	70	110	158
Scp6	40	73	115	165
Scp9	41	76	120	171
Total corresponding emissions in the base line scenario (incl. FSU)	5270	6846	8535	9814
Total abatement by OECD countries	829	1065	1235	1416

Source: Bernard and Vielle (2000).

Further analysis would require a model with a higher disaggregation by sector, where production functions need not resort to the Armington assumptions. Vintage production functions are promising candidates (B, p.16).

GREEN: General equilibrium environmental model by OECD

Sources

(A) Organisation for Economic Cooperation and Development. *GREEN: the Reference Manual.* Paris: OECD, 1994.
(B) Andrew Dean and Peter Hoeller. 'Costs of Reducing CO_2 Emissions: Evidence from Six Global Model.' Paris: OECD, 1992.
(C) Jean-Marc Burniaux, Giuseppe Nicoletti, and Joaquim Oliveira-Martins. 'GREEN: A Global Model for Quantifying the Costs of Policies to Curb CO_2 Emissions.' *OECD Economic Studies*, No.19, Winter 1992.

Time span: Five-year intervals up to 2010, twenty-year intervals to 2050 (C, p.51).
Regions: 12 (United States, Japan, EC, Other OECD, Central and Eastern Europe, the Former Soviet Union, Energy-Exporting LDCs, China, India, Dynamic Asian Economies, Brazil, Rest of the World) (C, p.52).
Branch classification: 11 (1. Agriculture; 2. Coal mining; 3. Crude oil; 4. Natural gas; 5. Refined oil; 6. Electricity, gas, and water distribution; 7. Energy-intensive industries; 8. Other industries and services; 9. Carbon-based back-stop; 10. Carbon-free back-stop; 11. Carbon-free electric back-stop) (C, p.52).
Energy sources: 5 (coal, crude oil, natural gas, refined oil products, electricity, gas and water distribution) and 3 back-stop, non-conventional energy sources (C, pp.52-53).
Scenarios: GREEN and five other global models compared based on output growth and population (B).
Technology: Back-stop energy technologies (C, p.53).
Energy intensity improvements: Sensitivity analysis conducted on autonomous energy efficiency parameter (AEEI), by assuming growth rate of AEEI of 1% per annum and 0.5% per annual (B, pp.26-27).
Lifestyle: Consumer sectors (1. Food, beverages and tobacco; 2. Fuel and power; 3. Transport and communication; 4. Other goods and services) (C, p.52).
Policy tools: Carbon tax, energy tax, tradable permits (constraint at the world level, initial quotas for each country, a single world price as the shadow price of carbon) (C, pp.61-62).
Environment linkage: CO_2 emissions.
Economic linkage: Full trade links in a trade sub-model with separate specification for each bilateral flow (C, pp.62-63).

Modelling logic: Econometric model with policy tools. Sensitivity analysis on energy efficiency.

Summary

The model focuses on energy use and CO_2 emissions. The production side of regional models describes the supply of fossil fuels and their use, as well as non-fossil energy inputs into the production process. Carbon-based back-stop includes liquid synthetic fuel derived from coal or shale; carbon-free back-stop includes liquid fuel such as derived from biomass; and carbon-free electric back-stop refers to carbon-free electricity not derived from hydro or nuclear fission sources (e.g. nuclear fusion, solar or wind). They become available at a given identical time period in all regions and their prices are exogenous and identical across regions.

The decisions by producers are described in several steps: (1) choose the mix between (a) intermediate inputs and (b) composite input including primary factors and energy; (2) divide among labour and other primary factors; (3) choose the mix between energy and the capital plus fixed factor bundle; (4) allocate the energy bundle among coal, oil, gas, refined oil products, and electricity; (5) decide the mix between conventional and back-stop technologies; (6) determine the mix between capital and the fixed factor; and (7) allocate intermediate and energy inputs among domestic supply and imports.

GREEN incorporates full trade linkages between all regions of the world. Country models are based on input-output data, providing sufficient sectoral detail. The model is suitable for the analysis of the international competitiveness issue and of the impact of different agreements to curb CO_2 emissions. Changes in government budget by the introduction of carbon tax are compensated by offsetting changes in the income tax rate, which amounts to lump-sum transfer to households.

The model is a simple recursive type. The dynamic feature comes from the saving decisions which affect the accumulation of productive capital and future economic growth. The capital stock is sector specific and distinction is made between old and new capital vintages. The model is closed by equating gross investment to net saving. The equilibrium search produces the world price of oil as well as prices of coal, gas, and the carbon-free resources. It is also possible to solve for the carbon or energy tax needed to satisfy the constraint on carbon emissions.

GTAP: Global trade analysis project model of global trade

Sources

(A) Thomas W. Hertel and Marinos E. Tsigas. 'Structure of GTAP.' In Thomas W. Hertel, ed. *Global Trade Analysis, Modeling and Applications*. Cambridge: Cambridge University Press, 1997.

Table 12-10. Average bilateral applied tariff (unit: percentage of cif values)

Exporting Countries	Importing Countries							
	AUS	NZL	CAN	US	JPN	KOR	EU	IDN
AUS	0.0	15.0	5.3	4.3	1.7	10.5	2.5	5.3
NZL	10.7	0.0	2.9	2.3	5.1	11.8	3.8	11.3
CAN	14.3	18.4	0.0	3.2	3.1	9.9	4.8	5.5
US	15.9	23.1	7.1	0.0	3.1	16.5	4.2	12.1
JPN	18.9	32.2	7.6	5.2	0.0	18.3	9.1	14.3
KOR	17.8	23.0	9.8	6.1	6.9	0.0	7.1	18.0
EU	15.4	21.8	8.9	4.5	7.1	18.0	0.0	16.2
IDN	8.9	17.3	9.2	7.5	3.2	9.3	7.9	0.0
MYS	13.8	15.6	7.6	4.4	2.6	12.2	7.1	12.3
PHL	14.8	25.7	11.9	8.7	5.1	14.1	10.0	10.8
SGP	16.1	42.3	6.3	3.8	5.2	20.1	9.7	11.0
THA	8.5	13.6	9.2	6.6	6.1	15.6	8.2	9.7
CHN	19.8	35.0	12.7	6.2	7.0	14.4	5.9	11.1
HKG	17.1	34.5	15.2	11.1	6.8	20.5	10.2	18.9
TWN	15.2	28.0	8.5	5.9	8.2	19.2	7.2	18.9
ARG	9.2	6.7	8.5	5.1	4.7	21.4	10.0	3.9
BRA	15.3	19.8	8.3	5.6	3.4	13.5	6.7	10.8
MEX	10.3	12.2	6.8	4.0	1.6	12.2	2.6	11.0
LAM	10.3	5.5	3.1	4.5	2.8	7.1	6.6	4.1
SSA	4.1	3.2	1.3	1.1	3.2	6.2	3.9	5.8
MNA	5.7	1.8	1.6	2.0	0.6	5.7	2.8	1.4
EIT	13.5	20.6	6.8	5.1	3.3	14.6	5.5	8.4
SAS	10.1	20.8	16.2	10.3	3.9	12.6	8.6	6.8
ROW	19.4	24.2	5.0	4.3	3.8	13.4	5.7	10.3

Source: Hertel and Tsigas (1997).

Note: This table represents part of the original table.

(B) Mark Gehlhar *et al.* 'Overview of the GTAP Data Base.' In Thomas W. Hertel, ed. *Global Trade Analysis, Modeling and Applications.* Cambridge: Cambridge University Press. 1997.
(C) Carlo Perroni and Randall Wigle. 'Environmental Policy Modeling.' In Thomas W. Hertel, ed. *Global Trade Analysis, Modeling and Applications.* Cambridge: Cambridge University Press. 1997.

Time span: 1992 as the base year. No time-series.

Regions: 24 (Australia, New Zealand, Canada, United States of America, Japan, Republic of Korea, EU-12, Indonesia, Malaysia, Philippines, Singapore, Thailand, People's Republic of China, Hong Kong, Taiwan, Argentina, Brazil, Mexico, Rest of Latin America, Sub-Saharan Africa, Middle East and North Africa, Eastern Europe and Former Soviet Union, South Asia, Regions not elsewhere classified.

Branch classification: 37 (Paddy rice; Wheat; Grains; Non-grain crops; Wool; Other livestock; Forestry; Fishing; Coal; Oil; Gas; Other minerals; Processed rice; Meat products; Milk products; Other food products; Beverages and tobacco; Textiles; Wearing apparel; Leather, etc.; Lumber and wood; Pulp, paper, etc.; Petroleum and coal products; Chemicals, rubber, and plastics;

Nonmetallic mineral products; Primary ferrous metals; Nonferrous metals; Fabricated metal products; Transport equipment; Machinery and equipment; Other manufacturing; Electricity, water, and gas; Construction; Trade and transport; Other services (private); Other services (government); Ownership of dwellings (A, p.75).

Energy sources: Coal, oil, gas, petroleum and coal products, electricity (A, p.75).

Scenarios: Various policy applications (A, pp.191-346).

Technology: Embedded in input-output structure of the base year (1992).

Energy intensity improvement: GTAP does not currently incorporate environment (C, p.305). Parameters to be specified by model users (A, p. 40).

Lifestyle: Embedded in consumer behaviour in final demand.

Policy tools: GTAP does not account for macroeconomic policies and monetary phenomena (A, p.52).

Economic linkage: Input-output tables (1992) linked by bilateral trade (A, pp.74-76).

Environment linkage: No. See (C) for possible extension in this direction.

Modelling logic: General equilibrium (A, p.29). Partial equilibrium closures can be examined by selectively omitting some equations (A, p.13).

Summary

The GTAP model describes the accounting for household, business firms, global transportation sector, and global banking sector (supply of investment goods and aggregate savings).

The database underlying the GTAP consists of bilateral trade, transport, and protection matrices that link 24 countries/regions (B, p.74). The regional databases are derived from country input-output tables. The input-output tables from the latest available year are updated to conform to the 1992 trade and macroeconomic data (B, p.75). Trade data are obtained from the United Nations and adjustments were made at the 4-digit SITC level prior to aggregation to 31 merchandise sectors (B, p.77).

The GTAP database, and a standard modelling framework built on it, is not in itself focused on particular policy issues including energy and environment. Rather, it is intended to be a collection of fully documented and publicly available data and a standard modelling tool.

It is suggested that at least four types of economy-environment interaction be made explicit in the model:

(1) Changes in the relative valuation of public goods compared to private goods.
(2) Changes in relative prices.
(3) Changes in consumption patterns caused by changes in environmental quality.
(4) Interactions between abatement activities taken by firms and private mitigating activities (C, p.307).

It is also pointed out that the models that lack abatement activities 'miss the interaction between abatement activities and other activities of the economy' (C, p.307). For this reason, a generalized algebraic modelling system was employed to 'calculate the "shocks" to GTAP that are necessary to achieve a given combination of emission fees and abatement.' These shocks are fed to GTAP, resulting changes in sectoral output, etc. are calculated, which are then used as inputs in the third step (C, p.309).

IEA energy model

Sources

(A) E. Lakis Vouyoukas. 'IEA Medium Term Energy Model.' OECD, *The Costs of Cutting Carbon emissions: Results from Global Models*. Paris. 1993.
(B) E. Lakis Vouyoukas. 'World Energy Outlooks.' In Jean-Baptiste Lasourd, Jacques Percebois, and François Valette, eds. *Models for Energy Policy.* London: Routledge. 1996.
(C) International Energy Agency. *World Energy Outlook to the Year 2010.* Paris: OECD, 1998b.

Time span: Annual to 2005 (A, p.107; B, p.127).
Regions: 10 (OECD North America, OECD Europe, OECD Pacific, Africa*, Asia*, Latin America*, Middle East*, East Europe*, Soviet Union*, China**)(*: less detailed. **: exogenous) (A, p.109).
Branch classification: 3 (Transportation, industry, other) (A, p.109). No description of industrial structure (A, p.108).
Energy sources: Eight oil products (gasoline, diesel, kerosene, gas, oil, heavy fuel oil, bunkers, other), seven other products (industrial gas, other gas, industrial coal, coke, other coal, industrial electricity, other electricity) (A, p.109).
Scenarios: Reference case, 1%, 2%, 3% reduction per annum in emission, stabilization case close to the 1990 level (A, pp.114-116). See Table 12-11.
Technology: Exogenous; slow introduction (A, p.108).
Energy intensity improvements: Detailed. Short- to medium-term rigidities of energy markets is the focus (A, p.108).
Lifestyle: No.
Policy tools: Carbon tax (A, p.109, pp.111-118).
Economic linkage: Macroeconomy exogenous, no economic cost estimates (A, p.107; B, p.127). Crude prices exogenous (A, p.108). Feedback between oil prices and GDP is possible (A, p.109). The data for non-OECD regions weak in end-use prices and their link to primary fuel prices (A, p.108).
Environment linkage: CO_2 emissions (A, p.110).
Modelling logic: Energy model with econometrically estimated parameters, with adjustment factors complementing econometric forecasts (A, p.107). Treats macroeconomy exogenously (B, p.127).

Table 12-11. Carbon tax and emission reduction due to substitution, total OECD

	1990	1995	2000	2005
Reference case:				
Energy (MTOE)	4184.6	4508.1	4776.3	5011.1
Emissions (million tons carbon)	3043.0	3274.7	3466.5	3631.9
One percent reduction case:				
Energy	4184.6	4327.3	4428.0	4520.9
Emissions	3043.0	3091.5	3113.1	3138.8
Reduction in emissions %		−5.8	−10.8	−14.6
Reduction in energy %		−4.1	−7.6	−10.3
Substitution %		1.7	3.2	4.3
as % of total reduction		28.9	29.6	29.4
Two percent reduction case:				
Energy	4184.6	4145.9	4088.2	4070.0
Emissions	3043.0	2923.1	2803.0	2736.3
Reduction in emissions %		−11.4	−21.2	−28.3
Reduction in energy %		−8.4	−15.6	−20.8
Substitution %		3.0	5.7	7.5
as % of total reduction		26.3	26.8	26.5
Three percent reduction case:				
Energy	4184.6	3859.1	3717.3	3657.1
Emissions	3043.0	2668.8	2480.7	2384.6
Reduction in emissions %		−20.5	−33.5	−42.1
Reduction in energy %		−15.5	−25.1	−31.5
Substitution %		4.9	8.4	10.6
as % of total reduction		24.0	25.1	25.1
Stabilization case:				
Energy	4184.6	4251.4	4333.6	4417.2
Emissions	3043.0	3019.7	3023.9	3043.2
Reduction in emissions %		−8.1	−13.7	−17.7
Reduction in energy %		−5.9	−9.7	−12.6
Substitution %		2.2	3.9	5.1
as % of total reduction		7.7	28.8	28.7

Source: Vouyukas (1993).

Summary

The model consists of four interdependent sub-models (final demand sub-model, transformation sub-model, supply sub-model, and price sub-model) and one self-contained sub-model (activity sub-model which converts exogenous assumptions on GDP etc. into sector-level activity variables).

The final demand sub-model solves for final energy demand based on the sector activity, the end-user price, and assumptions on sector-specific variables (saturation, technology, etc.). The transformation or power generation sub-model converts the demand for final electricity into primary fuel demand based on the structure of the electricity industry in each region, conversion efficiencies, and assumptions about non-fossil fuels (nuclear, hydroelectric, geothermal). The supply sub-model generates a set of primary fuel supplies, on the basis of output of the price sub-model and assumptions on reserves, discovery rates and other relevant variables (A, p.109).

It is claimed that the strength of the IEA model lies in (a) the detailed modelling of end-use consumer prices and their link to primary prices, and (b) the incorporation of rigidities in the current energy system of OECD regions (A, p.118).

Based on simulation results, the conclusion is reached that 'any reduction in emissions is likely to originate from a reduction in energy rather than from substitution in favor of less polluting fuels'. 'The model suggests that reducing emissions from current levels would require the imposition of implausibly high taxes and historical experience would suggest that this is likely to be very costly' (A, p.119).

IFs: International futures, the third generation

Source

(A) Barry B. Hughes. 'The International Futures (IFs) Modeling Project.' *Simulation and Gaming*, Vol. 30, No. 3, September 1999.

Time span: 1990s-2100 (A, p.305).
Regions: 20 in the model, 162 in the database (A, p.304).
Branch classification: 5 (agriculture, materials, energy, industry, services) (A, p.307).
Energy sources: 6 (oil, gas, coal, nuclear, hydroelectric, other renewables (e.g. photovoltaic, biomass, and wood) (A, p.307).
Scenarios: Variety of scenario interventions, including technological advance, multipliers on investment rates, tax rates, export levels, and other economic variables. Itis possible to intervene in systems concerning energy, agriculture, and environment (A, p.315). Change individual parameters. Change more than 20 functional relations within IFs by specifying analytic functions of choice (A, p.309). A parameter representing absorption of atmospheric CO_2 by oceanic and other sinks can be altered (A, p.319). Scenarios on feedback of CO_2 on agricultrue (A, p. 319).
Technology: Technological advance by assumption (A, p.315).
Energy intensity improvement: To be changed by scenarios (A, p.318).
Lifestyle: No.
Policy tools: Tax on fossil fuel use (A, p.317).
Economic linkage: Economic module encompasses the partial equilibrium analyses from the agriculture and energy modules (A, p.307).
Environment linkage: CO_2 emission (A, p.318).
Modelling logic: System dynamics. 'It is a general equilibrium model that does not assume exact equilibrium will exist in any given year; instead, the model chases equilibrium over time' (A, p.307).

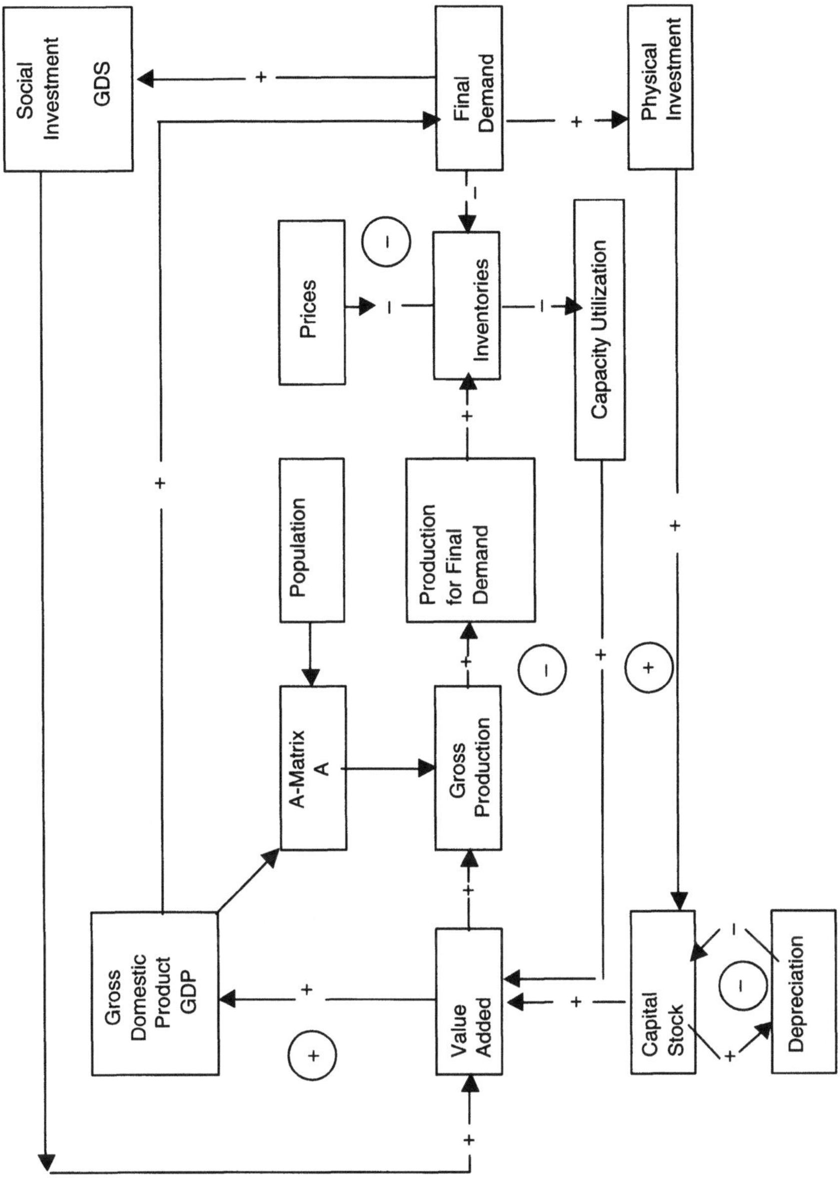

Figure 12-3. Economic module of International Futures.
Source: Hughes (1999).

Summary

The purpose of IFs is as a world model intended to analyse demographic, economic, energy, agricultural, environmental, and sociopolitical systems into the 21st century. It provides 200 variables across 20 countries and regions. In particular, the model looks at (1) the possible transition towards sustainability in the 21st century, and (2) aspects of sociopolitical change within countries and in the global system (A, p.304). It allows 'cross-sectional and limited longitudinal analysis of relationships among hundreds of variables across 162 countries' (A, p.305). IFs responds to the need to 'analyze long-term global change and human leverage with respect to that change' (A, p.310).

IFs consists of six modules (A, pp.306-308):

(1) Population module: represents 19 cohorts. Fertility, mortality, life expectancy are calculated.
(2) Agriculture module: production, consumption, and trade, price changes.
(3) Energy module: production of 6 energy types, consumption and trade at the aggregate level. Price changes. Known and ultimate reserves.
(4) Economic module: 5 sectors (see above). International trade uses the pooled rather than the bilateral approach.
(5) Sociopolitical module: fiscal policy including taxing and government spending. The spending categories are military, health, education, foreign aid, and a residual.
(6) Environment module: tracks the level of atmospheric CO_2, the area of forest land, the use of fresh water, and reserves of fossil fuels.

The model needs more detailed specification on the energy demand. 'It will be useful to elaborate the demand-side of the energy model by sector (household, transportation, etc.)' (A, p.424).

The model stresses 'feedback loops in reinforcing processes'. 'First, we can examine trends in population and economics … . Second, we can shift our perspective on … the specifics of agriculture, food, and environmental systems … . Those who have looked at human development with the first perspective are quite optimistic about the future, whereas those who adopted the second perspective are more skeptical.' IFs is able to integrate both perspectives (A, p.311).

'The IPCC itself builds two scenarios, a doubling and a quadrupling of global energy demand … . Even the 4-fold demand increase at the high end of IPCC forecasts for 2100 seems optimistically low, but IFs adopts something close to it for the base case.' Nuclear energy is described as becoming 'more expensive as the full costs of the nuclear fuel cycle, including long-term disposal of nuclear wastes, appropriately become part of costing'. The costs of renewable energy, while continuing to fall, will remain relatively high. 'Fossil fuels remain relatively plentiful and therefore not particularly expensive. There is an extreme global scarcity of political will to tax and discourage fossil fuel use' (A, p.317). As a result, 'We will have doubled atmospheric CO_2 (to about 600 ppmv) by 2050 and ... it will continue to rise slowly' (A, p.319).

IIASA and WEC's global energy perspective

Source

(A) Nebojsa Nakicenovic, Arnulf Grüber, and Alan McDonald. *Global Energy Perspectives.* Cambridge: Cambridge University Press, 1998.

Time span: 2020-2100 (A, p.5).
Regions: 11 world regions (North America, Latin America and the Caribbean, Sub-Saharan Africa, Middle East and North Africa, Western Europe, Central and Eastern Europe, Newly independent states of the former Soviet Union, Centrally planned Asia and China, South Asia, Other Pacific Asia, Pacific OECD (A, pp. 257-260).
Branch classification: No.
Energy sources: 5 (coal, oil, gas, nuclear, renewables) (A, pp.119-124).
Scenarios: See Table 12-12.
Technology: By assumption. Turnover of capital plays role (A, p.vii; pp.113-118).
Energy intensity improvements: By assumption, 0.8% to 1.4% (A, p.xiii).
Lifestyle: By assumption (floor space, appliances, passenger km) (A, p.95).
Policy tools: Environmental tax in case C (A, p.6). CO_2 emission constraint in case C (A, p.6).

Table 12-12. IIASA and WEC: three global energy scenarios in 2050 and 2100

	Case A High Growth	Case B Middle Course	Case C Ecologically Driven
Population (Billion)			
1990	5.3	5.3	5.3
2050	10.1	10.1	10.1
2100	11.7	11.7	11.7
Gross World Product (Trillion US$ in 1990 prices)			
1990	20	20	20
2050	100	75	75
2100	300	200	220
Global Primary Energy Intensity Improvement (% per year)			
1990-2050	−0.9	−0.8	−1.4
1990-2100	−1.0	−0.8	−1.4
Primary Energy Demand (gigatons of oil equivalent)			
1990	9	9	9
2050	25	20	14
2100	45	35	21
Net Carbon Emissions (giga tons of carbon)			
1990	6	6	6
2050	9 to 15	10	5
2100	6 to 20	11	2

Source: Nakicenovic *et al.* (1998).

Environment linkage: Endogenous carbon emissions.
Economic linkage: Gross world product by assumption (3 cases) (A, p.6). International trade by assumption (A, p.105). Prices and exchange rates outside the scope.
Modelling logic: Technology scenarios.

Summary

This work provides the future scenarios on world production, energy demand, and carbon emissions for the years 2050 and 2100 jointly conducted by the International Institute for Applied Systems Analysis (IIASA) and World Energy Council (Nakicenovic *et al.*, 1998).

Case A represents a scenario built around high rates of economic growth (2% per annum in OECD and double that pace in the developing countries) and rapid improvement in energy intensity (0.9% to 2050 and 1.0% during 2050-2100). Although there are a subset of assumptions under this scenario, it is suggested that technological change potential may make it possible to tap the vast oil and gas occurrences, an outcome which is contrary to the conventional wisdom which foresees exhaustion of fossil fuels. There is also a possibility of resorting to coal as the backstop fossil fuel, or to shift to large-scale introduction of renewables and a new generation of nuclear power plants, leading to a transition to a post-fossil-fuel age.

Case B is intended to describe a more pragmatic development, assuming more modest economic growth and technological change than Case A. In reality this reflects economic transition in Eastern Europe and the former Soviet Union and differentiated economic performance in the South.

Case C is labelled as 'ecologically driven'. In addition to an optimistic outlook as regard technology and geopolitics, it assumes 'unprecedented progressive international cooperation focused explicitly on environmental protection and international equity'. It assumes incentives to energy producers and consumers for more energy-efficient behaviour, green taxes, international agreements in environmental and economic spheres, and technology transfer. The growth in the South will be stimulated through substantial resource transfers from industrialized countries. As a result, CO_2 emission in the year 2100 will be contained to 2 GtC, or 1/3 of today's level. Scenario C incorporates two alternative scenarios concerning nuclear fuels: it either finds greater social acceptability through introduction of safe and small-scale new-generation reactors, or provides transient technology to be phased out by the end of the 21st century.

IMAGE 2.0: Integrated model to assess the greenhouse effect

Sources

(A) J. Alcamo. 'Modeling the Global Society-Biosphere-Climate System: Part 1: Model Description and Testing.' In Joseph Alcamo, ed. *IMAGE 2.0,*

Integrated Modeling of Global Climate Change. Dordrecht: Kluwer Academic Publishers, 1994a.
(B) J. Alcamo *et al.* 'Modeling the Global Society-Biosphere-Climate System: Part 2: Computed Scenarios.' In Joseph Alcamo, ed. *IMAGE 2.0, Integrated Modeling of Global Climate Change.* Dordrecht: Kluwer Academic Publishers, 1994b.
(C) Intergovernmental Panel on Climate Change. *Special Report on Emissions Scenarios*. Cambridge: Cambridge University Press, 2000.
(D) de Vries, "Greenhouse Gas Emissions in an Equity-, Envivonment- and Service-Oriented World: An IMAGE-Based Scenario for the 21st Century." *Technological Forecasting & Social Change.* 63 (2-3).

Time span: 1970-1990-2100 (A, p.1, p.18). Time steps vary (A, p.4).
Regions: 13 (Canada, USA, Latin America, Africa, OECD Europe, Eastern Europe, CIS (Former Soviet Union), Middle East, India (including Bangladesh, Bhutan, India, Myanmar, Nepal, Pakistan, Sri Lanka), China (including China, Korea DPR, Kampuchea, Laos, Mongolia, Vietnam), East Asia (Indonesia, Korea, Malaysia, Philippines, Thailand), Oceania, Japan) (A, p.5; C, p.341).
Branch classification: 5 (Industry, Transportation, Residential, Commercial, Other) (A, p.6).
Energy sources: 6 (coal, gas, oil, fuel wood, other biomass, electricity) (A, p.6).
Scenarios: Latest simulation scenarios as defined in the Special Report on Emissions Scenarios (SRES) (C, p.5). Earlier scenarios included:

(1) Conventional wisdom scenario, incorporating no assumptions about climate-related policies. Driving forces are partly based on the assumptions of the IS92 scenarios of the Intergovernmental Panel on Climate Change (IPCC).
(2) Biofuel crops scenario, differing from (1) above in their 'assumptions about how/where modern biofuels are grown'. Scenario (2) requires new cropland.
(3) No biofuels scenario, where biofuels are removed from (1) above and replaced by oil.
(4) Ocean realignment, investigating the consequences of a major change of the ocean's circulation pattern (B, pp.38-39).

Technology: Assumed efficiency of electricity and heat generation, improving from the 1990 level to 0.50 in 2100 in OECD regions, Eastern Europe, CIS, and Middle East; and to 0.45 in 2100 in other regions (B, p.45). Effectiveness of improved energy efficiency is examined (A, p.6).
Energy intensity improvement: Autonomous energy efficiency improvements, which are region specific. (B, p.45).
Lifestyle: No.
Policy tools: Introduction of biofuels (B, p.38).

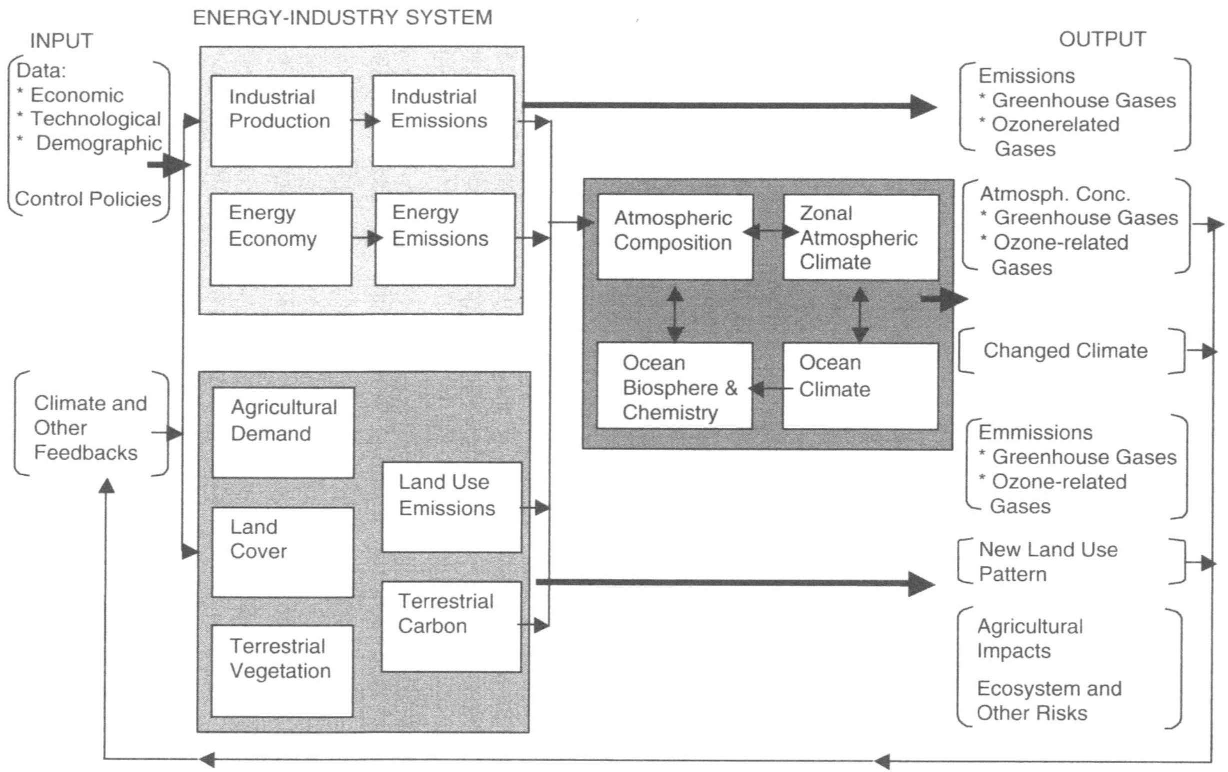

Figure 12-4. Flow chart of IMAGE 2.0.
Source: Alcamo *et al*. (1994).

Table 12-13. IMAGE 2.0 scenario results

Year, Scenario	Carbon Cycle (PgC)			Methane Emissions	Atmospheric Concentration					Average Surface Temperature (Degrees C)	
	CO_2-Emissions from Energy/Industry	Net Biosphere Flux*	Ocean Flux*	($TgCH_4$)	CO_2 (ppm)	CH_4 (ppm)	Trop. O_3	Chage of Agricultural Area (1000 km^2)	Change of Forest Area (1000 km^2)	Northern Hemisphere	Southern Hemisphere
1990	6.1	−1.2	−1.6	492	358	1.7	...	26.2	47.2	14.2	
							**	**	**	***	***
2050											
Conventional Wisdom	15.2	−7.2	−3.0	688	522	2.5	11.6	9%	−26%	plus 1.4	plus 1.0
Biofuel Crops	15.2	−6.0	−3.1	692	534	2.6	12.6	30%	−32%	1.5	1.0
No Biofuels	17.0	−7.5	−3.2	677	539	2.4	9.0	9%	−26%	1.4	1.0
Ocean Realign.	15.2	−4.5	−3.1	686	563	2.6	12.7	12%	−27%	0.0	1.1
2100											
Conventional Wisdom	24.0	−8.2	−4.2	778	777	2.3	10.0	14%	−27%	2.4	1.8
Biofuel Crops	24.0	−6.7	−4.5	793	821	2.4	12.0	65%	−31%	2.7	2.0
No Biofuels	29.2	−8.9	−4.8	746	857	1.7	(+ or −) 0.2	15%	−27%	2.4	1.9
Ocean Realign.	24.0	−6.7	−4.1	778	863	2.4	11.7	18%	−28%	1.2	2.0

* Minus sign indicates net sink.
** Percentages are relative to 1990.
*** 2050 and 2100 are absolute changes in degrees C relative to 1990.

Source: Alcamo *et al.* (1994).

Policy tools: Introduction of biofuels (B, p.38).

Economic linkage: Economic growth and population are exogenous (A, p.41). Regional population and GNP are the main driving forces of the model (A, p.7). Estimates of future population and economic growth are taken from the IPCC IS92s (B, p.39).

Environment linkage: Elasticity functions that relate activity levels of each sector with end-use energy consumption (A, p.7). Energy conservation function relates energy prices with the motivation for energy conservation (A, p.7). CO_2, CH_4, N_2O, and other greenhouse gas emissions are calculated by emission coefficients (A, p.7).

Modelling logic: The IMAGE Targets Image Energy Regional (TIMER) simulation model is a system dynamic model, with investment decisions in energy efficiency, electrical generation, and energy supply (based on anticipated demand, relative prices, institutional and informational delays) (C, p.340). Observed and assumed elasticity linking variables and modules (A, p.41).

Summary

IMAGE 2.0 is an integrated model designed to simulate the global society-biosphere-climate dynamics. The model consists of three subsystems:

(1) The Energy-Industry System (EIS): includes energy economy model, energy emissions model, industrial production and industrial emissions model. Computes the emissions of greenhouse gases in 13 world regions as a function of energy consumption and industrial production.

(2) The Terrestrial Environment System (TES): composed of agricultural demand model, terrestrial vegetation model, land cover model, terrestrial carbon model, and land-use emission model. Simulates the changes in global land cover on a grid-scale (0.5° $\times$ 0.5° latitude-longitude) and the flux of CO_2 and other greenhouse gases from the biosphere to the atmosphere.

(3) The Atmosphere-Ocean System (AOS): contains atmospheric composition model, zonal atmospheric climate model, ocean climate model, and ocean biosphere/chemistry model. Calculates the build-up of greenhouse gases in the atmosphere, the resulting zonal average temperature and precipitation patterns (A, p.3, pp.6-18; C, p.340).

It is expected that 'this approach can provide new scientific information and new policy information linking human activity with its consequences on the global biosphere and climate' (A, p.1).

Elasticity functions, energy conservation function, and other parameters are obtained from the 1970-1990 period (A, p.5, p.7). Energy-related scenario variables include value added of industrial output, value added of commercial services, public consumption, number of passenger vehicles, fuel mix, fuel prices, efficiency of primary energy conversion, autonomous efficiency improvements, and emission factors (B, p.40).

IPCC special report on emissions scenarios

Source

(A) Intergovernmental Panel on Climate Change. *Special Report on Emissions Scenarios*. Cambridge: Cambridge University Press, 2000.

Time span: 2100.
Regions: Mostly global aggregate; some tables provide figures by region (A, p.196, p.221) which distinguish 4 (OECD90, REF (Central and Eastern Europe and Former Soviet Union), ASIA, and ALM (Africa and Latin America), with totals for the world, developed countries, and developing countries. Detailed world region definition is given (A, pp.332-333). Underlying models typically distinguish around 10 regions (A, p.338).
Branch classification: No at the scenario level. Underlying models have branch disaggregation (A, pp.336-346).
Energy sources: 3 for resource use (coal, oil, gas), 6 primary sources (coal, oil, gas, nuclear, biomass, other renewables), 6 final energy (non-commercial, solids, liquids, gas, electricity, others) (A, pp.358-580).
Scenarios: As defined in the Special Report on Emissions Scenarios (SRES). Scenario 'families' are as follows (A, pp.4-5):

A1 story *line*: a world of very rapid economic growth, global population that peaks in mid-century and declines thereafter, and the rapid introduction of more efficient technologies. The A1 story line and scenarios assume 'convergence among regions, capacity building, and increased cultural and social interactions, with substantial reduction in regional differences in per capita income'.

A1F1: fossil fuel intensive.
A1T: predominantly non-fossil fuel.
A1B: balance across all sources.

A2 story line: a very heterogeneous world, with continuously increasing global population. Economic development is primarily regionally oriented and per-capita economic growth and technological change are fragmented and slower than in other story lines.

B1 story line: a convergent world, with the same population trends as in A1. Rapid changes in economic structure towards a service and information economy, with reductions in material intensity and introduction of resource-efficient technologies.

B2 story line: emphasis is on local solutions to economic, social, and environmental sustainability. Increasing global population at a lower rate than in A2, intermediate levels of economic development, and slower and diverse technological change than the A1 and B1.

Technology: By assumption in various scenarios.
Energy intensity improvement: Embedded in six underlying models.

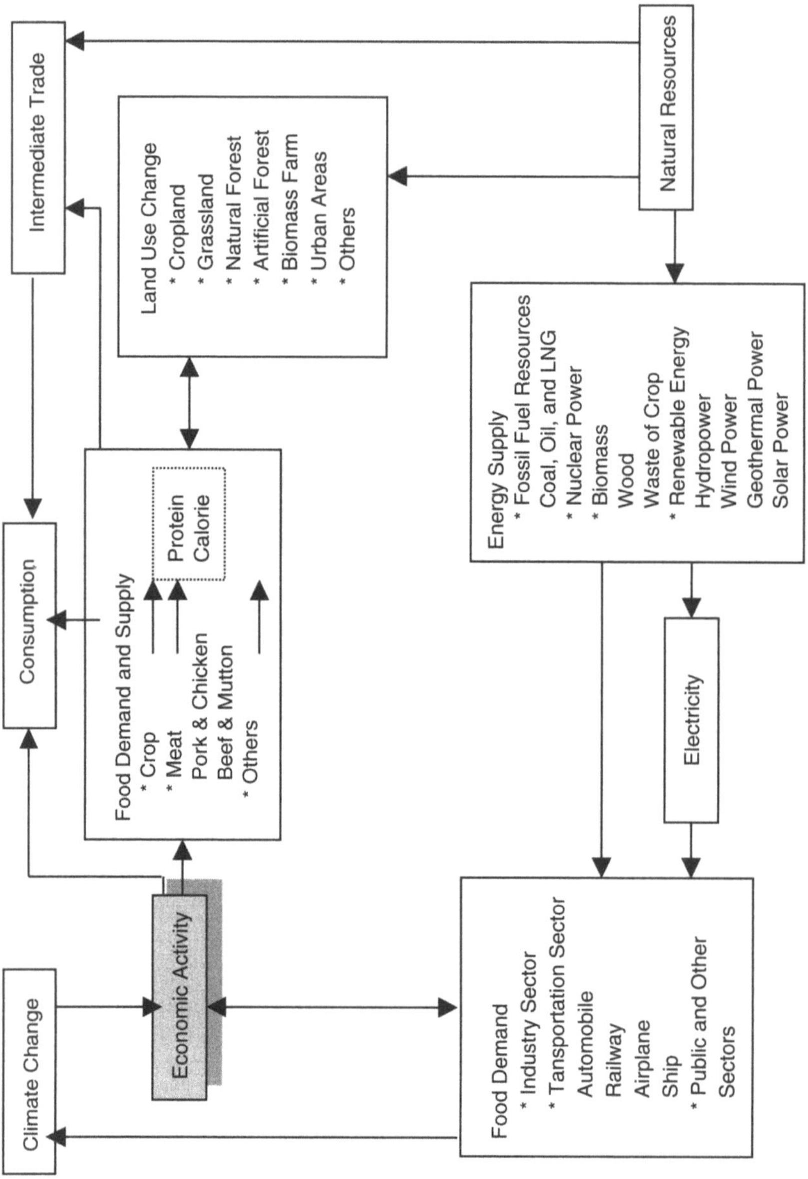

Figure 12-5. Structure of the MARIA model of one region.
Source: Intergovernmental Panel on Climate Change (2000).

Lifestyle: Embedded in some models.
Policy tools: No.
Economic linkage: Economic growth and population dynamics by exogenous assumptions.
Environment linkage: Detailed energy (resource use, primary energy, final use), cumulative CO_2 emissions, anthropogenic emissions of GHGs, land use.
Modelling logic: Scenarios (as above) based on six global models.

Summary

In 1992 the IPCC released emission scenarios to be used for driving global circulation models to develop climate change scenarios. The so-called IS92 scenarios were pathbreaking. They were the first global scenarios to provide estimates for the full suite of greenhouse gases. In response to a 1994 evaluation of the earlier IPCC IS92 emissions scenarios, the 1996 Plenary of the IPCC requested preparation of a *Special Report on Emissions Scenarios*, which was made available in 2000.

Table 12-14. IPCC emission outlook based on 26 harmonized scenarios

		Scenarios					
	Actual	A1			A2	B1	B2
Scenario Group		A1F1	A1B	A1T	A2	B1	B2
Final energy intensity (10^6J/US$)							
1990	16.7						
2020		9.4	9.4	8.7	12.1	8.8	8.5
2050		6.3	5.5	4.8	9.5	4.5	6.0
2100		3.0	3.3	2.3	5.9	1.4	4.0
Primary energy (10^8J/yr)							
1990	351						
2020		669	711	649	595	606	566
2050		1431	1347	1213	971	813	869
2100		2073	2226	2021	1717	514	1357
Share of coal in primary energy (%)							
1990	24						
2020		29	23	23	22	22	17
2050		33	14	10	30	21	10
2100		29	4	1	53	8	22
Share of zero carbon in primary energy (%)							
1990	18						
2020		15	16	21	8	21	18
2050		19	36	43	18	30	30
2100		31	65	85	28	52	49

Note: Overview of main secondary scenario driving forces in 1990, 2020, 2050, and 2100.
The range across different scenarios in the six scenario groups, provided in the original source, is omitted here.

Source: Intergovernmental Panel on Climate Change, *Special Report on Emissions Scenarios* (2000).

The new set of scenarios contained in this new publication is based on four sets of scenarios, which developed into 6 scenario groups (see above). Altogether 40 scenarios have been prepared (termed as SRES (Special Report on Emission Scenarios) scenarios. '26 scenarios were "harmonized" to share agreed common assumptions about global population and GDP development' (A, p.29).

Six global models have been employed to quantify story lines into 2100. Six models are 'representative of different approaches to modelling emissions scenarios and different integrated assessment frameworks in the literature and include so-called top-down and bottom-up models' (A, p.29). They are:

(1) Asian Pacific Integrated Model (AIM).
(2) Atmospheric Stabilization Framework Model (ASF).
(3) Integrated Model to Assess the Greenhouse Effect (IMAGE).
(4) Multiregional Approach for Resource and Industry Allocation (MARIA).
(5) Model for Energy Supply Strategy Alternatives and Industry Allocation (MESSAGE).
(6) Mini Climate Assessment Model (MiniCAM).

They are described individually in this survey.

MARIA: Multiregional approach for resource and industry allocation model

Source

(A) Intergovernmental Panel on Climate Change. *Special Report on Emissions Scenarios*. Cambridge: Cambridge University Press, 2000.

Time span: 2100.
Regions: 8 (NAM (USA and Canada), Japan, Other OECD countries, China, ASEAN countries (Indonesia, Malaysia, Philippines, Singapore, Korea, Thailand), SAS (India, Bangladesh, Pakistan, Sri-Lanka), EEFSU (Eastern Europe and the Former Soviet Union), and ALM (Africa and Latin America).
Branch classification: Energy demand consists of industry, transportation, and other public uses.
Energy sources: 9 (fossil (coal, natural gas, oil), biomass, nuclear power, renewable energy technologies (hydropower, solar, wind, geothermal)) (A, p.343).
Scenarios: As defined in the Special Report on Emissions Scenarios (SRES) (A, p.5).
Technology: Nuclear fuel recycling technologies, carbon sequestration technologies are taken into account (A, p.343).
Energy intensity improvement: No.
Lifestyle: No.
Policy tools: No.
Economic linkage: Future GDP growth is projected by potential growth rates (the products of two exogenous assumptions – population and per-capita

GDP growth) and endogenously determined energy costs and prices. Market prices are determined endogenously on the basis of model-calculated shadow prices (A, p.343).
Environment linkage: Energy (long-term profiles of fuel mix changes), food and land use, global warming.
Modelling logic: Intertemporal non-linear optimization model.

Summary

The origin of the model is the Dynamic Integrated Model of Climate and Economy (DICE) model (Nordhaus, 1994). The cost coefficients of energy conversion technologies are extracted from the GLOBAL 2100 model (Manne and Richels, 1992). The latest version of the model incorporates Rogner's estimates on fossil resource availability (Rogner, 1997) (A, p.343). The Global Warming Subsystem is based on Wigley's model for the emission-concentration mechanism (Wigley *et al.*, 1994).

The Food and Land Use module assesses the potential contribution of biomass. Forests are a source of biomass and wood products, but their function as a carbon sink is evaluated (A, p.344).

MARIA is designed for macro-level evaluation, and does not provide information on industrial structural change (A, p.344).

MARKAL models

Sources

(A) Richard Loulou and Amit Kanudia. 'Economic Indicators from a Multi-sector, Multi-region Bottom-up MARKAL Model'. Paper presented at the Joint EMF/IEA/IIASA Meeting, Stanford University, June 2000.
(B) Dolf Gielen and Tom Kram. 'Meeting UNFCCC Target via Materials Policies.' Paper presented at the Joint EMF/IEA/IIASA Meeting, Stanford University, June 2000.

Time span: Depends on the model versions.
Regions: More than 30 stand-alone models (A, p.5). Italy (country model), Matter (EU), Nordic, North America are among 15 participating models in ACROPOLIS (Assessing Climate Response Options through POLIcy Simulations) organized by IEA and EC collaboration project. MARKAL-Europe and MARKAL-US are among 14 models participating in Energy Modelling Forum study on Technology and Global Climate Change Policies.
Branch classification: Depends on the model versions.
Energy sources: Depends on the model versions.
Scenarios: Restricted or extended carbon trade, etc., depending on the model versions.
Technology: More than 5000 technologies in total (A, p.7).

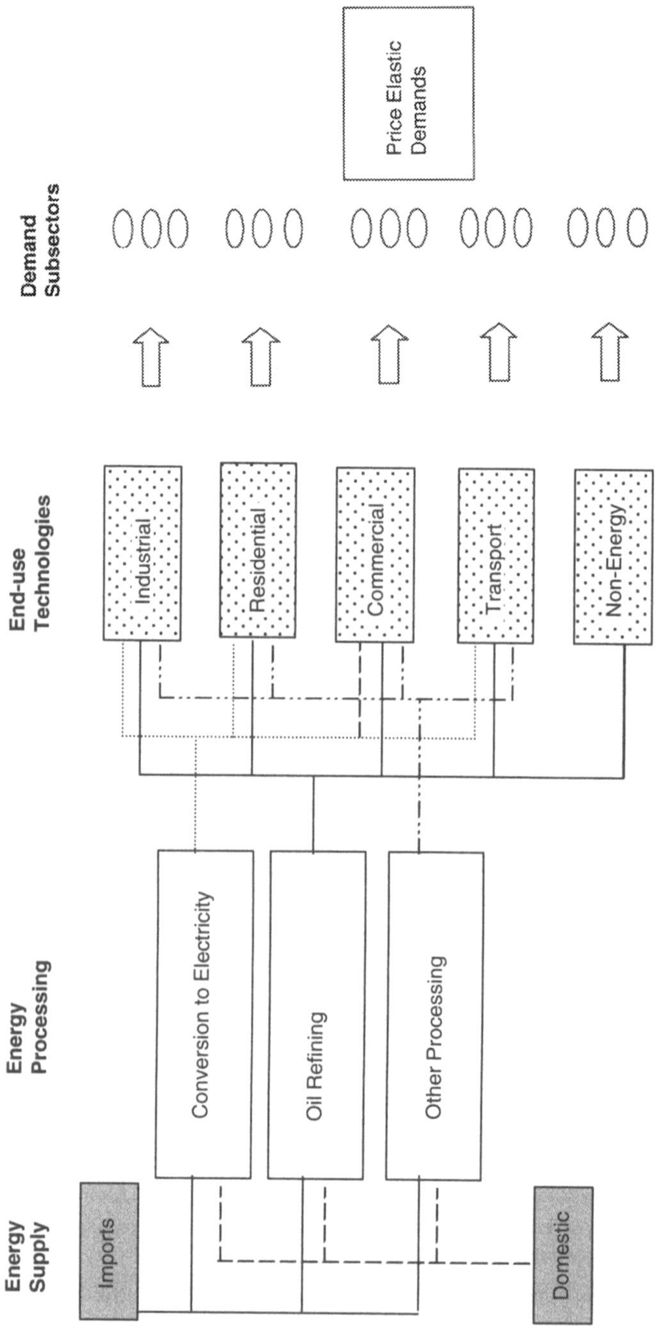

Figure 12-6. MARKAL reference system.
Source: Loulou and Kanudia (2000).

Energy intensity improvement: Embedded in choice of technology.
Lifestyle: No.
Policy tools: International permits (the case of North American MARKAL) (A, p.12).
Economic linkage: Commodity prices are derived based on marginal cost pricing. Equilibrium is based on least total cost (A, p.7).
Environment linkage: Through energy processing and use technologies. Scenarios include energy-related trade (the case of model for Canada)(A).
Modelling logic: Dynamic inter-temporal linear programming.

Summary

The MARKAL linear programming model was developed 20 years ago within the international IEA/ETSAP (International Energy Agency/Energy Technology Systems Analysis Programme) framework. More than 50 institutes in 27 countries use MARKAL as of today, it is reported (B, p.4). MARKAL consists of more than 30 stand-alone models. Most OECD countries and several LDCs are modelled, as well as several multi-regional models (A, p.5).

The economy is modelled as a system represented by processes and physical and monetary flows between these processes (B, p.4). Each piece of equipment is described by its technical parameters, cost parameters, and life duration. Capital turnover is tracked at the technology level (often at plant level) (A, p.7).

This is a technology rich, bottom-up model.
The last generation of the model is capable of producing economic information, such as:

(1) impact on sectors' output (physical and value),
(2) impact on sectors' value added (value),
(3) impact on product prices ($/ton),
(4) impact on trade (ton and value),
(5) impact on consumption (value), and
(6) impact on welfare (net social surplus) (A, p.3).

Future work includes TIMES (The Integrated Markal-Efom System, which is a world-wide, extended partial equilibrium model, with 15-20 regions, and endogenous trade (up to 10 commodities). This attempt, an ongoing project by ETSAP (above), is intended to embody all advances in MARKAL developments (A, p.26).

MARKAL MATTER

Source

(A) Dolf Gielen and Tom Kram. 'Meeting UNFCCC Target via Materials Policies.' Paper presented at the Joint EMF/IEA/IIASA Meeting, Stanford University, June 2000.

Time span: 1980-2000-2060.
Regions: Western Europe as a single region.
Branch classification: No; 150 materials, 100 products, and 30 categories of waste materials covered (A, pp.5-6).
Energy sources: 25 energy carriers. Final energy includes electricity, heat, and biomass (through incineration of waste) (A, p.5).
Scenarios: Constraints are set, representing the demand for products and services, the maximum introduction rate of new processes, the availability of resources, environmental policy goals for energy use and for emissions (A, p.4).
Technology: Processes are characterized by their physical inputs and outputs of energy and material, by their costs, and by their GHG emissions. The model contains a database of several hundred processes, covering the whole life cycle of both energy and materials (A, p.4).
Energy intensity improvement: No. Imbedded in selection of processes.
Lifestyle: No.
Policy tools: GHG policies are represented by emission penalties. The penalty levels have been selected in order to represent a broad spectrum of policy targets. The 20 EURO level corresponds with a 20% emission reduction compared to the base case, and the 200 EURO level corresponds with a 75% emission reduction.
Economic linkage: The demand is price dependent (A, p.4).
Environment linkage: The model is used to calculate the least-cost system configuration for the whole time period, meeting product and service demands and meeting emission reduction targets (A, p.4).
Modelling logic: Linear programming model (A, p.4).

Summary

'The relation between materials production and energy use has received a lot of attention because 60-80% of industrial energy use occurs during materials production. Apart from production, materials transportation represents another 5-10% of total energy use.' 'Little attention has been paid as of yet to energy and environmental strategies that focus on materials consumption. Changes in materials consumption can affect energy use and environmental impacts in other parts of the materials life cycle' (A, pp.1-2).

The MARKAL MATTER model has been built to capture the Western European energy and materials system in order to analyse the dematerialization potentials in detail. The work dates back five years, with participation of a large number of institutes and individuals. Detailed information regarding sub-modules for specific materials and products groups is made available.

In this study, materials include all substances excluding energy carriers and excluding food and fodder (A, p.1). Lifestyle changes can have a very significant impact on energy use and GHG emissions, but the analysis in this study is limited to engineering potentials for existing products. Three types of

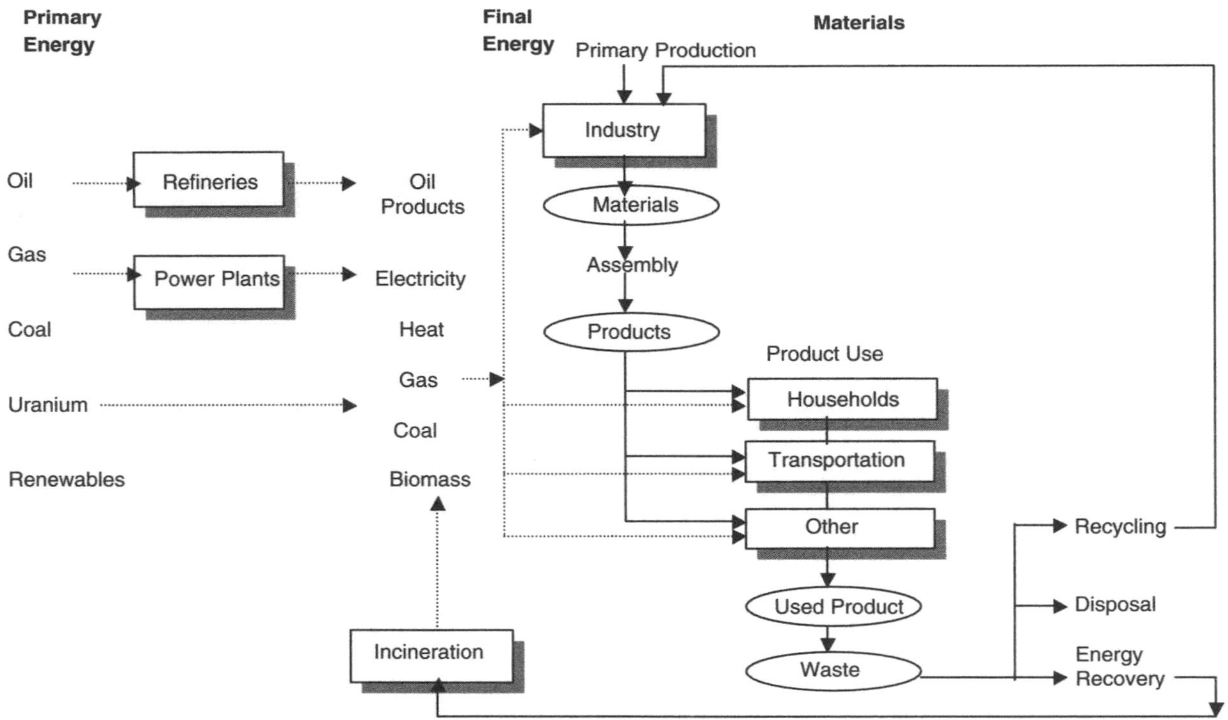

Figure 12-7. MARKAL MATTER energy and materials system model.

Note: Dotted lines represent energy flows; continuous lines represent materials flows.

Source: Gielen and Kram (2000); modified by the present author.

Table 12-15. Design-for-environment technology strategies

Strategy		Description
1	Increased resource efficiency	Reduced losses during materials production
2	Increased materials efficiency	Improvements of the chemical or physical materials quality
		Reduced variation in materials quality
		More diverse materials standardization
		Reduced materials losses during product manufacturing
		Less materials intensive product design
3	Increased product efficiency	Increased product life
		More efficient product use, e.g. shared car ownership
		Development of multi-functional products
4	Recycling/energy recovery of waste materials	Increased waste recovery
		Increased waste recycling (including cascading)
		Increased energy recovery
5	Reuse of waste products	Increased product recovery
		increased product reuse
6	Substitution of natural resources	Renewable organic feedstocks
7	Substitution of materials	Renewable materials (e.g. wood products)
		Less CO_2 intensive materials
		Materials with superior physical characteristics
		Recyclable materials
8	Substitution of products	Less material intensive products
		Products that require less maintenance
		Products with longer life, or multifunctional products

Source: Gielen and Kram (2000).

dematerialization strategies are considered explicitly; i.e. (1) increased efficiency of materials use; (2) materials substitution; and (3) recycling/energy recovery (A, p.3).

According to the calculation results, 'up to 800 Mt GHG emissions reduction (20% of 1990 emission levels) can be achieved through changes in materials consumption' (A, p.1). 'The modelling results suggest a 67% emission reduction potential in the materials system' (A, p.13).

'Dematerialization strategies result in emission reduction in other parts of the system than where the action takes place. As a consequence, the analysis must consider a number of linked processed simultaneously.' 'Two important differences exist between materials use and energy use. First, materials can be recycled and energy can be recovered from organic waste materials, while energy is used and released into the environment.... A life cycle approach is required for proper analysis, instead of a chain approach for energy. Second, materials are stored in products during the product life span. A time lag exists between materials consumption and waste release. As a consequence, proper analysis of materials strategies requires a dynamic approach, contrary to energy analysis.' 'The GHG impact of dematerialization depends on energy system characteristics, e.g. the GHG intensity of materials production and the GHG intensity of electricity production. As a consequence an integrated energy and materials systems approach is required, resulting in complex models' (A, p.2).

MERGE: Model for evaluating regional and global effects of GHG reduction policies

Sources

(A) Alan Manne, Robert Mendelsohn, and Richard Richels. 'MERGE: a Model for Evaluating Regional and Global Effects of GHG Reduction Policies.' *Energy Policy*. Vol. 23, No. 1 (1995).
(B) Alan Manne and Richard Richels. 'The Greenhouse Debate: Economic Efficiency, Burden Sharing and Hedging Strategies.' *The Energy Journal*, Vol. 16, No. 4 (1995).
(C) Alan Manne and Richard Richels. 'International Carbon Agreements, EIS Trade and Leakage.' Paper presented at the Joint EMF/IEA/IIASA Meeting, Stanford University, June 2000.

Time span: 1990 to 2200, with 10-year steps from 1990 to 2050, and 25-year steps during the rest of the projection period (A, p.19).
Regions: 5 (the USA, other OECD nations (Western Europe, Japan, Canada, Australia, and New Zealand), the Former USSR, China, and the rest of the world (A, p.19). 9 (USA, OECD-Europe, Japan, CANZ, EEFSU, China, India, Mexico plus OPEC, and the rest of the world) (C, Figure 2).
Branch classification: Energy (electric and non-electric), and an aggregate outside energy (A, p.20).
Energy sources: Electric, non-electric (A, p.20).
Scenarios: Five policy scenarios, i.e.:

(1) business as usual,
(2) a tax on carbon emissions starting at $1/tonne in 2000 and increasing at 5%/year,
(3) a tax on carbon emissions starting at $5/tonne in 2000 and increasing at 5%/year,
(4) stabilizing global CO_2 emissions at 1990 levels, and
(5) stabilizing concentrations of CO_2 in the atmosphere at near current levels.

Technology: Output depends on four inputs (capital, labour, electric and non-electric energy). There are constant returns to scale. There is a unit elasticity of substitution between (1) capital and labour, and (2) electric and non-electric energy. There is a constant elasticity of substitution between (1) and (2).

Energy intensity improvement: There are autonomous energy-efficiency improvements (AEEI) that are summarized by the scaling factor (A, p.21).
Lifestyle: There is a single representative producer-consumer. Savings decisions are made so as to maximize the sum of the discounted utility of consumption (A, p.19).
Policy tools: Tax on carbon emissions (A, p.26).
Economic linkage: Up to 2100 the growth rates represent the average of higher and lower rates in the IPCC's Working Group III. From 2100 onward it is assumed that per-capita growth rates will decline over time, but that the gap between the industrialized countries and the rest of the world will continue to be substantial (A, p.21).
Environment linkage: Energy-economy interactions, resource exhaustion, and the introduction of new technologies (A, p.20).
Modelling logic: General equilibrium model (A, p.19).

Summary

MERGE consists of three major submodels: (1) Global 2200 which is used to assess the economy-wide costs of alternative emission constraints; (2) the climate submodel with focus on greenhouse gases such as carbon dioxide (CO_2), methane (CH_4), and nitrous oxide (N_2O); and (3) the damage assessment submodel which carries out global climate cost-benefit analyses.

Global 2200: This CGE model is composed of 2800 linear and non-linear constraints and 3800 variables, many of which are subject to individual upper bounds (A, p.23). The GDP growth is a key determinant of energy demand. The rate of GDP growth depends on both population and per-capita productivity trends (A, p.21). Potential per-capita projections are made based on constant energy prices. 'Energy consumption need not grow at the same rate as the GDP. Over the long run, they may be decoupled' (A, p.21). These possibilities are represented in Global 2200 through two parameters, ESUB (the elasticity of price-induced substitution) and AEEI (autonomous energy efficiency improvements). ESUB is approximately equal to the absolute value of the price elasticity of demand. The values for ESUB are set at 0.40 for the United States and for the other OECD members. In the other regions of the world, it depends on long-term structural changes in the political and economic system. We have therefore set ESUB at 0.30 for these other three regions' (A, pp.21-22).

AEEI, which represents non-price efficiency improvements, may be brought about, for example, by deliberate changes in public policy, shifts in the economic mix away from manufactured goods and towards services (A, p.22). For most regions and time periods, AEEI of 0.5% per year is assumed regardless of GDP growth rate (A, p.22).

Climate model: The emission of anthropogenic greenhouse gases is divided into energy and non-energy. Energy-related emissions of CO_2, CH_4, and N_2O

Table 12-16. Identification of technologies by MERGE

A. Electricity Generation

Existing Technologies:	Description		
HYDRO	Hydroelectric, geothermal, and other renewables		
GAS-R	Remaining initial gas fired		
OIL-R	Remaining initial oil fired		
COAL-R	Remaining coal fired		
NUC-R	Remaining initial nuclear		
New Technologies:		Earliest possible introduction date (a)	Estimated US cost (b) (Mills.kWh)
GAS-N	Advanced combined cycle, gas fired	1995	34.3(c)
COAL-N	New coal fired	1990	51.0(d)
ADV-HC	High cost carbon free	2010	75.0
ADV-LC	Low cost carbon free	2020	50.0

a Estimated year when technology could provide 0.1 trillion kWh (approximately 20 GW of installed capacity at 60% capacity factor). For the other OECD region, we assume that the technology could provide 0.2 trillion kWh by this date.

b Based on 1990 dollar.

c Based on price of gas in 1990. Gas prices are projected to rise over time.

d Estimated costs are assumed to be 10 mills/kWh higher in the other OECD region due to higher fuel costs.

Table 12-16. continued

B. Non-electric Energy Supplies

Technology Name	Description	Carbon Emission Coefficient (Tons of Carbon per GJ of Crude Oil Equivalent)	Unit US Cost per GJ of Crude Oil Equivalent (1990 Dollars)
OIL-MX	Oil imports or exports	0.0199	3.70 in 1990 rising over time
CLDU	Coal – direct uses	0.0241	2.00
OIL-LC	Oil – low cost	0.0199	2.50
GAS-LC	Natural gas – low cost	0.0137	2.75(a)
OIL-HC	Oil – high cost	0.0199	3.00
GAS-HC	Natural gas – high cost	0.0137	4.25(a)
RNEW	Renewables	0.0000	6.00
SYNF	Synthetic fuels	0.0400	8.33
NE-BAK	Nonelectric backup	0.0000	16.67

a To allow for gas distribution costs, $1.25 per GJ is added to the wellhead price.

Source: Manne, Mendelsohm, Richels (1995).

are projected by fuel type for each period up to 2200, whereas emissions from other sources are exogenous to the model (A, pp.23-25).

Damage assessment submodel: Global cost-benefit analysis has to deal with damage in terms of (1) marketed goods and (2) non-marketed factors such as biodiversity, environmental quality, and human health. 'Many cost-benefit calculations ignore the issue altogether and confine their analyses to economically measurable consequences only' (A, p.25). The paper deals with this issue through a willingness-to-pay approach. 'The issue is framed in the context of how much the consumers in each region would be willing to pay to avoid ecological damages. Each region values ecological damages independently of where such damage occurs, whether they occur within or outside their own boundaries. The S-shaped function implies, however, that lower income regions will place a lower value on such losses than higher income regions' (A, p.26). 'Rather than advocate any particular set of numbers, we will explore the implications of alternative valuations' (A, p.26).

Changing location of production: MERGE has been modified to include the possibility of changes in the location of production of the energy-intensive sectors (EIS). EIS is an aggregate composed of ferrous and non-ferrous metals, chemicals, non-metallic minerals, paper, pulp and print. It does not include petroleum refining (C, p.1).

The GTAP database is employed to estimate initial EIS demand in each region (C, p.1).

'For projecting the impact of the Kyoto Protocol, each region is taken to be self-sufficient in EIS at base year energy prices. Each of the nine MERGE regions is sufficiently large to be nearly self-sufficient with respect to EIS. Changes in the location of production are attributed primarily to changes in the cost of energy. At base year prices in the USA about 85% of the cost of EIS consisted of non-energy inputs (labour and capital charges, shipping, iron ore, etc.), and 15% of the cost consisted of energy inputs. ... Under these conditions, a doubling of energy prices would imply only a 15% increase in the cost of EIS. This is why it is assumed that the demand for EIS is inelastic with respect to the price of energy. For projecting future demands, the income elasticity for EIS is taken to be 0.5' (C, p.1).

The analysis is constructed upon the Heckscher-Ohlin fraction, which is the relative price of imports at which domestic supplies drop to zero. 'The intercept of the non-energy supply curve for EIS in each region may be described as a Heckscher-Ohlin fraction. If this intercept is unity, EIS is viewed as a perfectly homogeneous commodity. Small change in energy cost will then lead to large changes in the international location of production. If the intercept is less than unity, the supply function is less elastic, and the changes in location will be less dramatic' (C, p.1).

The Kyoto Protocol 'exempts non-Annex B countries from any obligation to restrict greenhouse emissions. Accordingly, they will have an artificial incentive to produce EIS – even though their energy and carbon coefficients may be higher than in Annex B. This is bound to exacerbate the problem of leakage – that is, an increase in non-Annex B emissions associated with a reduction in those of Annex B.' 'Annex B reductions will impose downward pressure upon international oil and gas prices. This will lower the export earnings of two of our major regions: EFFSU (Eastern Europe + Former Soviet Union) and MOPEC (Mexico + OPEC). These lower prices would stimulate the importing countries to increase their consumption of oil and gas' (C, p.3).

Major findings are: (1) Leakage is a distinct possibility. (2) During the early decades of the 21st century, leakage would be associated primarily with EIS trade; during the middle decades, leakage would be associated primarily with trade in oil and gas. (3) Leakage ratios are sensitive to the numerical values of the EIS trade parameters (C, p. 4).

MESSAGE: Model for energy supply strategy alternatives and their general environmental impact

Sources

(A) Intergovernmental Panel on Climate Change. *Special Report on Emissions Scenarios*. Cambridge: Cambridge University Press, 2000.
(B) S. Messner and M. Strubegger. *User's Guide for MESSAGE III*, WP-95-69. International Institute for Applied Systems Analysis. Laxenburg, Austria, 1995.

Time span: 2100.
Regions: 10 (Former Soviet Union, Eastern Europe, Western Europe, North America, Pacific OECD, Pacific Asia, South Asia, Centrally Planned Asia, Middle East and North Africa, Sub-Saharan Africa, and Latin America) (A, p.338).
Branch classification: Energy sector and industrial emission sources only, and agricultural and land-use related emissions are not included.
Energy sources: Detailed.
Scenarios: As defined in the Special Report on Emissions Scenarios (SRES) (A, p.5). Scenario Generator (SG) model derives energy demand path consistent with historical dynamics and with the specific scenario features, such as high or moderate economic growth or rapid or gradual energy intensity improvements. Empirically estimated equations of trends are employed (A, p.341).
Technology: Menu of technologies is given. The model calculates cost-minimal supply structure under a set of constraints (A, p.341).
Energy intensity improvement: Incorporated in scenarios.
Lifestyle: No.
Policy tools: Not explicit.

Economic linkage: Exogenous assumptions about per-capita economic growth and population by region.

Environment linkage: Energy system structure, including energy demand, supply, and emissions patterns (A, p.341).

Modelling logic: Dynamic linear programming model.

Table 12-17. Selected energy technologies represented in MESSAGE

Technology Aggregates	Including
Centralized Electricity Generation:	
Coal conventional	Conventional coal power plants with DESOX (fuel gas desulfurization, FGD) and DENOX (flue-gas denitrification)
IGCC	Integrated coal Gasification Combined Cycle
Coal fuel cell	Coal-based high-temperature fuel cell (internal reforming)
Oil	New standard oil power plant (Rankine cycle, low Nox, and with FGD); existing crude oil and light oil engine-plants; light oil combined cycle power plants
Gas standard	Standard gas power plant (Rankine cycle, potential for cogeneration)
NGCC	Natural Gas-fired Combined cycle power plant with DENOX
NGFC	Natural Gas-powered high-temperature Fuel Cell, cogeneration possibilities
Bio	New biomass-fired power plant (Rankine cycle, cogeneration possibilities), advanced biomass plants (gasified biomass is burned in combined gas turbines)
Nuclear	Conventional, existing nuclear power plants
Advanced nuclear/other	Nuclear high-temperature reactors for electricity and hydrogen co-production, future inherently safe nuclear reactor designs, and other future zero-carbon electricity-generating technologies for base load
Hydro	Hydropower plants (low and high cost)
Wind	Wind power plant
Other renewables	Geothermal power plant (cogeneration potential); grid-connected solar photo-voltaic power plant (no storage); solar thermal power plants with storage, and solar thermal power plant for hydrogen production
Decentralized Electricity Generation:	
Hydrogen fuel cell	Decentralized stationary and mobile hydrogen fuel cells (cogeneration systems or off-hours electricity generation)
Photo-voltaics	On-site solar photo-voltaic power plant in the residential and/or commercial sectors, and in the industrial sector
Synfuels:	
Coal synliquids	Light oil and methanol production from coal
Biomass synliquids	Ethanol production from biomass
Gas synliquids	Methanol production from natural gas
Syngases	Syngases from various sources, including biomass and coal gasification
Hydrogen H2 (1)	Hydrogen production from fossil fuels (coal or gas)
Hydrogen H2(2), (3)	Non-fossil hydrogen production: H2(2): from biomass and electricity H2(3): from nuclear and solar

Source: IPCC (2000).

Summary

MESSAGE is an energy systems engineering model, and is one of the six models that constitute IIASA's integrated modelling framework, which includes:

(1) Scenario Generator (SG) model,
(2) Linked MESSAGE (bottom-up type) and MACRO (top-down type),
(3) Regional Air Pollution Impacts Model (RAIMS),
(4) Basic Linked System of National Agricultural Models (BLS),
(5) Model for Assessment of GHG-Included Climate Change (MAGICC), and
(6) General Circulation Model (GCM) (A, p.342).

MESSAGE is a bottom-up type model that calculates cost-minimal supply structure under the constraints of resource availability, the set of given technologies, and the demand for useful energy. It generates energy demand, supply, and emission patterns that are consistent with the development produced by the SG. MACRO, on the other hand, is a top-down model, built on the Global 2100 model (Manne and Richels, 1992). Solution is obtained by maximizing the inter-temporal utility function in each world region. It establishes the relationship between macroeconomic development and energy use. MESSAGE and MACRO are linked and used in tandem (A, p.341).

In one of the scenarios (B2 market), it is reported that the price-induced energy demand savings are 8% by 2020, 23% by 2050, and 30% by 2100 (A, p.341). In addition, the MESSAGE model was used to quantify four scenario groups of the A1 story line and a number of scenarios of the B1 story line of the SRES.

In the B2 market, shadow prices for international trade are calculated as follows (A, p.342) (1990 US$/GJ):

Year	Gas	Coal	Oil
2020	0.4	0.3	0.5
2050	0.7	0.4	1.1
2100	0.7	1.1	2.3

MIDAS: Multinational integrated demand and supply

Source

(A) P. Capros, P. Karadeloglou, L. Mantzos, and G. Mentzas. 'The Energy Model MIDAS.' In Jean-Baptiste Lasourd, Jacques Percebois, and François Valette, eds. *Models for Energy Policy*. London: Routledge, 1996.

Time span: 10 years (A, p.41).
Regions: United Kingdom, Italy, France, and Germany (A, p.43).

Table 12-18. Policy analysis that can be studied with MIDAS

Crude-oil prices	Impacts of international crude-oil price shocks
Fiscal policy	Changes of tax-rates on energy products and sectors
Investment programs	Impact on energy demand and supply and on equilibrium energy prices of alternative technological and utilities investment programs; impact of nuclear programs
Demand policy	Direct and secondary impacts of incentives and measures in the energy demand side
Impact of growth	Energy system impacts and secondary effects of macro-economic policy and forecasts
Sectoral policy	Energy system impact evaluation of specific energy sectoral policies (natural gas, mining, refineries, etc.)
Environmental policy	Carbon tax policy, impacts evaluation and potential assessment of pollution abatement equipment and reduced emission technologies
Forecasts	Consistent construction of future EUROSTAT balances and price-lists; e.g. Energy 2010 studies of CEC
EC issues	Tax harmonization, EC unified market analysis, elimination of trade barriers

Source: Capros, Karadeloglou, Mantzos, and Mentzas (1996).

Branch classification: Electricity use of the MIDAS Demand model has iron and steel, non-ferrous metals, chemicals, building materials, paper and pulp, other industries, space heating, cooking and water, specific uses, transport, exports (A, pp.49-50).

Energy sources: Electricity, petroleum, coal, natural gas, synthetic gas (A, p.44).

Scenarios: Technological characteristics. As a single-country model it has no international linkages.

Technology: Described in detail in sub-models on electricity production, petroleum refining, coal mining, natural gas distribution, synthetic gas and coal production.

Energy intensity improvement: Not explicit.

Lifestyle: As reflected in energy use.

Policy tools: Fiscal policy, investment programme, demand policy (A, p.46).

Economic linkage: Crude oil prices, demand policy (A, p.46).

Environment linkage: Environmental policy, e.g. carbon-tax policy, pollution abatement, reduced emission technologies (A, p.46).

Modelling logic: Bottom-up technology model. Macroeconomic variables are exogenous.

Summary

MIDAS is developed under the auspices of CEC/DG-XII (A, p.41). The complete MIDAS model is available for the United Kingdom, Italy, France, and Germany.

MIDAS is linked to the macroeconomic model HERMES and a sub-model that evaluates CO_2 emissions (A, p.41, p.43).

MIDAS is a country-specific energy model for European countries covering both demand and supply of energy. The energy demand sub-model is an econometric system using relative prices of energy to estimate energy demand and energy substitution by sector. Energy supply consists of five sub-models; namely electricity generation, petroleum refining, natural gas and synthetic gas, coal mining and coke-oven plants (A, p.41).

The paper offers interesting observations about price determination. 'Models built for the USA energy system generally retain a price-adjustment formulation. ... On the contrary, European energy models ... such as MIDAS retain quantity adjustment formulation. This difference reflects the diverse conditions that prevail in energy markets' (A, p.43). 'If the market clearing price p^* is calculated such that $D = S$, then the market is formulated as a price adjusted one. Such a formulation corresponds to a perfectly competitive market. ... If the market has an oligopolistic or monopolistic character and competiton does not operate perfectly, supply firms may adopt a cost mark-up pricing policy. In this case, prices are a function of supply, i.e. $p = p(S)$ while demand and supply are rationed by each other at the given level of prices so that $D = S = min\{D(p), S(p)\}$. ... This market clearing formulation is called quantity-adjustment. If we further assume excess supply in the market, ... we may calculate this case, prices are a function of supply, i.e. $p = p(S)$ while demand and supply are rationed by each other at the given level of prices so that $D = S = min\{D(p), S(p)\}$. ... This market clearing formulation is called quantity-adjustment. If we further assume excess supply in the market, ... we may calculate a rate of capacity utilization as $U = S/S^*$ where $S^* = S(P)$. This rate may influence prices, i.e. $p = p(S, U)$, representing the disequilibrium pressures exerted on price formulation' (A, p.43).

MiniCAM: The mini climate assessment model

Source

(A) Intergovernmental Panel on Climate Change. *Special Report on Emissions Scenarios*. Cambridge: Cambridge University Press, 2000.

Time span: 2100.
Regions: 11 (Centrally Planned Europe, OECD-Europe, Canada, USA, Oceania, Japan, South East Asia, Centrally Planned Asia, Middle East, Africa, Latin America) (A, p.338). A 14-region version is being prepared (A, p.344).
Branch classification: 3 (industrial, transportation, residential/commercial).
Energy sources: 7 (coal, oil, gas, nuclear, hydro, solar, and biomass) (A, p.344).
Scenarios: As defined in the Special Report on Emissions Scenarios (SRES) (A, p.5).
Technology: Labour productivity increase in the long term as part of scenario exercise (A, p.344).
Energy intensity improvement: Represented by the improvement in end-use energy efficiency (A, p.345). 'Fuel-specific rates of technical change are

available for primary fuel production and conversion, as are technical change coefficients for each category of electricity production' (A, p.344).

Lifestyle: No.

Policy tools: No.

Economic linkage: Prices adjust energy supply and demand. Population times aggregate labour productivity exogenous. The resultant estimate of GNP is corrected for changes in energy prices using GNP/energy elasticity. Detailed age breakdown and labour force participation rates to deal with population scenarios (A, p.344).

Environment linkage: Global GHG emissions; agriculture, forestry and land use; climate change; impacts of climate change through damage functions.

Modelling logic: FRB is a partial equilibrium model.

Summary

MiniCAM is being developed by the Global Change Group at Pacific Northwest Laboratory. It has recently expanded to deal with emissions of GHG gases (A, p.344). The MiniCAM framework integrates ERB model for global GHG emission estimates (Edmonds *et al.*, 1994, 1996), MAGICC model for climate change assessment (Wigley and Raper, 1993), the SCENGEN tool to deal with regional climate changes (Hulme *et al.*, 1995), and the Manne *et al.* (1995) damage functions to examine the impacts of climate change (A, p.344).

Energy demand is estimated for 3 categories of activities (residential/ commercial, industrial, and transportation) as a function of price and income. Energy services are provided by 4 secondary fuels (solids, liquids, gases, and electricity), demand for which depends on their relative costs and the improvement in the end-use energy efficiency. Demand for primary fuels is determined by the relative costs of transforming them into the secondary fuels (A, p.344).

The energy supply sector provides both renewable and non-renewable resources. The cost of fossil resources relates to the resource base by grade. This practice allows the model to test the importance of fossil fuel constraints. Unconventional fuels such as oil shale and methane hydrates are potentially available at high cost, or after extensive technology development. The cost of production includes both technical and environmental factors (A, p.345).

Biomass is supplied by the agriculture sector, and provides the linkage with the land-use model and the energy model.

MS-MRT: Multi-sector, Multi-region trade model

Sources

(A) Paul M. Bernstein, W. David Montgomery, Thomas F. Rutherford, and Gui-Fang Yang. 'Effects of Restrictions on International Permit Trading.' *Energy Journal*, May 1999.

Table 12-19. Carbon permit trade

Countries/Regions	2010						
	Permit Price (1995$ / Ton)			Permit Purchases (Million of Tons of Carbon)		Permit Purchases (Billion of 1995 Dollars)	
	NT	A1	GL	A1	GL	A1	GL
USA	274	89	30	305.0	461.8	27.3	13.2
JPN	482	89	30	57.4	81.1	5.1	2.4
EUR	211	89	30	103.5	165.1	9.2	5.0
OOE (other OECD)	246	89	30	40.1	61.9	3.6	1.9
SEA (south-east Asia)	0	0	30	0.0	−49.0	NA	−1.5
OAS (other Asia)	0	0	30	0.0	−12.0	NA	−0.4
CHI (China & India)	0	0	30	0.0	−274.6	NA	−7.5
FSU (Eastern Europe & former SU)	0	89	30	−506.0	−311.8	−45.2	−9.4
MPC (Mexico & OPEC)	0	0	30	0.0	−58.3	NA	−1.8
ROW	0	0	30	0.0	−91.3	NA	−2.8

Countries/Regions	2030						
	Permit Price (1995$ / Ton)			Permit Purchases (Million of Tons of Carbon)		Permit Purchases (Billion of 1995 Dollars)	
	NT	A1	GL	A1	GL	A1	GL
USA	353	224	32	180.4	709.40	40.4	22.5
JPN	526	224	32	108.5	183.90	24.3	5.8
EUR (EU of 15)	431	224	32	184.8	429.00	41.4	13.6
OOE (other OECD)	505	224	32	59.1	141.20	13	4.5
SEA (south-east Asia)	0	0	32	0	−123.30	NA	−3.9
OAS (other Asia)	0	0	32	0	−37.60	NA	−1.2
CHI (China & India)	0	0	32	0	−901.30	NA	−28.6
FSU (Eastern Europe & former SU)	50	224	32	−532.8	−6.10	−119.5	−0.2
MPC (Mexico & OPEC)	0	0	32	0	−153.80	NA	−4.9
ROW	0	0	32	0	−241.40	NA	−7.7

Source: Bernstein, Montgomery, Rutherford, and Yang (1999).

(B) Paul M. Bernstein, W. David Montgomery, and Thomas Rutherford. 'Global Impacts of the Kyoto Agreement: Results from the MS-MRT Model.' Paper presented at the Joint Meeting organized by the International Energy Agency, Energy Modeling Forum, and the International Energy Workshop (IIASA), 16-18 June 1999b, Paris.
(C) Paul Bernstein. 'MS-MRT Model Enhancements and Results.' Paper presented at the Joint EMF/IEA/IIASA Meeting, Stanford University, June 2000.

Time span: 2000 to 2030 with 5-year intervals (A, p.223).
Regions: 10 (United States, Japan, Europe Union of 15, Other OECD, Eastern Europe and Former Soviet Union, China and India, South East Asia (Korea, Singapore, Taiwan, Thailand, and Malaysia), Other Asia, Mexico and OPEC, Rest of the world) (A, p.224). New version incorporates 9 global regions, 35 individual countries (C, p.2).
Branch classification: 6, including 4 energy forms (oil, coal, natural gas, electricity) and 2 non-energy goods (energy intensive sectors and all other goods) (A, p.224).
Energy sources: 4 (oil, coal, natural gas, and electricity) (A, p.224).
Scenarios: (1) No-trade scenario where the emissions permits are tradable within each Annex I region but not across regions; (2) the Annex I trading scenarios where permits are tradable across Annex I countries; and (3) the global trading scenarios where the permits are tradable throughout the world (A, p.232, p. 234). 10%, 30%, and 50% ceiling on satisfying country's emissions obligation through purchase of emissions permits (A, p.241).
Technology: Represented by key elasticities. See below.
Energy intensity improvement: Assumed (A, p.233).
Lifestyle: Infinitely-lived representative agent allocates region's entire lifetime income in different time periods in order to maximize welfare (A, p.227). The consumer faces the choice between current consumption and future consumption purchased via savings (A, p.227).
Policy tools: Emissions trading along three scenarios. See above.
Economic linkage: The business-as-usual growth rates from MERGE by Alan Manne and Richard Richels. The energy production and consumption forecasts as well as regional emissions from the Department of Energy Information Center's.
International Energy Outlook, 1998 (A, p.233). Key elasticities are assumed (oil supply price elasticity, Armington elasticity, demand elasticity, interfuel elasticity of substitution, and the cost of the carbon-free backstop technology) (A, p.233). New version includes fully dynamic treatment of investment, saving, and capital flows (C, p.2).
Environment linkage: Emissions trading.
Modelling logic: Dynamic, general equilibrium model (A, p.222).

Summary

The work is intended to respond to some of the key questions concerning the Kyoto Protocol.

Distribution of economic impacts across regions: how will developing countries be affected by emission limits adopted by the industrial countries? How will the cost of reducing emissions be distributed among countries?

Carbon leakage and effects on international competition: how will energy-intensive industries be affected when some countries agree to binding emission limits and others do not? How large will leakage of carbon emissions from participating to non-participating countries be?

Effects of international emission permit trading: how will emission limits and international permit trading affect the terms of trade and capital flows between countries? Who will be the buyers and sellers of permits in an international permit trading system? Will developing countries be made better or worse off by participating in a system of international permit trading? (B, pp.2-3).

'Countries have argued extensively over the equitable structuring of an international permit trading protocol. The United States has advocated unrestricted emissions trading, extended as rapidly as possible to include key non-Annex I countries. The European Union and a number of developing countries have proposed tight restrictions on emissions trading and oppose any efforts to include non-Annex I countries. Russia has made it clear that its participation in the Kyoto Protocol is contingent on its unrestricted ability to sell permits to other Annex I counties' (A, p.222).

The paper considers several different Annex I trading regimes (A, pp.241-242):

(1) Ceilings on percentage of a country's emissions obligation that may be satisfied through the purchase of emissions permits. 10%, 30%, and 50% limits are suggested.
(2) Impacts of prohibiting the East European and Former Soviet Union region from selling its 'hot air'. 'Hot air' or paper tons refer to the positive difference between the Kyoto emissions targets for the EE/FSU and its actual forecasted emissions. This resulted from its economic decline since 1990.
(3) The case where EE/FSU exercise some market power and charge a monopoly price for its emissions permits.
(4) Restrictions on sales where countries are allowed to sell only 30% of the amount they would sell under unrestricted Annex I trading.
(5) Comparisons of full global trading with clean development mechanism (CDM).

Of the alternative measures of economic impacts (including change in consumers surplus measured by the equivalent valuation, the discounted present value of consumption, GDP, and direct costs), the paper employs the intertemporal equivalent variation (A, p.237).

MS-MRT is based on the GTAP database (C, p.2).

New Earth 21 (Sometimes referred to as DNE21 or Dynamic New Earth)

Source

(A) Kenichiro Nishio, Yasumasa Fujii, and Kenji Yamaji, 'Analysis of the Kyoto Mechanisms using a Global System Model DNE21.' Paper presented at the Joint EMF/IEA/IIASA Meeting, Stanford University, June 2000.

Time span: Up to 2020 with a two-year step.
Regions: 10 (North America, Latin America, Western Europe, Former USSR and Eastern Europe, Middle East and Northern Africa, Subsaharan and Southern Africa, South and East Asia, China, Japan, Oceania) (A, p.2).
Branch classification: No.
Energy sources: 9 primary (natural gas, oil, coal, biomass, photovoltaics, wind and other renewables, hydro, geothermal, nuclear). 4 secondary (gaseous fuel, liquid fuel, solid fuel, electricity), where liquid fuel is decomposed into 3 (gasoline, light fuel oil, and heavy fuel oil) (A, p.3).
Scenarios: Constraints represent supply-demand balances, energy and CO_2 balances in the various types of energy processing plants, several inter-temporal dynamics such as depletion of fossil fuel resources and limitations on the maximum growth rates of annual fuel production.
Technology: Detailed description of conversion from primary energy sources to secondary energy carriers.
Energy intensity improvement: No. Embedded in technology selection.
Lifestyle: No.
Policy tools: Emission trading, clean development mechanism, banking and borrowing of emission reduction (A, pp.4-5).
Economic linkage: Future energy demands are exogenously given (A, p.3).
Environment linkage: CO_2 emission reductions.
Modelling logic: Multi-period inter-temporal nonlinear optimization. Objective function is defined as the sum of the discounted total energy system costs distributed over time (5% discount rate) (A, p.2).

Summary

New Earth 21 is an engineering process model.

Two sets of analysis are provided based on mathematical formulation of the Kyoto Mechanisms. Formulations correspond to:

Case 0: without the Kyoto Mechanism with Kyoto Mechanism
Case 1: No trading
Case 2: Emissions trading

Case 3: Emissions trading + clean development mechanism
Case 4: Emissions trading + clean development mechanism (including the banking)

The first analysis concerns the potential of the Kyoto Mechanism in the first commitment period, i.e. 2000 to 2008-2012 (6 representative time points). The whole annual emission in case 0 (BAU) reaches 9000 Mt-C/year by 2012, implying 1.5 times as much as that of 1990. The average emission of Annex I regions during the 1st commitment period is 4873 Mt-C/year (21% over 1990 levels). In Case 2 the emission increases a little by the so-called hot air of FSSU and Eastern European countries. In Cases 3 and 4 a noticeable amount of emission credits will be transferred from non-Annex I regions to Annex I regions. Furthermore, the banking of CDM stimulates the reduction of non-Annex I regions during the interim period (A, pp.5-6).

The second analysis includes both the 1st and the 2nd commitment period, and the time horizon is extended to 2000 to 2018 (9 representative time points). Here, CDM action during the interim period enables Annex I regions to bank the emission reductions from the 1st commitment period to the 2nd commitment period. Various cases are simulated (A, pp.7-8).

The PRIMES project, European Commission

Sources

(A) European Commission, Directorate-General XII Science, Research and Development, *The PRIMES Project*. EUR 16713 EN. Brussels and Luxembourg: European Commission. 1995.

Time span: 1990-2030 (A, p.11).
Regions: 12 European Union countries (Belgium, Denmark, France, Germany, Greece, Ireland, Italy, Luxembourg, the Netherlands, Portugal, Spain, and the United Kingdom) (A, p.11).
Branch classification: Residential, commercial, industry (6 sub-sectors: iron and steel, non-ferrous metals, building materials, chemicals, paper and pulp, and other industries), transport (4 transport modes) (A, p.14).
Energy sources: Coal (3 types), oil (4 different refineries), natural gas (regional detail), electricity (A, p.14).
Scenarios: Operate on decision variables. See below.
Technology: 50 thermal generation technologies, all renewables. 2 technology vintages for residential sector, 2 technology vintages for commercial sector, 4 fuels for industry sector, 4 transport modes and different car technologies in transport sector (A, p.14).
Energy intensity improvement: Embedded in technology penetration structure by fuel use.
Lifestyle: Embedded in 'individual agents' behavior'.

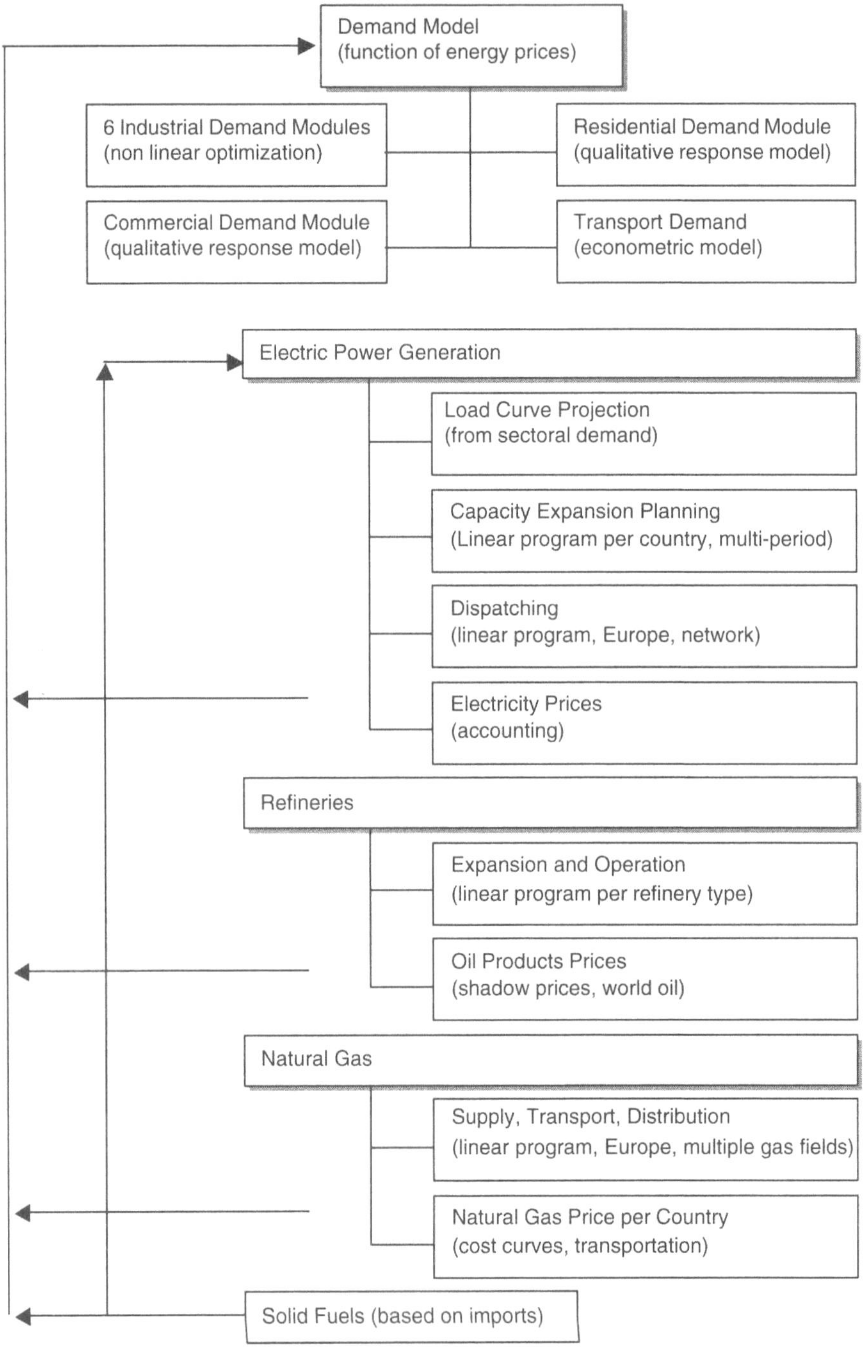

Figure 12-8. General model structure of PRIME.
Source: European Commission (1995).

Policy tools: Standards, taxes, tradables (A, p.67). Third-party access for electricity (A, p.12). Decision variables introduced in the optimization routine (A, p.18).
Economic linkage: Market clearing energy prices, individual agents' behaviour (A, p.12).
Environment linkage: Emissions of CO_2, SO_x, and NO_x (A, p.14, p.67).
Modelling logic: Partial equilibrium (A, p.11). Non-linear optimization (A, p.18).

Summary

'PRIMES is a partial equilibrium model for the European energy system and market, including environmental applications' (A, p.7). 'The PRIMES model combines the detailed "bottom-up" technological approach ... with an accurate presentation of the different energy markets and the feedback between supply and demands as determined by market prices' (A, p.12).

'PRIMES is developed under JOULE II, which is a programme on non-nuclear energy (1991-1994). The general objective of the JOULE II is 'to contribute to the development of (1) economically viable and environmentally safer energy options, and (2) energy saving technologies. The programme covers four areas, namely: (1) analysis of strategies and modelling, (2) minimum emission power reduction from fossil fuels, (3) renewable energy sources, and (4) energy utilization and conservation. The sub-programme, 'analysis of strategies and modelling', focuses on 'the utilization of Community-wide models to analyze the different energy, environmental and related economic policies.' (A, pp.5-6).

Under the JOULE II programme, other models such as GEM-E3, E3-ME, WARM, MEGAPESTES, POLES, and SAFIRE have been developed besides PRIMES (A, p.7).

Compared to previous models available to the European Commission, PRIMES differs in the following points.

(1) Formation of energy prices is explicitly considered. It calculates price equilibria in each energy market.
(2) Market clearing mechanisms are the main explanatory force in the model, making it suitable for providing realistic trends towards the liberalization of European energy markets.
(3) The behaviour of individual agents is recognized, removing the central planning paradigm.
(4) Details about technologies are contained in both supply and demand (A, pp.11-12).

The decision variables (fuel switching, cogeneration, exploitation of renewable energy, use of recycled material and scrap, consumption of by-products, and heat recovery) represent a number of options in satisfying demand while minimizing costs. 'The optimisation process is also constrained by upper limits on pollutant emissions, but measures can be taken to alleviate this constraint by either investing in pollution abatement equipment and/or by purchasing pollution permits' (A, p.18).

The prototype model focuses on: (1) global warming and (2) acid deposition and ambient quality. For this purpose, energy-related emissions of CO_2, NO_x, and SO_2 are investigated (A, p.67).

The following policy instruments are considered in the prototype.

(1) A minimum abatement percentage compared to a reference level.
(2) Prescriptive technological standards or fuel quality.
(3) Emission or product charges.
(4) Tradable permits (A, p.69).

This is in contrast to the main instruments actually used at EU or national levels which are 'command and control' type, such as emission standards, technology standards for motor vehicles (imposition of catalytic converter), and fuel quality standards (regulation of sulphur content) (A, p.69).

RAINS-Asia

Sources

(A) Resource Management Association. *RESGEN: Regional Energy Scenario Generator for Asia*. Madison, 1996.
(B) Markus Amman and Jhuzer Dhoondia. *RAINS-Asia User's Manual*. Washington, D.C.: International Bank for Reconstruction and Development, 1994.

Time span: 2000-2010-2020.
Regions: 23 plus sea lanes (Bangladesh, Bhutan, Brunei, Cambodia, China, Hong Kong, India, Indonesia, Japan, Korea (North), Korea (South), Laos, Malaysia, Mongolia, Myanmar, Nepal, Pakistan, Philippines, Sea lanes, Singapore, Sri Lanka, Sri Lanka, Taiwan, Thailand, Vietnam). 94 sub-national regions.
Branch classification: 6 (fuel production and conversion, power plants and district heating, household and other consumers, transport, industry, non-energy use) with sub-divisions by process.
Energy sources: 17 (brown coal/lignite (high grade), brown coal/lignite (low grade), hard coal (high grade), hard coal (medium), hard coal (low quality), derived coal, other solid (low S), other solid (high S), heavy fuel oil, medium distillates (diesel, light fuel oil), light fraction (gasoline, kerosene, naphtha, LPG), gas, renewable (solar, wind, small hydro), hydro, nuclear, electricity, heat).
Scenarios: Default base case energy scenario is based on country projections of energy consumption as much as possible (A, p.47). Low energy scenario assumes the overall energy intensity improvement by 26%.

Table 12-20. RAINS-Asia energy consumption projection by country (unit: petajoules)

	1990	2000	2000	2010	2010	2020	2020
Country		Base	Low	Base	Low	Base	Low
Bangladesh	805	1062	1019	1554	1427	2363	1965
Bhutan	22	31	30	41	40	58	56
Brunei	62	130	122	171	149	229	178
Cambodia	101	183	161	354	258	757	454
China	30255	50527	433438	73518	59132	101016	75348
Hong Kong	422	662	619	912	762	1211	980
India	15812	22908	20192	36932	29764	63622	48368
Indonesia	3739	6247	5879	9250	7769	13684	10139
Japan	17866	22606	20188	25679	20914	28809	21536
North Korea	1764	3186	2826	4980	3923	7910	5455
South Korea	3594	6264	5814	9497	7793	13449	9744
Laos	38	56	55	82	79	121	108
Malaysia	1060	1953	1928	3457	3186	6031	4686
Mongolia	151	193	178	260	212	361	266
Myanmar	309	426	421	577	567	737	719
Nepal	270	357	336	467	425	615	514
Pakistan	1436	2531	2359	4353	3910	7361	6391
Philippines	585	1082	948	1921	1820	3113	2401
Sea Lanes	163	208	208	266	266	343	343
Singapore	342	645	641	1130	1061	1731	1522
Sri Lanka	190	399	285	502	372	632	432
Taiwan	2094	3334	2910	4796	4241	6843	5771
Thailand	1688	3292	2903	5693	4482	9326	6697
Vietnam	772	1186	1176	2118	1926	3801	3358
Total	83594	129468	114669	188528	154476	274121	207431

Note: Base refers to base case scenario, and Low to efficiency/low energy scenario.

Source: Resource Management Association (1996).

Technology: Energy intensities, fuels and power plant specifications, and locations.

Energy intensity improvement: For each of the 23 countries, base case (business-as-usual) and energy efficiency scenarios are provided (A, p.32). The efficiency improvements through 2020 are based on the experiences of industrialized countries during the period 1973-1983 as they responded to the sharp rise in energy prices (A, p.48).

Lifestyle: No.

Policy tools: Technology-driven cost-curves, free and competitive market for the exchange of emission control technology.

Economic linkage: No feedback. Country GDP and sub-national GDP needed (A, p.34). GDP is disaggregated into end-use sectors (industrial, commercial, agriculture, residential, non-energy, transportation) (A, p.35).

Environment linkage: Energy, sulphur.

Modelling logic: Based on socioeconomic assumptions (population and economic activity), energy and sulphur deposition details are generated (A, p.31).

Summary

The Regional Air Pollution INformation and Simulation (RAINS) model was developed as a tool for the integrated assessment of alternative strategies to reduce acid deposition in Europe (B, p.3) The RAINS-Asia, developed under collaboration of the World Bank, the Asian Development Bank, and the IIASA, contains technical and economic details of about 250 (B, p.4) or 355 (A, p.7) of the largest point sources in Asia. The model consists of three modules (energy-emissions (ENEM), acid deposition (ATMOS), and ecosystems impact (IMPACT) (B, p.8).

Although the model as a whole is designed to examine sulphur deposition, the energy-emissions module has general applicability to energy projection models.

'Regional and national potentials for emission control and the associated costs are estimated on the basis of detailed data on the most commonly used emission control technologies. The cost evaluation is based on the international operating experiences of pollution-control equipment, by extrapolating them to the country-specific situation. ... Some important country-specific factors that have a strong impact on abatement costs include the sulfur content of fuels, plant capacity utilization regime, and boiler sizes.' 'A free and competitive market for the exchange of emission control technology is assumed' (B, p.5). 'The RAINS model also computes national cost curves that rank the abatement measures by their cost-effectiveness. Due to country-specific factors, such as energy-use patterns and technical infrastructure, national cost curves show significant differences among countries' (B, p.5).

Sulphur deposition is calculated on a 1 degree by 1 degree grid (B, p.6).

RAINS-Asia is accompanied by a regional scenario-developing database tool called Regional Energy Scenario GENerator (RESGEN) for Asia.

In the base case scenario the average energy consumption per capita increases by a factor of 2.25 by 2020, or about 50% of the current average for Europe and 20% of the USA. The energy intensity of the Asian economy as a whole decreases 3.5%. Under the low energy scenario the overall energy intensity improves by about 26%, to a level close to the industrialized countries in the 1980s (A, p.50).

RICE: Regional integrated model of climate and the economy

Sources

(A) William Nordhaus and Zili Yang. 'A Regional Dynamic General Equilibrium Model of Alternative Climate-Change Strategies.' *The American Economic Review*, September 1996.
(B) William Nordhaus. *Managing the Global Commons: The Economics of the Greenhouse Effects*. Cambridge, MA: MIT Press, 1994.

Time span: 1990-2100-2200. Solutions are provided with 10-year intervals up to 2100.

Table 12-21. Major input parameters for the RICE Model

	Cost Intercept	Climate Damage Intercept	CO_2 Emissions, 1990		Population	Per Capita Output	CO_2 Ratio, 2100
			Land-use	Industrial			
Region	(a)	(b)	(c)	(d)	(e)	(f)	(g)
United States	0.07	0.01102	0.010	1.360	0.294	68.8	0.119
Japan	0.05	0.01174	0.000	0.292	0.125	89.1	0.063
China	0.15	0.01523	0.136	0.669	1.656	9.9	0.512
European Union	0.05	0.01174	0.100	0.872	0.427	63.0	0.074
Former Soviet Union	0.15	0.00857	0.000	1.066	0.366	18.9	0.322
Rest of World	0.10	0.02093	1.730	1.700	6.738	18.1	1.185

a The intercept of cost function equals the fraction of annual output required to reduce net CO_2 emissions to 0.
b The intercept of climate-damage function equals the reduction in annual net output from an increase of 2.5 degres C in global mean temperature.
c Emissions are measured in billion of tons of carbon per year. Land-use emissions are primarily from deforestation.
d Emissions are measured in billion of tons of carbon per year. Industrial uses primarily from burning fossil fuels.
e Population is in billions of people.
f Gross domestic product (GDP) is measured at 1990 market exchange rates in thousands of 1990 US dollars.
g The ratio is industrial CO_2 emission to GDP (tons of carbon per $1000 of output in 1990 US dollars).

Source: Nordhaus and Yang (1996).

Regions: 10 (the United States, Japan, China, the European Union, Former Soviet Union (FSU), India, Brazil and Indonesia, 11 large countries, 38 medium-sized countries, and 137 small countries). Each of the first 5 regions are treated as single decision maker. Regions 6 through 10 are sometimes treated as the rest of the world (A, p.743).

Branch classification: No.

Energy sources: No breakdown. Expressed in terms of CO_2 emissions.

Scenarios: Three strategies are compared. They are: (1) a market approach, (2) a global cooperation approach, and (3) a noncooperative or nationalistic approach (A, p.743, p.749, p.761).

Technology: Level of technology is exogenous. Emissions-reduction cost function, marginal atmospheric retention ratio of CO_2 emissions, elasticity of output with respect to capital, etc. are parameters (A, p.763). Hicks-neutral technological change, approximately at the observed rates for 1960-1990 generates each region's income. After 1990, growth rates are assumed to decline exponentially (A, p.741).

Energy intensity improvement: Expressed in terms of CO_2 emissions reduction, which is assumed.

Lifestyle: No.

Policy tools: Control rates for CO_2 emissions, carbon taxes (A, p.751). Policy variables include gross investment and rate of emissions reduction (A, p.763).

Economic linkage: Endogenous variables include consumption (per-capita and total), damage from greenhouse warming, atmospheric temperature relative to pre-industrial levels, among other things, are endogenous (A, p.762). Population is exogenous (A, p.763), taken from the United Nations projections (A, p.744). *Environment linkage*: The economic choices faced by countries are: (1) to consume goods and services, (2) to invest in productive capital, and (3) to slow climate change through reducing CO_2 emissions (A, p.743).

Modelling logic: Optimization-type dynamic general-equilibrium model (A, p.761). The model solves for the intertemporal general equilibrium. For algorithm for finding the cooperative solution, see A, pp.745-748, and for noncooperative solution, see A, p.748.

Summary

'Most existing models of global climate change take the vantage point of the Global Commoner engaged in determining how nations *should* design sensible strategies. The RICE model takes a positive point of view by asking how nations would in practice choose climate change policies in light of economic trade-offs and national self-interest' (A, p.742). The model 'emphasizes the implications of the fact that while climate change is a global externality, the decision makers are national and relatively small' (A, p.762). Countries may not undertake costly efforts today for three reasons: (1) the benefits are 'so conjectural'; (2) they occur so far in the future; and (3) no individual countries can have significant impact upon the pace of global warming (A, p.762).

Three alternative solution concepts are developed. (1) Market approach: this strategy assumes that there is no correction for the climate change. Consequently, there is no abatement of CO_2 emissions. (2) Cooperative approach: countries undertake policies that reduce greenhouse-gas emissions efficiently. (3) Noncooperative approach: this case corresponds to a situation where each country sets its CO_2 emissions controls to maximize its own economic welfare assuming that other countries' strategies are invariant to a country's policies (A, p.749).

'There are major gains to taking an efficient cooperative approach to coping with global warming as opposed to the noncooperative approach. The net economic gains from an efficient policy has a discounted value of $433 billion relative to the market scenario, while noncooperative policy has a gain of only $43 billion' (A, p.762). In the efficient solution, carbon taxes are identical in all regions, which is calculated to be about $6 per ton of carbon in the year 2000. The control rates will differ, however, because of different costs of reducing CO_2 emissions. ... These results indicate that there will be substantial inefficiencies in any policy (such as that currently in force under the Framework Convention) that equalizes emissions control rates across countries or does not allow trading of emissions permits' (A, p.761).

DICE (the Dynamic Integrated model of Climate and the Economy) is an earlier model whereby policies are taken so as to maximize the generalized level of consumption now and in the future (Nordshaus, 1994). Output is explained by Cobb-Douglas production function, where the elasticity of output with respect to capital is assumed to be 0.25 (B, p.12). Total factor productivity term starts from an empirically observed value during 1965-1975, but is assumed to decline by half every six decades in the years to come (B, pp.13-14). The model specifies in sequence the linkage between the GHG emissions to economic activity, accumulation of GHG in the atmosphere, the relationship accumulation of GHGs and increased radiative forcing, radiative forcing and climate change, the impact of climate change on human and natural systems, and the cost of reducing GHG emissions (B, pp.15-19). The rate of reduction of emissions relative to uncontrolled emissions is determined by the optimization (B, p.15). Series of coefficients are employed representing the impact of CO_2 doubling on output by sector, including agriculture, energy, coastal activities, and other (B, p.51). No change in relative prices is assumed.

WEPS: World energy projection system by EIA

Sources

(A) Energy Information Administration, US Department of Energy. *World Energy Projection System Model Documentation.* DOE/EIA-M050(97). Washington, DC, 1997.

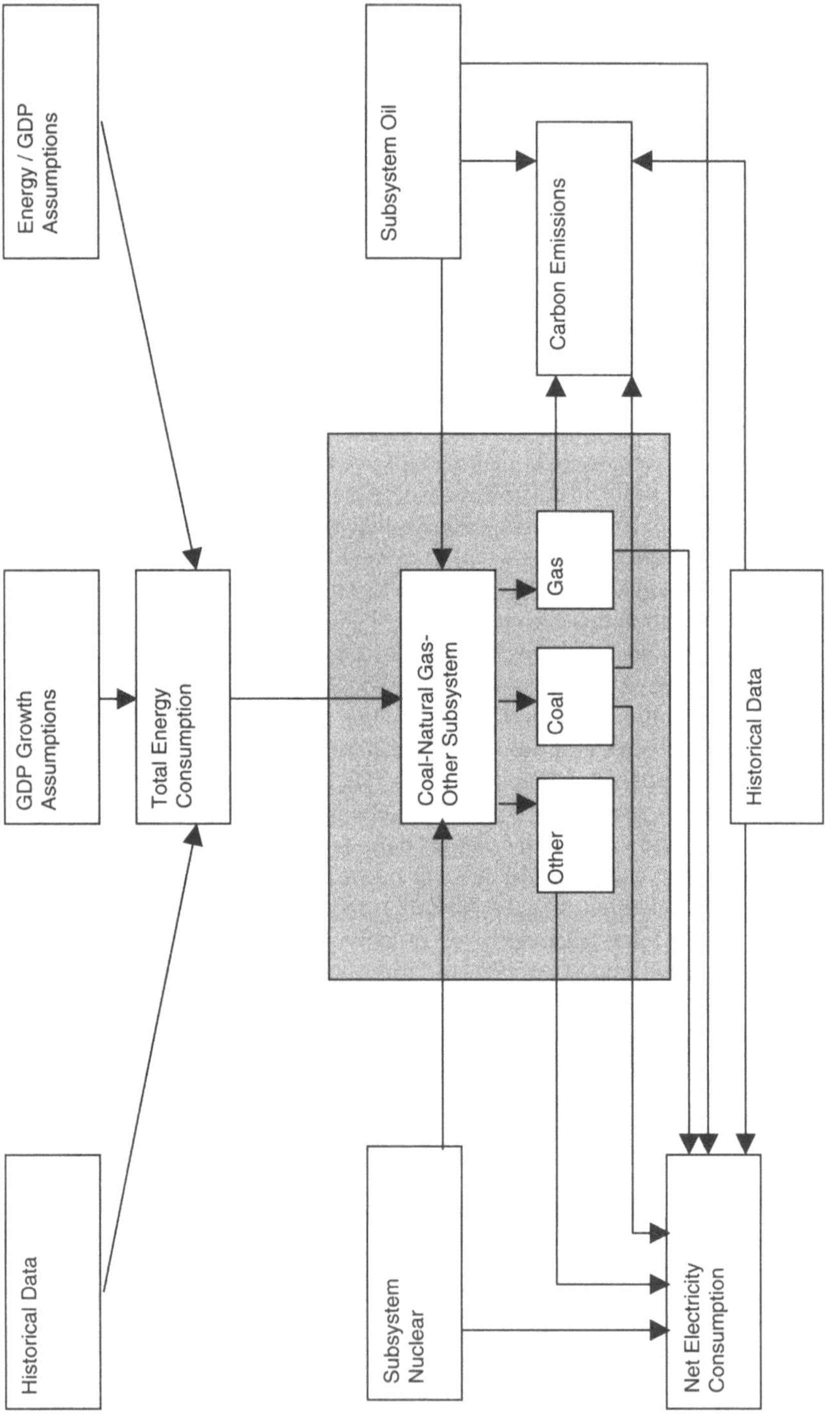

Figure 12-9. World energy consumption system interaction in WEPS.
Source: Energy Information Administration, US Department of Energy (1997a).

(B) Energy Information Administration, US Department of Energy. *International Energy Outlook 1999*. DOE/EIA-0484(99). Washington, DC, 1999.

Time span: 1990-2020 with 5-year intervals (A, p.26).

Regions: 7 individual countries (United States, Canada, Mexico, Japan, China, India, and Brazil) and 9 regions including North America (Canada, Mexico, and the United States), Western Europe, Pacific (Japan, Australia, New Zealand, etc.), developing Asia (China, India, and other Asia), Middle East, Africa, and Central and South America (Brazil and other Latin America), Eastern Europe, and the former Soviet Union (A, p.1, p.4).

Branch classification: No. Embedded in sub-models.

Energy sources: Oil, natural gas, coal, nuclear, hydroelectricity and other renewable sources, electricity (A, p.26).

Scenarios: High and low economic growth, high and low electricity consumption (A, p.27).

Technology: Embedded in assumptions on energy intensity, conversion factors, etc. (A, p.15, p.26).

Energy intensity improvement: By independent models and assumptions (A, p.26; B. p.14, pp.131-134).

Lifestyle: No.

Policy tools: No.

Economic linkage: Future elasticity assumptions (A, p.15).

Environment linkage: Carbon emissions (A, p.26).

Modelling logic: Scenarios incorporating high and low economic growth.

Table 12-22. Projected 2010 emission and required reduction of Annex I parties

Region and Country	1990 Emission (MMTC)	2010 Reference Case (MMTC)	2010 Kyoto Target (MMTC)	Change from Reference Case, 2010 (%)
Annex I Industrialized				
United States	1384	1790	1252	−30
Canada	126	162	118	−27
Western Europe	936	1021	862	−16
Japan	274	322	258	−20
Australia	90	133	97	−14
Total Annex I Industrialized	2772	3408	2586	−24
Transitional (EE/FSU)				
Former Soviet Union	991	666	990	49
Eastern Europe	299	270	320	18
Total EE/FSU	1290	935	1309	40
Total	4062	4344	3895	−10

Source: Energy Information Administration (1999).

Summary

The WEPS provides projection of total energy consumption by world regions based on an assumed relationship to gross domestic product. 'Econometric techniques are not used to determine changes in relationships over the projection period' (A, p.28). The model is used to provide an accounting framework that produces consistent world energy consumption balances from diverse sources of information. The data are in terms of physical units (such as tons and kilowatt hours).

WEPS is described as 'an integrated set of personal-computer-based spreadsheets containing data compilations, assumption specifications, descriptive analysis procedures, and projection models'. 'WEPS provides projections of total world primary energy consumption, as well as projections of energy consumption by permit type (oil, natural gas, coal, nuclear, and hydroelectric and other renewable sources), and projections of net electricity consumption. Carbon emissions resulting from fossil fuel use are derived from the energy consumption projections' (A, p.1, p.26). The files are updated annually.

The system is composed of total energy consumption subsystem, high and low economic growth subsystem, oil subsystem, nuclear energy subsystem, coal-natural gas-other allocation system, carbon emissions subsystem, high and low carbon emissions subsystem, electricity consumption subsystem, and high and low electricity subsystem. Oil consumption projections are from the International Energy Module of the National Energy Modeling System (NEMS). Nuclear generation projections are supplied by the International Nuclear Model. International coal trade is accepted from NEMS Coal Export Sub module (A, pp.26-27).

IEA's *Energy Balances of OECD Countries* and *Energy Balances and Statistics of Non-OECD Countries* are among the input data. WEFA's *World Economic Outlook* is employed as the economic projections. Projections produced by the WEPS are published in EIA's annual report *International Energy Outlook*. See B. The results are reported in Table 6-2 of this volume.

About the energy prospect in relation to the Kyoto Protocol, the EIA's annual report writes as follows. 'Additional uncertainty arises from the commitments being made by developed countries under the Kyoto Protocol of the Framework Convention on Climate Change. If those commitments are realized, energy demand could actually be reduced over the next decade. Such a development could lower total world energy consumption in 2010 by more than 10 percent relative to the reference case projection presented here. ... The protocol identifies stringent targets for reduced greenhouse gas emissions in developed countries; however, neither policies nor technologies to achieve the targets have been identified for implementation, nor have countries ratified the agreement. Thus, the Kyoto Protocol is viewed in this report as a factor heightening uncertainties and the need for collateral analysis of the IEO99 reference case rather than one that *per se* alters it (B, p.8).

World energy outlook by the international energy agency

Source

(A) International Energy Agency. *World Energy Outlook*. Paris: OECD. 1998.

Time span: 1995-2010-2020.
Regions: 10 (OECD: North America, Europe, Pacific; Non-OECD: Transition Economies, Africa, China, East Asia, South Asia, Latin America, Middle East) (A, p.26, p.147).
Branch classification: Electrical services (final consumers), mobility (transport), stationary services (heating in buildings and industrial processes), power generation (A, p.25).
Energy sources: 8 (A, p.26).
Scenarios: Continued trends and policies.
Technology: Detailed description for individual energy sources (A, pp.63-174).
Energy intensity improvement: Embedded in analyses.
Lifestyle: Embedded in 'electrical services' and 'mobility'.
Policy tools: Scale of regulation and 'carbon value' that need to be included into fossil fuel prices in order to meet Kyoto commitment (A, p.26).
Economic linkage: Energy prices (A, p.26).
Environment linkage: Limitations imposed by finite recoverable reserves (A, p.26).
Modelling logic: Scenarios on energy supply (including reserves) and demand. There is an underlying world energy model (A, p.3) Fuel prices not formally modelled (A, p.26).

Summary

'Our aim is not to foretell the future. ... Instead, we see this publication as an opportunity to identify and discuss the main energy issues that could arise over this period. ... Because the *Outlook* now projects twenty-five years ahead, we have paid particular attention to the relationship between cumulative oil production to date and estimates of oil reserves. Our analysis of the current evidence suggests that world oil production from conventional sources could peak during the period 2010 to 2020. There will be no shortage of liquid fuels if this happens because reserves of unconventional oil are large' (A, p.3).

The business-as-usual case assumes energy policies existing before the Kyoto Conference will be continued and no account is taken for many flexibility mechanisms in the Protocol. 'This can be described as an illustration of how energy demand, supply, and prices are likely to develop if recent trends and current policies continue' (A, p.24).

Without such new policies the *Outlook* projects that, between 1995 and 2020, world energy demand will grow by 65% and CO_2 emission by 70%. The underlying assumption is the growth of world economy at 3.1% per annum, which is

Table 12-23. World oil and gas supply and prices, mid-1990s-2020

A. Oil Supply 1996-2020

Conventional Oil Reserves of 2.3 trillion barrels	(Million barrels per day)		
	1996	2010	2020
Total Demand for Liquid Fuels	72.0	94.8	111.5
Total Natural Gas Liquids, Processing Gains and Identified Unconventional Oil	9.3	15.9	20.1
Conventional Crude Oil			
Middle East OPEC	17.2	40.9	45.2
World excluding Middle East OPEC	45.5	38.0	27.0
Total Crude Oil	**62.7**	**79.0**	**72.2**
World Liquids Supply excluding Unidentified Unconventional Oil	72.0	94.8	92.3
Balancing Item – Unidentified Unconventional Oil	0.0	0.0	19.1

B. Natural Gas Production and Net Imports	(Mtoe)		
Indigenous Production	1995	2010	2020
OECD North America (including Mexico)	592	759	764
OECD Europe	199	276	238
OECD Pacific	31	77	68
Transition Economies	585	809	1116
China	17	57	81
Rest of World (excluding Mexico)	395	750	1208
World	**1818**	**2727**	**3474**
Net Imports	1995	2010	2020
OECD North America (including Mexico)	−2	−2	−2
OECD Europe	104	230	387
OECD Pacific	42	42	64
Transition Economies	−74	−162	−281
China	0	0	0
Rest of World (excluding Mexico)	−76	−114	−174

C. World Fossil Fuel Prices, BAU Assumptions	1995	1998-2010	2015-2020
IEA Crude Oil Import Price ($ 1990/bbl)	15	17	25
OECD Steam Coal Import Price ($ 1990/ton)	40.3	42	46
US Natural Gas Wellhead Price ($ 1990/thousand cubic ft.)	1.35	1.7*	3.5
Natural Gas Import Price into Europe ($ 1990/toe)	89.9	103	150
Japan LNG Import Price ($ 1990/toe)	125.6	141	210

Note: * refers to 1998-2005.

Source: International Energy Agency (1998b).

close to the actual record since 1971 (1990 US$, purchasing power parity). It is suggested that 'two-thirds of the increase in energy demand over the period 1995-2020 comes from China and other developing countries' (A, p.19).

The dependence on oil supplies from the Middle East will increase and oil prices could rise during the period. The *Outlook* says that 'the probability of supply disruptions and price shocks could rise'. As 2020 approaches, it is predicted that unconventional sources (such as shale oil, tar sand, and conversion from coal, biomass, and gas) will start to play their roles (A, p.19).

The report identifies several sources of uncertainties including the following.

(1) Economic output and structure, population growth.
(2) Technical change and capital stock turnover.
(3) Human attitudes and behaviour.
(4) Fossil fuel supplies and extraction costs.
(5) Energy market developments.
(6) Energy subsidies.
(7) Changing environmental objectives and policies.

Regarding energy subsidies, the report writes: 'In many developing countries, energy prices are set at levels well below the full cost of supply. ... Over time, prices are expected to rise. ... This change permit the privatization of state-owned facilities and encourage foreign investment in energy supplies.' On environmental aspects, the report adds that 'Perhaps the greatest unceertainty ... is that surrounding the policy choices that governments will make to meet the commitments they entered into in Kyoto' (A, p.24).

As for the Kyoto Protocol, the *Outlook* says that 'it is not possible to prepare a projection that meets the Kyoto targets because the countries involved have not yet announced the policies they intend to use'. Two sets of analyses are presented in this regard: (1) the scale of regulation that would be required to meet the Kyoto commitments; and (2) the carbon value that needs to be built into fossil fuel prices to meet the Kyoto targets (A, p.26).

World model for the future of the environment

Source

(A) Faye Duchin and Glenn-Marie Lange. *The Future of the Environment. Ecological Economics and Technological Change.* New York and Oxford: Oxford University Press, 1994.

Time span: 2020 at decade intervals.
Regions: 16 (1. High-income North America; 2. Newly industrializing Latin America; 3. Low-income Latin America; 4. High-income Western Europe; 5. Medium-income Western Europe; 6. Eastern Europe; 7. Former Soviet Union; 8. Centrally Planned Asia; 9. Japan; 10. Newly industrializing Asia;

11. Low-income Asia; 12. Major oil producers; 13. Northern Africa and other Middle East; 14. Sub-Sahara Africa; 15. Southern Africa; 16. Oceania (A, p. 6, pp.199-201).

Branch classification: 44 industrial (A, p.6, pp.203-204), 39 trade (A, pp.205-206).

Energy sources: Fossil fuels (oil, gas, coal) (A, p.32).

Scenarios: *Our Common Future* (OCF) scenario constructed from scattered information contained in the Brundland Report, vs. reference scenario (no technological change after 1990), alternative technological scenarios.

Technology: Details in Table 12-24.

Energy efficiency improvements: Details in Table 12-24.

Lifestyle: Reflected in the percentage of the population in living in cities (A, p.192).

Policy tools: Alternative technologies, rather than carbon tax (A, p.7).

Economic linkage: GDP and population are exogenous. Imports are from a single world trade pool rather than being specified by bilateral flows. A balance of payments deficit is assumed to be covered by an inflow of short-term capital. Price vector is used to sum up GDP. Relative price change of petroleum can be captured. Net inflow of economic aid, long-term capital, and foreign earnings are specified (A, p.193).

Environment linkage: 3 types of emissions (carbon dioxide, sulphur oxides, nitrogen oxides) based on sulphur contents, types of production processes, etc. (A, pp.51-71).

Modelling logic: Static input-output model, extended by the explicit representation of investment and international exchanges (A, p.191).

Summary

On the methodological issue the authors write: ‘A conventional textbook might describe the economist’s objective as identifying those alternatives, and in the process determining the corresponding “right” prices, that ensure optimal allocation of resources and maximum utility for consumers in a global economy in equilibrium. This theoretical position, and the actual models that are based on it (general equilibrium models), are appealing for their comprehensiveness and simplicity of objective. However, any theory or model of an economy has its shortcomings, and some of the major ones of this perspective are that a real economy is never in equilibrium; that social welfare is not the same as the “utility” of consumers; and that most environmental problems cannot reasonably be associated with prices that would make it possible to simply add them to calculations of utility because any such prices would be arbitrary. The challenge we have tried to face is to go beyond criticism of mainstream economic models, and beyond qualitative description alone, to build a formal framework that is capable of representing the activities of a real economy and to arrive at quantitative results which avoid these shortcomings’ (A, p.6).

Table 12-24. Technology scenarios in 2020 relative to 1990

1. Petroleum price	$44 by 2020
2. GDP and population	
3. Household energy:	
A. In-house use	Small increase in investment requirements
B. Developed	50 percent energy savings
C. Developing	200-400 percent increase in energy per capita (including fuel for transportation); greater increases in electricity per capita
4. Transportation:	
A. Motor vehicles	Changing material inputs, 20 percent increase in fuel efficiency
B. Transportation services	50 percent less fuel per mile
C. Households	50 percent less fuel per mile, 16-450 percent expansion in vehicles per capita in developing countries
5. Electricity:	
A. Fuel mix	Displacement of 15 percent of coal and oil by gas in most regions
	Unchanged share of non-thermal electricity in most regions after 2000
B. Conversion efficiency:	
Developed (including pollution control)	15 percent improvement
	20 percent incease in capital
Eastern Europe, former Soviet Union, and other regions	25 percent improvement
	20-25 percent increase in capital
6. Emissions:	
A. Carbon	Reduction of emissions only through lower energy use
B. Sulfur and nitrogen:	
Developed	Scrubbers provide 90 percent removal after combustion; 10 percent increase in electricity use and 10 percent increase in capital; installation in 100 percent of power plants in Japan, 50 percent in Europe, 20 percent in Eastern Europe, 25 percent in former Soviet Union, 33 percent in North America; installation in 50 percent of industries in Western Europe and Japan; 30 percent reduction of coal sulfur in Eastern Europe and former Soviet Union through coal washing; Sulfur content of petroleum products

Table 12-24. continued

	reduced 33-50 percent; 33-50 percent reductions of nitrogen with 100 percent installation of catalytic converters in motor vehicles
Developing	No use of scrubbers; 35 percent decrease in coal sulfur content in centrally planned Asia and low-income Asia due to increased coal washing with increase in capital requirements; little use of catalytic converters
7. Industrial conservation:	Small increase in investment requirements
Developed	20 percent energy saving on average
Eastern Europe and former Soviet Union	20-50 percent energy savings (depends on sector)
Developing	20-50 percent energy savings (depends on sector); 10-30 percent of fossil-fuel use displaced by electricity
8. Construction:	
A. Maintenance vs. new construction	Share increase of 50 percent
B. Materials	Copper inputs decrease by 33 percent; aluminum inputs increase 15-33 percent; wood inputs unchanged; cement inputs decrease 15-25 percent; finished chemical inputs (mainly paint) increase 35 percent
Developed	Steel inputs decrease 25 percent
Developing	Steel inputs decrease 50 percent, more highly processed inputs
9. Metal-fabricating sectors	Virgin material inputs decrease 15 percent
10. Paper	Recycled content reaches 60 percent in developed regions
11. Electronics:	
A. Investment	
Most regions	Value share in total investment goods rises to 56 percent
Poorest regions	Share remains at 1990 value
B. Production	
Developed and newly industrializing	65 percent decrease in inputs
Other regions	45 percent decrease in inputs
12. Capital coefficients	
Developing regions	Increase in various sectors

Source: Duchin and Glenn-Lange (1994).

'The World Model does not attempt to determine a unique and optimal path to sustainable development, but simply to evaluate the implications of a set of technical and organizational choices that are made outside the model (by the Brundtland Report, in this case). 'The first stage of analysis needs to focus on the implications of the technological options for achieving specific targets. If the objectives still appear desirable, the economic incentive needed to make the shift voluntary can be explored at a later stage' (A, p.7). What the authors have in mind is introduction of carbon tax, for example.

Thus, the authors provide details on technological potentials of main activities of industry, including electric power generation, industrial energy conservation, processing and fabrication of metals, construction, cement, pulp and paper, chemicals, household energy conservation, and motor vehicles.

'The *Our Common Future* (OCF) (World Commission on Environment and Development, 1987) scenario reduces emissions of the three pollutants that are tracked below what they would be under the Reference scenario. But the problem is that these emissions still increase substantially between 1990 and 2020; in particular, annual carbon emissions rise by 60 percent' (A, p.8).

'In addition, the locus of pollution shifts decisively from the industrial countries, where historically most of it has originated, to the developing countries, where most of the world's people live and where most of the future increase in population will take place' (A, p.8).

If the objectives of sustainable development spelled out in the Brundland Report are to be realized, authors argue, the need for motorized transport has to be reduced through new design of communities, and reliance on renewable sources for energy and materials would have to be promoted.

WorldScan

Source

(A) Johannes Bollen, Ton Manders, and Hans Timmer. 'The IPCC Stabilisation Scenarios.' Paper presented at the Joint Meeting organized by the International Energy Agency, Energy Modeling Forum, and the International Energy Workshop (IIASA), 16-18 June 1999; Paris, 1999.

Time span: 1995-2100.
Regions: 12 (United States, Western Europe, Japan, Rest of the OECD, Eastern Europe, Former Soviet Union, Middle East and North Africa, Sub-Saharan Africa, Latin America, China, South-East Asia, South Asia and Rest).
Branch classification: 11 in WorldScan core (agriculture, services, trade and transport, electricity, intermediate goods, consumer goods, capital goods, oil, natural gas, coal, other raw materials (A, p.4, p.5), with further disaggregation of 'raw materials' and 'utility' in carbon extension of the model.
Energy sources: Coal, gas, oil, electricity (A, p.4).

Scenarios: Scenario A1 is the case with emphasis on global technological convergence and strongly intensified trade linkages (A, p.10).
Technology: Fixed technology trend. Overall technology differs across countries. Catch up is possible by adopting foreign technology (A, p.5).
Energy intensity improvement: Sharp decarbonization after 2050.
Lifestyle: Scenario A1 assumes income maximization (A, p.10). Dualistic economy is assumed (A, p.5).
Policy tools: (1) immediate action (global permit market by 2030) vs (2) delayed response (by 2050) (A, pp.10-11).
Economic linkage: Permit price, which drops sharper in the early action scenario compared to the delayed response (A, p.18). Technology transfers related to the relocation of energy-intensive industries.
Environment linkage: Timing of new entrants into permit trading altering the cost-benefit of participating countries (A, pp.13-15).
Modelling logic: Applied general equilibrium model (A, p.1).

Summary

WorldScan is based on neoclassical theories of growth and international trade. The emphasis is placed on long-term analysis (A, p.4). The model is benchmarked to the GTAP database.

In addition to neoclassical specification, the model has the following extensions:

(1) An Armington trade specification, explaining two-way trade and allowing market power to determine trade patterns in the medium run, while allowing Hecksher-Ohlin mechanisms in the long run;
(2) Imperfect financial capital mobility;
(3) Consumption patterns depending upon per-capita income, and developing towards a universal pattern;
(4) A Lewis-type low-productivity sector in developing regions, from which the high-productivity sector can draw labour, enabling high growth for a long period (A, p.5).

As for the policy experiments, the authors write: 'We show how a specific burden sharing rule yields different regional target profiles, if the global policy either calls for an early action or a delayed response. It is shown that by applying system of tradable permits to achieve a global constraint, new partners can be attracted to join the abatement coalition. But it is also shown that some countries within the abatement coalition might be against entry of any new partners' (A, p.1). Welfare triggers are used to force entrance of new members to the Annex I group of countries. The welfare target is set at $10,000 per capita (A, p.10).

The overall conclusion by the authors is that 'stabilization of greenhouse gases in the atmosphere at 550 ppmv can be achieved at moderate costs'. 'Model simulations show that there might be conflicting interests due to income transfers

Table 12-25. Economy–energy–environment in the 21st century: simulation/scenario results from main projects

Model / Scenario Names	Sub Title	References	Item	Unit	Year							
					1990	1995	2000	2005	2010	2020	2050	2100
6 model results on C emission (AEEI 1% Case)		Dean & Hoeller (1992)										
CRTM	Carbon Rights Trade Model	Rutherford (1992)	Carbon	billion tons	6.003		6.931		8.031	9.327	11.337	35.863
ERM(1)	Edmonds-Reilly Model	Barns *et al.* (1992)			5.767			6.709		8.180	11.838	22.579 (2095 figure)
GREEN(1)	OECD Model	Oliveira-Martins (1992)			5.815		7.071	7.704	8.705	10.806	18.998	
IEA	IEA Model	Vouyoukas (1992)			5.919		7.316	7.932				
MR(2)	Global 2100	Manne (1992)			6.003		6.748		7.581	8.681	11.356	26.039
WW	Whalley-Wigle Model	Whalley & Rigle (1992)			Average 1990-2100 25.2							65.5
APEC Energy Outlook	APEC Energy Demand	APERC (1998)	Final Energy Consumption	Mtoe								
	APEC Total		B98 Baseline			3130.2	3486.8	3901.2	4368.5			
			Env.Friendly Scenario				3412.8	3699.2	4039.4			
	United States		B98 Baseline			1507.3	1651.0	1766.3	1880.9			
			Env.Friendly Scenario				1632.8	1700.9	1767.3			
	Other Americas		B98 Baseline			284.5	316.3	368.6	435.4			
			Env.Friendly Scenario				307.9	345.9	392.2			
	China		B98 Baseline			623.4	750.9	882.1	1014.3			
			Env.Friendly Scenario				716.6	808.9	922.3			
	Other East Asia		B98 Baseline			505.0	538.3	601.3	675.1			
			Env.Friendly Scenario				530.1	577.6	629.6			
	Oceania		B98 Baseline			83.9	95.4	110.8	130.1			
			Env.Friendly Scenario				93.5	104.3	116.8			
	Southeast Asia		B98 Baseline			126.1	134.8	172.0	232.8			
			Env.Friendly Scenario				131.8	161.7	211.3			

Table 12-25. continued

Model / Scenario Names	Sub Title	References	Item	Unit	Year 1990	1995	2000	2005	2010	2020	2050	2100
			Env.Friendly Scenario				656.5	654.2	647.0			
	United States		B98 Baseline			439.4	531.3	629.0	708.4			
			Env.Friendly Scenario				490.4	498.7	482.5			
	Other Americas		B98 Baseline			−168.6	−185.3	−158.1	−108.6			
			Env.Friendly Scenario				−198.0	−193.5	−179.9			
	China		B98 Baseline			−12.0	49.4	99.5	121.4			
			Env.Friendly Scenario				−31.3	−46.5	−68.0			
	Other East Asia		B98 Baseline			580.4	634.3	708.4	788.3			
			Env.Friendly Scenario				609.4	645.3	676.0			
	Oceania		B98 Baseline			−94.7	−120.2	−139.8	−148.4			
			Env.Friendly Scenario				−126.4	−154.2	−181.7			
	Southeast Asia		B98 Baseline			−53.7	−79.9	−74.4	−33.2			
			Env.Friendly Scenario				−87.6	−95.5	−81.7			
ECN Study CO_2 Emissions	Western Annex I	Sijm *et al.* (2000)	CO_2	Mt	9920				11702			
	CEE/FSU Annex I				4124				3962			
	Total Annex I				14044				15664			
	Latin America				1021				1284			
	Africa				767				1876			
	Asia				4051				6337			
	FSU non Annex I				341				442			
	Middle East				838				1317			
	Oceania				4				4			
	Total non-Annex I				7022				11260			
	World				21065				26924			

ECN Study		Sijm *et al.* (2000)	Greenhouse gases	Mt GHG		
Emissions Trends	Netherlands		Emissions		225	258
			Kyoto Target			212
	Germany		Emissions		1203	976
			Kyoto Target			951
	Italy		Emissions		511	592
			Kyoto Target			478
	France		Emissions		501	515
			Kyoto Target			501
	United Kingdom		Emissions		752	699
			Kyoto Target			658
	Japan		Emissions		1333	1587
			Kyoto Target			1253
	USA		Emissions		6187	7751
			Kyoto Target			5754
	Australia		Emissions		423	496
			Kyoto Target			456
	Canada		Emissions		340	402
			Kyoto Target			320
				Mt CO2		
	Western Annex I				9920	11702
	CEE/FSU Annex I				4124	3962
	Total Annex I				14044	15664
	Latin America				1021	1284
	Africa				767	1876
	Asia				4051	6337
	FSU non Annex I				341	442
	Middle East				838	1317
	Oceania				4	4
	Total non-Annex I				7022	11260
	World				21065	26924

Table 12-25. continued

Model / Scenario Names	Sub Title	References	Item	Unit	Year 1990	1995	2000	2005	2010	2020	2050	2100
G-cubed	Emission Stabilization Taxes	Bagnoli, McKibbin, Wilcoxen (1996)	Emission Stabilization Taxes	US$/t Carbon								
	Scenario 1 (Base Line)											
	USA					2.80	16.80		44.80			
	Japan					10.50	63.00		168.00			
	Australia					3.80	22.80		60.80			
	Other OECD					6.80	40.80		108.80			
	China					1.15	6.90		18.40			
	Developing Countries					2.60	15.60		41.60			
	Eastern Europe & Former USSR					1.15	6.90		18.40			
	Scenario 2 (Energy Scarcity)											
	USA					1.40	8.40		22.40			
	Japan					5.50	33.00		88.00			
	Australia					2.60	15.60		41.60			
	Other OECD					3.70	22.20		59.20			
	China					0.24	1.44		3.84			
	Developing Countries					1.15	6.90		18.40			
	Eastern Europe & Former USSR					0.35	2.10		5.60			
IEA Energy Model	Medium Term Energy Model	Vouyukas (1993)										
	Energy		Reference Case	MTOE	4184.6	4508.1	4776.3	5011.1				
			Stabilization Case	MTOE	4184.6	4251.4	4333.6	4417.2				
	Emissions		Reference Case	Mil. Tons C	3043.0	3274.7	3466.5	3631.9				
			Stabilization Case	Mil. Tons C	3043.0	3019.7	3023.9	3043.2				
IIASA-WEC	Global Energy Scenarios											
	Gross World Product	Nakicenovic *et al.* (1998)		Trillion US$, 1990 prices								
			High Growth		20						100	300
			Middle Course		20						75	200
			Ecologically Driven		20						75	220

Primary Energy Demand	Nakicenovic *et al.* (1998)		Giga tons of oil equivalent				
		High Growth		9		25	45
		Middle Course		9		20	35
		Ecologically Driven		9		14	21
Carbon Emissions, Net	Nakicenovic *et al.* (1998)		Giga tons of Carbon				
		High Growth		6		9 to 15	6 to 20
		Middle Course		6		10	11
		Ecologically Driven		6		5	2
International Energy Outlook	Energy Information Admin., US Dept. of Energy (1997b)	Carbon Emissions	Mil. Tons				
Annex I Industrialized Countries		IOE Reference Case		2772	3408		
		Kyoto Protocol Target			2586		
North America		IOE Reference Case		1472	1952		
		Kyoto Protocol Target			1370		
United States		IOE Reference Case		1346	1790		
		Kyoto Protocol Target			1252		
Canada		IOE Reference Case		126	162		
		Kyoto Protocol Target			118		
Western Europe		IOE Reference Case		936	1021		
		Kyoto Protocol Target			862		
Industrialized Asia		IOE Reference Case		364	435		
		Kyoto Protocol Target			354		
Japan		IOE Reference Case		274	322		
		Kyoto Protocol Target			258		
Australasia		IOE Reference Case		90	113		
		Kyoto Protocol Target			97		

Table 12-25. continued

Model / Scenario Names	Sub Title	References	Item	Unit	Year 1990	1995	2000	2005	2010	2020	2050	2100
	Transitional (EE/FSU)		IOE Reference Case		1290				935			
			Kyoto Protocol Target						1309			
	Former Soviet Union		IOE Reference Case		991				666			
			Kyoto Protocol Target						990			
	Eastern Europe		IOE Reference Case		299				270			
			Kyoto Protocol Target						320			
	Total		IOE Reference Case		4062				4344			
			Kyoto Protocol Target						3895			
IMAGE 2.0	Scenario Results	Alcamo *et al.* (1994)										
	CO_2 Emissions from Industry		Conventional Wisdom	Pg C	6.1						15.2	24.0
			Biofuel Crops								15.2	24.0
			No Biofuels								17.0	29.2
			Ocean Realignment								15.2	24.0
	Atmospheric Concentration		Conventional Wisdom	ppm	358						522	777
			Biofuel Crops								534	821
			No Biofuels								539	857
			Ocean Realignment								563	863
	Average Surface Temperature, Northern Hemisphere		Conventional Wisdom	Centigrade							plus 1.4	2.4
			Biofuel Crops								1.5	2.7
			No Biofuels								1.4	2.4
			Ocean Realignment								0.0	1.2
IPCC Special Report	A1F1 Fossil fuel intensive case	IPCC (2000)	Primary energy	10**18 J/yr	351					669	1431	2073
	A1B Balance across sources				351					771	1347	2226
	A1T Predominantly non-fossil				351					649	1213	2021
	A2 Heterogenous world				351					595	971	1717
	B1 Convergent world				351					606	813	514
	B2 Local solution, diverse tech.				351					566	869	1357

MESSAGE	Scenario A1B, World	IPCC (2000)								
	Population			Million	5262	6117	6888	7617	8704	7056
	Final Energy									
	Non-commercial			EJ	38	28	22	16	5	0
	Solids			EJ	42	58	59	67	48	15
	Liquids			EJ	111	125	160	204	357	254
	Gas			EJ	41	48	67	85	164	96
	Electricity			EJ	35	47	70	107	311	859
	Others			EJ	8	10	20	38	122	525
	Total				275	316	398	517	1005	1749
	Primary Energy			EJ						
	Coal			EJ	91	105	120	157	210	192
	Oil			EJ	128	155	172	198	281	138
	Gas			EJ	71	85	128	178	378	350
	Nuclear			EJ	7	9	12	19	84	358
	Biomass			EJ	46	47	63	87	204	475
	Other Renewables			EJ	8	13	29	52	240	1172
	Total			EJ	352	415	524	689	1397	2683
	Anthropogenic Emissions									
	Fossil Fuel CO_2			GtC	5.99	6.90	8.31	10.56	16.47	13.83
	Other CO_2 (land-use changes)			GtC	1.11	1.07	1.04	0.26	−0.02	0.00
	Total CO_2			GtC	7.1	7.97	9.36	10.81	16.45	13.83
	Cumulative Resources Use									
	Coal			ZJ	0.0	0.9	2.0	3.2	9.0	19.0
	Oil			ZJ	0.0	1.4	3.0	4.7	11.4	25.2
	Gas			ZJ	0.0	0.7	1.6	2.9	10.3	31.9
	Cumulative CO_2 Emissions			GtC	7.1	82.4	169.1	269.9	695.1	1544.4
MS-MRT	Carbon Permit Trade	Bernstein *et al.* (1999b)								
	Permit Price			1995$ / ton						
	USA		No trade				274			
			A1 trade				89			
			Global trade				30			

Table 12-25. continued

Model / Scenario Names	Sub Title	References	Item	Unit	Year 1990	1995	2000	2005	2010	2020	2050	2100
	Japan		No trade						482			
			A1 trade						89			
			Global trade						30			
	Europe		No trade						211			
			A1 trade						89			
			Global trade						30			
	Other OECD		No trade						246			
			A1 trade						89			
			Global trade						30			
	South East Asia		No trade						0			
			A1 trade						0			
			Global trade						30			
	Other Asia		No trade						0			
			A1 trade						0			
			Global trade						30			
	China		No trade						0			
			A1 trade						0			
			Global trade						30			
	Former Soviet Union		No trade						0			
			A1 trade						89			
			Global trade						30			
	Mexico and Oil Producers		No trade						0			
			A1 trade						0			
			Global trade						30			
	Rest of the World		No trade						0			
			A1 trade						0			
			Global trade						30			
	Permit Purchases	Bernstein *et al.* (1999b)										
	USA		A1 trade						305.0			
			Global trade						461.8			

	Japan	A1 trade				57.4	
		Global trade				81.1	
	Europe	A1 trade				103.5	
		Global trade				165.1	
	Other OECD	A1 trade				40.1	
		Global trade				61.9	
	South East Asia	A1 trade				0.0	
		Global trade				−49.0	
	Other Asia	A1 trade				0.0	
		Global trade				−12.0	
	China	No trade				0.0	
		A1 trade				−274.6	
	Former Soviet Union	A1 trade				−506.0	
		Global trade				−311.8	
	Mexico and Oil Producers	A1 trade				0.0	
		Global trade				−58.3	
	Rest of the World	A1 trade				0.0	
		Global trade				−91.3	
RAINS-Asia	Energy Consumption	Resource Management Association	Petajoules				
	China	Base		30255	50527	73518	101016
		Low			433438	59132	75348
	India	Base		15812	22908	36932	63622
		Low			20192	29764	48368
	Indonesia	Base		3739	6247	9250	13684
		Low			5879	7769	10139
	Japan	Base		17866	22606	25679	28809
		Low			20188	20914	21536

Table 12-25. continued

Model / Scenario Names	Sub Title	References	Item	Unit	Year 1990	1995	2000	2005	2010	2020	2050	2100
	South Korea		Base		3594		6264		9497	13449		
			Low				5814		7793	9744		
	Malaysia		Base		1060		1953		3457	6031		
			Low				1928		3186	4686		
	Philippines		Base		585		1082		1921	3113		
			Low				948		1820	2401		
	Singapore		Base		342		645		1130	1731		
			Low				641		1061	1522		
	Taiwan		Base		2094		3334		4796	6843		
			low				2910		4241	5771		
	Thailand		Base		1688		3292		5693	9326		
			Low				2903		4482	6697		
	Asia Total		Base		83594		129468		188528	274121		
			Low				114669		154476	207431		
World Energy Outlook Oil Supply		IEA (1998)		Mil. Barrels/day		1996						
	Total Demand for Liquid Fuels					72.0			94.8	111.5		
	Natural Gas Liquids, etc.					9.3			15.9	20.1		
	Crude Oil					62.7			79.0	72.2		
	Middle East OPEC					17.2			40.9	45.2		
	World excl. Middle East OPEC					45.5			38.0	27.0		
	Unidentified Unconventional Oil					0.0			0.0	19.1		
Natural Gas	Indigenous Production	IEA (1998)		Mtoe		1995						
	OECD North America incl. Mexico					592			759	764		
	OECD Europe					199			276	238		
	OECD Pacific					31			77	68		

Transition Economies				585		809	1116
China				17		57	81
Rest of World excl. Mexico				395		750	1208
World				1818		2727	3474
Net Imports							
OECD North America incl. Mexico				−2		−2	−2
OECD Europe				104		230	387
OECD Pacific				42		42	64
Transition Economies				−74		−162	−281
China				0		0	0
Rest of World excl. Mexico				−76		−114	−174
Crude Oil Import Price, IEA	IEA (1998)		\$ 1990 / bbl	15		17	25
Natural Gas Import Price into Europe	IEA (1998)		\$ 1990 / toe	89.9		103	150
CO_2 Emissions by non Annex I and FSU	Bernard and Vielle (2000)	Base line scenario	Mt C			5270	6846

from permit trade'. 'The entrance of new members, e.g. the Middle East, has a negative impact on the position of some of the Annex I countries (Eastern Europe and Former Soviet Union). We see that EEFSU will become a net importer of permits after entrance of Latin America and the Dynamic ASEAN economies (A, pp.18-19).

In a recent paper the focus is on carbon leakage. 'Leakages occur as a result of trade in energy carriers and trade in energy-intensive products. ... Trade in energy leads to lower energy prices in developing countries, making production there more energy-intensive. Energy-intensive industries may reallocate from Annex-B countries to developing countries, making imports in Annex-B countries more energy-intensive.' 'We analyze technology transfers related to the relocation of energy-intensive industries. It is shown that these transfers mitigate the leakage and increase the efficiency in global energy use'.

'Leakage occurs mainly within the developing countries. However, we show that carbon leakage can also occur in the former Soviet Union, which is part of the Annex-B group, because their agreed emission limits are above their emissions in a business-as-usual scenario. Typical value for the leakage rate turns out to be around 20 percent, if one analyzes the impact of the Kyoto Protocol with a time horizon of 20 years.'

4. What do we know about the future?

Table 12-25 summarizes the numerical results for the future from various studies on economy–energy–environment spheres. The results are cited directly from the original sources that we examine in this chapter in order to show similarity and divergence of experts' opinion. It is noted that not all studies provide numerical data for the future.

APPENDIX: A very long-term view of the global community

DIRK VANWYNSBERGHE and KIMIO UNO

1. Introduction

A 50-year time span is nearly the maximum at the detailed level of econometric analysis. However, the economy–energy–environment interaction would have to be placed under a much longer perspective. Apart from a concerted effort towards database construction and model-building based on the data from 1970 to 2000 that comprise earlier chapters of this volume, the COMPASS project has produced a series of research memos concerning long-term performance of the global community. Taken together, the data set covers the period from 1820 to 2100.

A long-term historical view on the performance of individual economies from 1820 to the present is available from Maddison (1995), together with demographic trends. His work has the advantage that per-capita income is PPP (purchasing power parity) adjusted, so that cross-section comparisons across countries are possible for any time in the 180 years of time span, together with time-series analysis within a single economy. The results from our study up to 2020 are fitted onto Maddison's long-term trend. This reveals the long-term economic trends, with their implication on energy consumption and GHG emissions.

2. The needs for a long-term perspective

Apart from a concerted effort towards database construction and model-building based on the data from 1970 to 2000, the COMPASS project has produced the following research memo concerning long-term performance of the global community. Taken together, the data set covers the period from 1820 to 2100, based both on data compilation by Angus Maddison from various country sources (1820-1990) and the updates by the COMPASS project (1991-1997/1998), those generated by econometric modelling and simulation (1995-2010) and scenarios (2000-2100).

It is clear that the econometric method, which is restricted by the availability of historical data as the basis of determining the functional forms and parameters of individual components of the vast feedback system running through the global community, is not sufficient to provide a global view in the longer term. First,

K. Uno (ed.), Economy–Energy–Environment Simulation, 299–305.

there are countries/regions for which historical data are lacking, or do not provide guidelines for the future. China and Russia, for example, have a long history of central economic planning, which came to an end abruptly, and are now in a long process towards a free market. Their performance, together with that of many others, exerts a profound impact on the global community in the future, but their future is full of uncertainties. Various future scenarios would have to be examined. Second, while more rigorous econometric modelling and simulation can be applied for the recent past, and probably 10 to 20 years into the future, such short- to medium-term analysis would have to be put into a longer-term perspective. Third, relative positioning in the global community may provide additional insights into the potentials of individual countries/regions. Here, catching up and saturation of industrial growth, among others, seem to be relevant considerations (Fei and Ranis, 1964; Rosenberg and Birdzell, 1986; Rostow, 1990; Schmalensee *et al.*, 1998; Sathaye and Ketoff, 1991, among others).

The basis of the work is Maddison (1995), which discusses the economic performance (PPP-adjusted GDP, demography, and PPP-adjusted per capita-GDP) of individual countries (56 representative sample countries and 143 non-sample countries) between 1820 and 1992. Recently, this work has been extended into the last millennium and updated to the most recent year (Maddison, 2001). Prior to the start of the project COMPASS, Keio University Uno Laboratory had obtained permission from Angus Maddison for the use of his data set for educational and research purposes on campus, and was supplied with data and text on electronic media. Keio University also invited Angus Maddison as a Special Invited Professor for lecture and collaborative work. Based on the above, a country/regional classification has been modified in order to be consistent with the one used in the COMPASS project. This step is important for the COMPASS approach because it covers the world exhaustively by including non-individually distinguished economies in Rest of Europe, Rest of Asia, etc. so that the total of countries/regions would truly be a global total. Subsequently, the time span has been extended to cover the period 1991 to 1997/1998 based on GDP indexes published in *International Financial Statistics* (IMF, various years), supplemented by GDP data from the IEA (originally from the World Bank) (IEA, various years b). For the period from around 1995 to 2010 we refer to actual records, short-term projections by various international organizations (ADB, various years a; APEC; 1999; EBRD, 1998 and 2000; OECD, various years c; United Nations ESCAP, various years; World Bank, various years a, among others), and the COMPASS simulation results. For 2010 to 2100 the projection is generated by a Scenario Writer. This is a graphic user interface (GUI) based on a widely used spreadsheet that allows users to operate on key parameters. The COMPASS database for this purpose provides data annually for 199 countries/regions, with an explicit classification scheme that leads to the COMPASS classification of about 60 countries/regions. Data are then aggregated to 13 countries/regions in order to facilitate presentation of the global perspective.

3. Brief examination of the data

As an illustration, data sheets have been compiled covering the period 1950-2100. They are:

Population
PPP-adjusted GDP
PPP-adjusted per-capita GDP.

The table provides two sums, one representing the trend value and the other based on a scenario. The trend values do not depend on subjective judgement but are the extension of an actual trend that prevailed up to the present. On the other hand, since scenario parameters are user-defined, as explained above, the projection figures listed in the following tables are for illustration only, and not a point estimate of the future. Two kinds of sums are compared in terms of percentage difference from the trend. Looking at the population projections for the year 2100, the scenario value (11,274 million) is 11.3% lower than the trend value (12,709 million). This view is consistent with the recent downward revision of world population projections by major international organizations (United Nations, 1991 and 1998; United States Bureau of Census, 1996; Wexler, 1996; World Bank, 1996; Zaba and Clarke, 1994, among others). However, in the case of Japan, it is assumed in the scenario that the present population growth rate would be maintained throughout the 21st century. This of course is not possible from current birth rates. A large amount of population inflow from abroad is implicitly assumed.

Turning to PPP-adjusted GDP per-capita projection, the scenario value for the year 2100 (8993 US$) is 31% higher than the trend value (6825 US$). Needless to say, these numbers are nothing but supposition. However, the supposition comes from various external information items regarding individual country/ region performance in the future, as well as in the recent past. For example, Japan's growth rate in terms of per-capita GDP will be low throughout the projection period, reflecting the difficulty in shifting the top-down planning on the real side of the economy to the bottom-up decision making and resource allocation on the financial side. However, in the latter half of the 21st century growth is assumed to resume at a faster pace when the institutional set-up is altered from the one fitted for catching up to the one where innovation is the key to the success. It is also noteworthy to examine the relative performance of China and Russia. At present the per-capita GDP of China is about one-half of that of Russia. By the end of the 21st century China will enjoy per-capita GDP that slightly exceeds that of Russia. This is attributable to China's success in shifting to a market-oriented development pattern. China has declared its intention to double the size of its economy in the coming decade (*People's Daily*, 13 March 2001). Whether this is possible cannot be judged simply by an econometric method because the historical record is not long enough and/or detailed enough to enable empirical examination. It is helpful to

Appendix Table 1. Population trends, 1950-2100

			1950	1960	1970	1980	1990	2000	2010	2020	2050	2100
A. Population size (unit: 1000)												
RAFN	1	Africa	173,484	217,939	281,089	367,950	492,935	582,213	686,087	763,962	994,085	1,542,537
CNHN	2	China	548,725	670,070	822,274	986,298	1,139,380	1,261,284	1,385,396	1,435,947	1,549,581	1,830,569
EASN	3	South East Asia	74,794	98,981	132,263	166,899	200,774	232,040	267,021	290,310	356,280	509,777
RASN	4	Other Asia	182,318	225,633	283,781	360,925	459,158	500,694	544,674	575,762	663,826	834,451
EURN	5	Europe	413,494	453,470	495,788	529,693	558,056	594,112	633,236	675,268	809,615	1,076,350
FSUN	6	Former Soviet Union	180,050	214,350	242,757	265,542	288,734	325,067	365,869	399,845	502,568	718,672
INDN	7	India	359,000	434,000	541,000	679,000	848,000	959,136	1,078,442	1,144,852	1,321,119	1,719,393
JAPN	8	Japan	83,563	94,053	104,334	116,800	123,540	127,984	132,607	137,486	152,848	181,224
NAMN	9	North America	166,047	198,625	226,432	251,883	276,600	301,889	328,491	347,061	400,364	502,599
OCEN	10	Oceania	13,899	17,598	21,661	25,614	30,079	35,858	42,757	48,389	65,552	107,999
OPRN	11	Oil Producers	246,324	312,910	402,666	524,191	668,213	772,159	888,856	966,830	1,188,631	1,697,683
SAMN	12	Other South America	71,990	91,669	115,808	143,509	175,281	207,051	243,959	271,783	353,659	550,432
UNIN	13	Unidentified	0	0	0	0	0	0	0	0	0	0
		Total	2,513,688	3,029,298	3,669,852	4,418,304	5,260,750	5,899,488	6,597,392	7,057,493	8,358,127	11,271,688
		COMPASS trend	2,513,688	3,029,298	3,669,852	4,418,304	5,260,750	5,813,302	6,502,917	7,192,531	9,261,375	12,709,447
		Percentage off from trend						1.5%	1.5%	−1.9%	−9.8%	−11.3%
B. Population shares (unit: percent of world total)												
RAFN	1	Africa (%)	6.90	7.19	7.66	8.33	9.37	9.87	10.40	10.82	11.89	13.69
CNHN	2	China (%)	21.83	22.12	22.41	22.32	21.66	21.38	21.00	20.35	18.54	16.24
EASN	3	South East Asia (%)	2.98	3.27	3.60	3.78	3.82	3.93	4.05	4.11	4.26	4.52
RASN	4	Other Asia (%)	7.25	7.45	7.73	8.17	8.73	8.49	8.26	8.16	7.94	7.40
EURN	5	Europe (%)	16.45	14.97	13.51	11.99	10.61	10.07	9.60	9.57	9.69	9.55
FSUN	6	Former Soviet Union (%)	7.16	7.08	6.61	6.01	5.49	5.51	5.55	5.67	6.01	6.38
INDN	7	India (%)	14.28	14.33	14.74	15.37	16.12	16.26	16.35	16.22	15.81	15.25
JAPN	8	Japan (%)	3.32	3.10	2.84	2.64	2.35	2.17	2.01	1.95	1.83	1.61
NAMN	9	North America (%)	6.61	6.56	6.17	5.70	5.26	5.12	4.98	4.92	4.79	4.46
OCEN	10	Oceania (%)	0.55	0.58	0.59	0.58	0.57	0.61	0.65	0.69	0.78	0.96
OPRN	11	Oil Producers (%)	9.80	10.33	10.97	11.86	12.70	13.09	13.47	13.70	14.22	15.06
SAMN	12	Other South America (%)	2.86	3.03	3.16	3.25	3.33	3.51	3.70	3.85	4.23	4.88
UNIN	13	Unidentified (%)	0.00	0.00	0.00	0.00	0.00	0.00	0.00	0.00	0.00	0.00
		Total (%)	100.00	100.00	100.00	100.00	100.00	100.00	100.00	100.00	100.00	100.00

Appendix Table 2. PPP-adjusted GDP trends, 1950-2100

			1950	1960	1970	1980	1990	2000	2010	2020	2050	2100
A. GDP size (unit: million 1990 US$)												
RAFN	1	Africa	148,945	221,796	345,893	489,039	640,438	961,342	1,485,263	2,185,114	3,305,810	5,971,580
CNHN	2	China	339,278	595,012	914,681	1,485,014	3,158,584	4,591,094	6,828,857	9,666,577	13,230,693	19,373,900
EASN	3	South East Asia	79,668	133,760	267,034	564,847	1,107,139	1,439,990	1,880,907	2,330,600	3,176,603	4,626,809
RASN	4	Other Asia	120,650	182,089	288,798	457,985	668,280	836,043	1,052,133	1,306,657	1,661,231	2,418,440
EURN	5	Europe	1,605,502	2,583,386	4,122,351	5,602,413	6,880,591	7,573,061	8,354,521	9,216,940	10,754,040	14,838,993
FSUN	6	Former Soviet Union	510,243	843,434	1,351,818	1,709,174	1,987,995	2,487,467	3,130,777	3,886,534	4,945,518	7,185,033
INDN	7	India	214,288	318,928	474,804	637,189	1,115,999	1,675,136	2,587,390	3,805,636	5,754,684	9,809,482
JAPN	8	Japan	156,546	364,791	985,736	1,531,612	2,291,456	2,408,457	2,532,351	2,663,519	3,107,292	4,754,846
NAMN	9	North America	1,554,502	2,173,848	3,296,743	4,553,212	5,986,653	6,592,379	7,224,326	7,804,908	9,941,286	14,473,055
OCEN	10	Oceania	79,641	117,044	189,073	257,935	345,206	448,773	586,062	725,830	986,406	1,438,489
OPRN	11	Oil Producers	363,590	637,136	1,144,297	2,098,329	2,504,906	3,514,624	5,014,503	6,799,527	9,057,917	12,727,253
SAMN	12	Other South America	199,308	285,647	444,682	651,474	706,805	989,719	1,408,785	1,904,501	2,595,454	3,779,228
UNIN	13	Unidentified	0	0	0	0	0	0	0	0	0	0
		Total	5,372,159	8,456,871	13,825,910	20,038,224	27,394,051	33,518,084	42,085,875	52,296,341	68,516,934	101,397,110
		COMPASS trend	5,372,159	8,456,871	13,825,910	20,038,224	27,394,051	31,214,683	36,767,020	42,319,357	58,976,368	86,738,053
		Percentage off from trend						7.38%	14.47%	23.58%	16.18%	16.90%
B. GDP shares (unit: percent of world total)												
RAFN	1	Africa (%)	2.77	2.62	2.50	2.44	2.34	2.87	3.53	4.18	4.82	5.89
CNHN	2	China (%)	6.32	7.04	6.62	7.41	11.53	13.70	16.23	18.48	19.31	19.11
EASN	3	South East Asia (%)	1.48	1.58	1.93	2.82	4.04	4.30	4.47	4.46	4.64	4.56
RASN	4	Other Asia (%)	2.25	2.15	2.09	2.29	2.44	2.49	2.50	2.50	2.42	2.39
EURN	5	Europe (%)	29.89	30.55	29.82	27.96	25.12	22.59	19.85	17.62	15.70	14.63
FSUN	6	Former Soviet Union (%)	9.50	9.97	9.78	8.53	7.26	7.42	7.44	7.43	7.22	7.09
INDN	7	India (%)	3.99	3.77	3.43	3.18	4.07	5.00	6.15	7.28	8.40	9.67
JAPN	8	Japan (%)	2.91	4.31	7.13	7.64	8.36	7.19	6.02	5.09	4.54	4.69
NAMN	9	North America (%)	28.94	25.71	23.84	22.72	21.85	19.67	17.17	14.92	14.51	14.27
OCEN	10	Oceania (%)	1.48	1.38	1.37	1.29	1.26	1.34	1.39	1.39	1.44	1.42
OPRN	11	Oil Producers (%)	6.77	7.53	8.28	10.47	9.14	10.49	11.91	13.00	13.22	12.55
SAMN	12	Other South America (%)	3.71	3.38	3.22	3.25	2.58	2.95	3.35	3.64	3.79	3.73
UNIN	13	Unidentified (%)	0.00	0.00	0.00	0.00	0.00	0.00	0.00	0.00	0.00	0.00
		Total (%)	100.00	100.00	100.00	100.00	100.00	100.00	100.00	100.00	100.00	100.00

Annex Table 3. PPP-adjusted per-capita GDP, 1950-2100

			1950	1960	1970	1980	1990	2000	2010	2020	2050	2100
A. PPP-adjusted per capita GDP (unit: 1990 US$)												
RAFN	1	Africa	859	1,018	1,231	1,329	1,299	1,651	2,165	2,860	3,325	3,871
CNHN	2	China	618	888	1,112	1,506	2,772	3,640	4,929	6,732	8,538	10,584
EASN	3	South East Asia	1,065	1,351	2,019	3,384	5,514	6,206	7,044	8,028	8,916	9,076
RASN	4	Other Asia	662	807	1,018	1,269	1,455	1,670	1,932	2,269	2,503	2,898
EURN	5	Europe	3,883	5,697	8,315	10,577	12,330	12,747	13,193	13,649	13,283	13,786
FSUN	6	Former Soviet Union	2,834	3,935	5,569	6,437	6,885	7,652	8,557	9,720	9,840	9,998
INDN	7	India	597	735	878	938	1,316	1,747	2,399	3,324	4,356	5,705
JAPN	8	Japan	1,873	3,879	9,448	13,113	18,548	18,818	19,097	19,373	20,329	26,237
NAMN	9	North America	9,362	10,944	14,560	18,077	21,644	21,837	21,992	22,489	24,831	28,796
OCEN	10	Oceania	5,730	6,651	8,729	10,070	11,477	12,515	13,707	15,000	15,048	13,320
OPRN	11	Oil Producers	1,476	2,036	2,842	4,003	3,749	4,552	5,642	7,033	7,620	7,497
SAMN	12	Other South America	2,769	3,116	3,840	4,540	4,032	4,780	5,775	7,007	7,339	6,866
UNIN	13	Unidentified	0	0	0	0	0	0	0	0	0	0
		Total	2,137	2,792	3,767	4,535	5,207	5,682	6,379	7,410	8,198	8,996
		COMPASS trend	2,137	2,792	3,767	4,535	5,207	5,370	5,654	5,884	6,368	6,825
		Percentage off from trend						5.81%	12.83%	25.94%	28.73%	31.81%
B. Per capita GDP levels (unit: percent of world average)												
RAFN	1	Africa (%)	40.17	36.45	32.66	29.31	24.95	29.06	33.94	38.60	40.57	43.03
CNHN	2	China (%)	28.93	31.81	29.53	33.20	53.24	64.07	77.27	90.85	104.15	117.65
EASN	3	South East Asia (%)	49.84	48.41	53.59	74.62	105.90	109.23	110.42	108.34	108.76	100.89
RASN	4	Other Asia (%)	30.96	28.91	27.01	27.98	27.95	29.39	30.28	30.63	30.53	32.22
EURN	5	Europe (%)	181.68	204.07	220.70	233.21	236.78	224.36	206.82	184.20	162.03	153.25
FSUN	6	Former Soviet Union (%)	132.60	140.95	147.81	141.92	132.22	134.69	134.14	131.17	120.04	111.14
INDN	7	India (%)	27.93	26.32	23.30	20.69	25.27	30.74	37.61	44.86	53.14	63.42
JAPN	8	Japan (%)	87.66	138.93	250.78	289.14	356.20	331.22	299.36	261.44	247.99	291.66
NAMN	9	North America (%)	438.05	392.04	386.46	398.58	415.65	384.35	344.75	303.49	302.90	320.11
OCEN	10	Oceania (%)	268.11	238.25	231.70	222.04	220.40	220.28	214.87	202.43	183.56	148.06
OPRN	11	Oil Producers (%)	69.07	72.94	75.43	88.26	71.99	80.11	88.44	94.91	92.96	83.34
SAMN	12	Other South America (%)	129.54	111.62	101.92	100.10	77.44	84.13	90.52	94.57	89.52	76.32
UNIN	13	Unidentified (%)	0.00	0.00	0.00	0.00	0.00	0.00	0.00	0.00	0.00	0.00
		Total (%)	100.00	100.00	100.00	100.00	100.00	100.00	100.00	100.00	100.00	100.00

learn from other countries' experience in the similar stage of economic development, and to be able to judge China's relative size in the global community and/or its per-capita income in the future is actually tenable. In the meantime, Russian income growth is assumed to recover from the current negative or very low level growth to positive and sustained growth. Like Japan the future success will depend on fundamental changes in the institutional set-up.

The scenario adopted here is based on the view that population growth will be contained in the future, whereas per-capita income will grow faster than the record in the recent past. Thus, in terms of aggregate GDP, higher-than-trend expansion is implied. At this point in research no deeper examination of scenario parameters has been conducted, and care should be taken not to take the figures provided as the COMPASS projection. It should be stressed that they are the test result of the Scenario Writer software.

4. The long-term view as the reference for medium-term simulation runs

Projection from 2000 to 2100 is therefore based on the historical trends of individual countries/regions, adjusted by the most plausible new development in the future. In order to be able to go back into history and to look far into the future, we have to adopt a suitable methodology focusing on key parameters such as per-capita GDP and population trends. Methodologically, this can be considered as a scenario type. In fact, the COMPASS Scenario Writer allows users to insert their own scenarios on global perspectives. This work has a clear advantage in providing a global view, even when the user's scenarios are confined to specific countries/regions or to a specific time horizon.

COMPASS simulation is based on an econometric method covering the economy–energy–environment interaction at detailed industrial description disaggregated to 36 branches, with country/region resolution of about 60. Its observation period is from 1970 to mid-1990, and its projection period is from mid-1990 to 2010/2020.

The Scenario Writer gives a long-term perspective of the global community for the period 1950 to 2100. Angus Maddison's original work goes back to 1820. The very-long term view of the global community running from 1820 to 2100 provides a reference baseline for judging the sustainability of the global environment while satisfying human aspirations.

Bibliography

Abramovitz, M. (1986). 'The Search for the Sources of Growth: Areas of Ignorance, Old and New.' *Journal of Economic History*, 52(2), pp.217-243.

Adelman, M.A. and Lynch, M.C. (1997). 'Fixed View of Resource Limits Creates Undue Pessimism.' *Oil and Gas Journal*, April, pp.50-60.

Ahmed, K. (1994). *Renewable Energy Technologies. A Review of the Status and Costs of Selected Technologies*. Washington, DC: World Bank.

Alcamo, J., ed. (1994). *IMAGE 2.0: Integrated Modelling of Global Change*. Dordrecht: Kluwer Academic Publishers.

Alcamo, J., Kreileman, E., and Leemans, R., eds. (1998). *Global Change Scenarios of the 21st Century. Results from the IMAGE 2.1 Model*. London: Elsevier Science.

Alcamo, J., Shaw, R., and Hordijk, L. (1990). *The RAINS Model of Acidification: Science and Strategies in Europe*. Dordrecht: Kluwer Academic Publishers.

Allen, R.G.D. (1968). *Economic Theory: A Mathematical Treatment*. New York: St. Martin's Press.

Amman, M. and Dhoondia, J. (1994). *RAINS-Asia User's Manual*. Washington, DC: World Bank.

Andersen, E.S. (1994). *Evolutionary Economics: Post-Schumpeterian Contributions*. London: Pinter.

Armington, P.S. (1969). *A Theory of Demand for Products Distinguished by Place of Production*. International Monetary Fund Staff Papers, Vol. 16, pp.159-176.

Asian Development Bank (ADB) (various years a). *Asian Development Outlook*. Hong Kong: Oxford University Press.

Asian Development Bank (ADB) (various years b). *Key Indicators of Developing Asian and Pacific Countries*. Hong Kong: Oxford University Press (China) Ltd.

Asian Development Bank (ADB) (1994). *Energy Indicators of Developing Member Countries of Asian Development Bank*. Manila.

Asian Policy Forum (2000). *Policy Recommendations for Preventing Another Capital Account Crisis*. Tokyo: Asian Development Bank Institute.

Asia Pacific Economic Cooperation (APEC) Secretariat (various issues). *APEC Energy Statistics*. Tokyo: Institute of Energy Economics, Energy Data and Modelling Center.

Asia-Pacific Economic Cooperation (APEC) (1999). *APEC Economic Outlook*. Singapore: APEC Secretariat.

Asia Pacific Energy Research Centre (APERC) (1998). *APEC Energy Demand and Supply Outlook*. Tokyo.

Australian Bureau of Agricultural and Resource Economics (1996). *The MEGARARE Model: Interim Documentation*. Canberra: ABARE.

Ayres, R.U. (1994). 'On Economic Disequilibrium and Free Lunch.' *Environmental and Resource Economics*, 4, pp.435-454.

Azar, C. (1997). 'Managing the Global Commons: The Economics of Climate Change by W.D. Nordhaus.' *Environmental Values*, 6(1), pp.106-108.

Azar, C. and Dowlatabadi, H. (1999). 'A Review of Technical Change in Assessment of Climate Policy.' Prepared for the *Annual Review of Energy & Environment*, 24.

Azar, C. and Rodhe, H. (1997). 'Targets for Stabilization of Atmospheric CO_2.' *Science*, Vol. 276.

Bach, S., Bork, C., Kohlhaas, M., Lutz, C., Meyer, B., Praetorius, B., and Welsch, H. (2001). *Die ökologische Steuerreform in Deutschland. Eine modellgestützte Analyse ihrer Wirkungen auf Wirtschaft und Umwelt*. Heidelberg: Physica-Verlag.

Bagnoli, P., McKibbin, W.J., and Wilcoxen, P.J. (1996). 'Future Projections and Structural Change.' In Nakicenovic, N., Nordhaus, W.D., Richels, R., and Toth, F.L., eds. *Climate Change: Integrating Science, Economics, and Policy*. CP096-1. Laxenbourg, Austria: International Institute for Applied Systems Analysis.

Baranzini, M. and Scazzieri, R., eds. (1990). *The Economic Theory of Structure and Change*. Cambridge: Cambridge University Press.

Barker, T. (1998). 'The Use of National Accounts in Modelling Greenhouse Gas Mitigation.' Paper presented at the 25th General Conference of the International Association for Research in Income and Wealth, Cambridge. August.

Barker, T. (1999). 'Achieving a 10% Cut in Europe's Carbon Dioxide Emissions Using Additional Excise Duties: Coordinated, Uncoordinated and Unilateral Action Using the Econometric Model E3ME.' *Economic Systems Research*, 11, pp.401-422.

Barker, T., Gardiner, B., Chao-Dong, H., Jennings, N., and Cshurich, C. (1999). *E3ME Version 2.2 (E3ME22) User's Manual*. Cambridge: Cambridge Econometrics.

Barns, D.W., Edmonds, J.A., and Reilly, J.M. (1992). 'Use of the Edmonds-Reilly Model to Model Energy-related Greenhouse Gas Emissions.' OECD Economics Department Working Paper No.113. Paris: OECD.

Baron, R., Bosi, M., Lanza, A., and Pershing, J. (1999). 'A Preliminary Analysis of the EU Proposals on the Kyoto Mechanisms.' Paper presented at the IEA-EMF-IEW workshop. Paris.

Baumol, W.J. (1995). 'Environmental Industries with Substantial Start-up Costs as Contributors to Trade Competitiveness.' *Annual Review of Energy and the Environment*, 20, pp.71-81.

Benhaïm, J. and Schembri, P. (1996). 'Technical Change: An Essential Variable in the Choice of Sustainable Development Trajectory.' In Faucheux, S., Pearce, D., and Proops, J., eds. *Models of Sustainable Development*. Cheltenham: Edward Elgar.

Bernard, A.L. and Vielle, M. (1999). 'Efficient Allocation of a Global Environment Cost between Countries: Tradable Permits VERSUS Taxes or Tradable Permits AND Taxes? An Appraisal with a World General Equilibrium Model.' Paper presented at the Joint Meeting organized by the International Energy Agency, Energy Modeling Forum, and the International Energy Workshop (IIASA), June, Paris.

Bernstein, P. (2000). 'MS-MRT Model Enhancements and Results.' Paper presented at the Joint EMF/IEA/IIASA Meeting, Stanford University, June.

Bernstein, P.M., Montgomery, W.D., Rutherford, T.F., and Gui-Fang Yang (1999a). 'The Effects of Restrictions on International Permit Trading: the MS-MRT Model.' *Energy Journal*, Special Issue, pp.221-256.

Bernstein, P.M., Montgomery, W.D., and Rutherford, T.F. (1999b). 'Global Impacts of the Kyoto Agreement: Results from the MS-MRT Model.' Paper presented at the Joint Meeting organized by the International Energy Agency, Energy Modeling Forum, and the International Energy Workshop (IIASA), June, Paris.

Bohm, P. (1999). 'An Emission Quota Trade Experiment among Four Nordic Countries.' In Sorrell, S. and Skea, J., eds. *Pollution for Sale: Emissions Trading and Joint Implementation*. Cheltenham: Edward Elgar.

Bollen, J., Manders, T., and Timmer, H. (1999). 'The IPCC Stabilisation Scenarios.' Paper presented at the Joint Meeting organized by the International Energy Agency, Energy Modeling Forum, and the International Energy Workshop (IIASA), 16-18 June, 1999; Paris.

Bos, E. and Vu, M.T. (1994). *World Population Projections: Estimates and Projections with Related Demographic Statistics*, 1994-1995 edition. Washington, DC: World Bank.

Boulding, K.E. (1968). *Beyond Economics, Essays on Society, Religion, and Ethics*. Ann Arbor: University of Michigan Press.

Brenton, P.A. (1989). 'Modeling Bilateral Trade Flows: An Empirical Analysis Using Disaggregate Commodity Data.' *Journal of Policy Modeling*, No. 11, pp.547-567.

Brown, L.R. *et al.*, eds. (various years). *State of the World*. London: Earthscan.

Burniaux, J.-M. (1999). 'Burden Sharing Rules in Post-Kyoto Strategies: A General Equilibrium Evaluation Based on the GREEN Model.' Paper presented at the IEA-EMF-IEW workshop, Paris.

Burniaux, J.M., Nicoletti, G., and Oliveira-Martins, J. (1992). 'GREEN. A Global Model for Quantifying the Costs of Policies to Curb CO_2 Emissions.' *OECD Economic Studies*, No. 19/Winter.

Capros, P. *et al.* (1995). 'GEM-E3 Computable General Equilibrium Model for Studying Economy-Energy-Environment Interactions for Europe.' DG-XII. The European Commission.

Capros, P., Karadeloglou, P., Mantzos, L., and Mentzas, G. (1996). 'The Energy Model MIDAS.' In Lasourd, J.-B., Percebois, J., and Valette, F., eds. *Models for Energy Policy*. London: Routledge.

Chasek, P.S. (2001). *Earth Negotiations, Analyzing Thirty Years of Environmental Diplomacy*. Tokyo, New York, Paris: United Nations University Press.

Chenery, H., Robison, S., and Syrquin, M., eds. (1986). *Industrialization and Growth: A Comparative Study*. Oxford: Oxford University Press.
Cline, W.R. (1999). 'Discounting the Very Long Run.' In Portney, P.R. and Weyant, J.P., eds. *Discounting and Intergenerational Equity*. Washington, DC: Resources for the Future.
Cohen, W.M. and Levintal, D.A. (1989). 'Innovation and Learning: Two Faces of R&D.' *Economic Journal*, 99, pp.569-596.
Constanza, R., ed. (1991). *Ecological Economics: The Science and Management of Sustainability*. New York: Columbia University Press.
Cooper, A., Livermore, S., Rossi, V., Wilson, A., and Walker, J. (1999). 'A Cross-Country Quantitative Investigation using the Oxford Global Macroeconomic and Energy Model.' *Energy Journal*, Kyoto Special Issue, pp.335-364.
Darmstadter, J., Dunkerley, J., and Alterman, J. (1977). *How Industrial Societies Use Energy – A Comparative Analysis*. Resources for the Future. Baltimore: Johns Hopkins University Press.
Dean, A. and Hoeller, P. (1992). 'Costs of Reducing CO_2 Emissions: Evidence from Six Global Models.' Economics Department Working Paper No. 122. GD(92)140. Paris: Organisation for Economic Co-operation and Development.
De Jong, A. and Zalm, G. (1991). 'Scanning the Future: A Long-term Scenario Study of the World Economy 1990-2015.' OECD: *Long-term Prospects of the World Economy*, Paris, pp.27-74.
Demeny, P. (1990). 'Population.' In Turner II, B.L. *et al.*, eds. *The Earth as Transformed by Human Action*. Cambridge: Cambridge University Press.
De Vries, B., Janssen, M., and Beusen, A. (1999). 'Perspectives on Global Energy Futures – Simulations with the TIME Model.' *Energy Policy*, 27, pp.477-494.
De Vries, B., Bollen, J., Bouwman, L., den Elzen, M., Janssen, M., and Kreileman, E. (2000). 'Greenhouse Gas Emissions in an Equity-, Environment- and Service-Oriented World: An IMAGE-based Scenario for the 21st Century.' *Technological Forecasting & Social Change*, 63(2-3).
De Vries, H.J.M., Olivier, J.G.J., van den Wijngaart, R.A., Kreileman, G.J.J., and Toet, A.M.C. (1994). 'Model for Calculating Regional Energy Use, Industrial Production and Greenhouse Gas Emissions for Evaluating Global Climate Scenarios.' *Water, Air, Soil Pollution*, 76, pp.79-131.
Dicken, P. (1986). *Global Shift, Industrial Change in a Turbulent World*. London: Paul Chapman.
Dornbusch, R. and Poterba, J.M. (1991). *Global Warming: Economic Policy Responses*. Cambridge, MA: MIT Press.
Douglas, M., Gasper, D., and Thompson, M. (1998). 'Human Needs and Wants.' In Rayner, S. and Malone, E.L., eds. *The Societal Framework*. Columbus, OH: Battelle Press.
Dowlatabadi, H. (1995). 'Integrated Assessment Models of Climate Change: An Incomplete Overview.' *Energy Policy*, 23(4/5), pp.289-296.
Downing, R.J., Ramankutty, R., and Shah, J.J. (1997). RAINS-ASIA, An Assessment Model for Acid Deposition in Asia. Washington, DC: World Bank.
Duchin, F. and Lange, G.-M. (1994). *The Future of the Environment: Ecological Economics and Technological Change*. New York and Oxford: Oxford University Press.
Duchin, F. and Smyshlyaev, A. (1995). 'Development and the Environment: Extending the Global Input-Output-Modelling System'. In Klein, L.R. and Fu Chen Lo, eds. *Modelling Global Change*. Tokyo: United Nations University Press.
Economic and Social Commission for Asia and the Pacific (various years). *Economic and Social Survey of Asia and the Pacific*. New York: United Nations.
Economic and Social Commission for Asia and the Pacific (ESCAP) (1991). *Energy Policy Implications of the Climate Effects of Fossil Fuel Use in the Asia-Pacific Region*. ST/ESCAP/1007. Bangkok: ESCAP.
Economic Commission for Europe (1992). *East-West Energy Efficiency*. New York: United Nations.
Edmonds, J.A. and Barns, D.W. (1992). 'Factors Affecting the Long-term Cost of Fossil Fuel CO_2 Emissions Reduction.' *International Journal of Global Energy Issues*, 4(3), pp.140-166.
Edmonds, J., Wise, M., and MacCracken, C. (1994). *Advanced Energy Technologies and Climate Change. An Analysis Using the Global Change Assessment Model (GCAM)*. PNL-9798, UC-402. Richland, WA: Pacific Northwest Laboratory.
Edmonds, J., Wise, M., Pitcher, H., Richels, R., Wigley, T., and MacCracken, C. (1996). 'An Integrated Assessment of Climate Change and the Accelerated Introduction of Advanced

Energy Technologies: An Application of MiniCAM 1.0.' *Mitigation and Adaptation Strategies for Global Change*, 1(4), pp.311-339.
Energy Information Administration, US Department of Energy (1997a). *World Energy Projection System Model Documentation*. DOE/EIA-M050(97). Washington, DC.
Energy Information Administration, US Department of Energy (1997b). *Annual Energy Outlook 1998 with Projections to 2020*. DOE/EIA-0383 (98). Washington, DC: EIA.
Energy Information Administration, US Department of Energy (1999). *International Energy Outlook 1999*. DOE/EIA-0484(99). Washington, DC.
Engelman, R. (1994). *Stabilizing the Atmosphere: Population, Consumption and Greenhouse Gases*. Washington, DC: Population Action International.
EUREC Agency (1996). *The Future of Renewable Energy. Prospects and Directions*. London: James & James.
European Bank for Reconstruction and Development (EBRD) (1998). *Transition Report 1998*. London: EBRD.
European Bank for Reconstruction and Development (EBRD) (2000). *Transition Report Update*. London: EBRD.
European Commission, Directorate-General XII Science, Research and Development (1995). *The PRIMES Project*. EUR 16713 EN. Brussels and Luxembourg: European Commission.
European Commission (1997). *Methodological Problems in the Calculation of Environmentally Adjusted National Income Figures*, Vol. One and Vol. Two. Studies for the European Commission Directorate General XII.
EUROSTAT (Statistical Office of the European Communities) (1997). *Material Flow Accounting. Experience of Statistical Institutes in Europe*. Luxembourg.
Fankhauser, S. (1995). *Valuing Climate Change. The Economics of the Greenhouse*. London: Earthscan.
Faucheux, S., Pearce, D., and Proops, J. (1996). *Models of Sustainable Development*. Cheltenham: Edward Elgar.
Fei, J.C.H. and Ranis, G. (1964). *Development of the Labor Surplus Economy, Theory and Policy*. Homewood, IL: Richard D. Irwin.
Fischer, G. and Rosenzweig, C. (1996). 'The Impacts of Climate change, CO_2 and SO_2 on Agricultural Supply and Trade.' WP-96-5. Laxenburg, Austria: International Institute for Applied Systems Analysis.
Fischer, G., Frohberg, K., Keyzer, M.A., and Parikh, K.S. (1988). *Linked National Models: A Tool for International Policy Analysis*. Dordrecht: Kluwer Academic Publishers.
Fiksel, J. (1996). *Design for Environment: Creating Eco-efficient Products and Processes*. New York: McGraw-Hill.
Førsund, F.R. (1995). 'Input-Output Models, National Economic Models, and the Environment.' *Handbook of Natural Resource and Energy Economics*, Vol. I. Amsterdam: Elsevier Science Publishers.
Freeman, C. (1982). *The Economics of Industrial Innovation*. Cambridge, MA: MIT Press.
Freeman, C. (1990). 'The Economics of Technical Change.' *Cambridge Journal of Economics*, 18, pp.463-514.
Freeman, C., ed. (1996). *Long Wave Theory*. Cheltenham: Elgar Reference.
Gaffin, S.R. (1998). 'World Population Projections for Greenhouse Gas Emissions Scenarios.' *Mitigation and Adaptation Strategies for Global Change*, 3(2-4), pp.133-170.
Gana, J.L. *et al.* (1979). 'Alternative Approaches to Linkage of National Econometric Models.' In Sawyer, J.A. ed., *Modelling the International Transmission Mechanism*. Amsterdam: North Holland.
Gehlhar, M. *et al.* (1997). 'Overview of the GTAP Data Base.' In Hertel, T.W., ed. *Global Trade Analysis, Modeling and Applications*. Cambridge: Cambridge University Press.
Gielen, D. and Kram, T. (2000). 'Meeting UNFCCC Target via Materials Policies.' Paper presented at the Joint EMF/IEA/IIASA Meeting, Stanford University, June.
Giovannini, B. and Baranzini, A., eds. (1997). *Energy Modelling: Beyond Economics and Technology*. Geneva: International Academy of the Environment and Geneva: Centre for Energy Studies of the University of Geneva.
Glenn, J.C. and Gordon, T.J., eds. (1997). *State of the Future: Implications for Actions Today*. The Millennium Project. Washington, DC: American Council for the United Nations University.

Grassini, M. (1995). 'Comments on Chapter 5.' In Klein, L.R. and Fu Chen Lo, eds. *Modelling Global Change*. Tokyo: United Nations Press.

Greene, D.L. and Santini, D.J., eds. (1993). *Transportation and Global Climate Change*. Washington, DC: American Council for an Energy-Efficient Economy.

Greening, L.A., Davis, W.B., Schipper, L.J., and Khrushch, M. (1997). 'Comparison of Six Decomposition Methods: Application to Aggregate Energy Intensity for Manufacturing in Ten OECD Countries.' *Energy Economics*, 19(3), pp.375-390.

Gregory, K. and Rogner, H.H. (1998). 'Energy Resources and Conversion Technologies for the 21st Century.' *Mitigation and Adaptation Strategies for Global Change*, 3(2-4), pp.171-229.

Grubb, M., Vrolijk, C., and Brack, D. (1999). *The Kyoto Protocol: A Guide and Assessment*. London: Royal Institute of International Affairs, Energy and Environmental Programme.

Grueber, A. (1998). *Technology and Global Change*. Cambridge: Cambridge University Press.

Gürer, N. and Ban, J. (1997). 'Factors Affecting Energy-Related CO_2 Emissions: Past Levels and Present Trends.' *OPEC Review*, XXI(4), pp.309-350.

Hall, G. and Howell, S. (1985). 'The Experience Curve from the Economist's Perspective.' *Strategic Management Journal*, 6, p.197.

Hertel, T.W., ed. (1997). *Global Trade Analysis, Modeling and Applications*. Cambridge: Cambridge University Press.

Hertel, T.W. and Tsigas, M.E. (1997). 'Structure of GTAP.' In Hertel, T.W., ed. *Global Trade Analysis, Modeling and Applications*. Cambridge: Cambridge University Press.

Hickman, B.G. and Lau, L.J. (1973). 'Elasticities of Substitution and Export Demand in a World Trade Model.' *European Economic Review*, 4, pp.347-380.

Higgins, M. and Williamson, J.G. (1997). 'Age Structure Dynamics in Asia and Dependence on Foreign Capital.' *Population and Development Review*, 23(2), pp.261-293.

Hogan, W.W. and Jorgenson, D.W. (1991). 'Productivity Trends and the Cost of Reducing CO_2 Emissions.' *Energy Journal*, 12, pp.68-85.

Hufschmidt, M.M., James, D.E., Meister, A.D., Bower, B.T., and Doxon, J.A. (1983). *Environment, Natural Systems, and Development. An Economic Valuation Guide*. Baltimore and London: Johns Hopkins University Press.

Hughes, B.B. (1999). 'The International Futures (IFs) Modeling Project.' *Simulation and Gaming*, 30(3), September.

Hulme, M., Jiang, T., and Wigley, T. (1995). *SCENGEN: A Climate Change SCENario GENerator: Software User Manual*, Version 1.0. Climate Change Research Unit, School of Environmental Sciences. Norwich: University of Anglia.

Institute of Developing Economies (IDE) (1982). *International Input-Output Table for ASEAN Countries, 1975*. Tokyo.

Institute of Developing Economies (IDE) (1992). *Asian International Input-Output Table 1985*. Tokyo.

Institute of Developing Economies (IDE) (1998). *Asian International Input-Output, 1990*. Tokyo.

Intergovernmental Panel on the Climate Change (IPCC) (1992a). *Climate Change 1992 – The Supplementary Report to the IPCC Scientific Assessment*. Houghton, J.T., Callandar, B.A., and Varney, S.K., eds. Cambridge: Cambridge University Press.

Intergovernmental Panel on the Climate Change (IPCC) (1992b). *Climate Change 1992 – The Supplementary Report to the IPCC Impacts Assessment*.

Intergovernmental Panel on the Climate Change (IPCC) (1992c). *Climate Change: The IPCC 1990 and 1992 Assessment*.

Intergovernmental Panel on the Climate Change (IPCC) (1994). *IPCC Guidelines for National Greenhouse Gas Inventories*, 3 volumes. London: IPCC WGI Technical Support Unit.

Intergovernmental Panel on the Climate Change (IPCC) (1995). *Climate Change 1994 – Radiative Forcing of Climate Change and an Evaluation of the IPCC IS92 Emission Scenarios*. Cambridge: Cambridge University Press.

Intergovernmental Panel on the Climate Change (IPCC) (1996a). *Climate Change 1995 – The Science of Climate Change*. Contribution of Working Group I to the Second Assessment Report. Cambridge: Cambridge University Press.

Intergovernmental Panel on the Climate Change (IPCC) (1996b). *Climate Change 1995 – Impacts, Adaptations, and Mitigation of Climate Change: Scientific-Technical Analyses – Contribution of Working Group II to the Second Working Group II to the Second Assessment Report*.

Intergovernmental Panel on the Climate Change (IPCC) (1996c). *Climate Change 1995 – Economic and Social Dimensions of Climate Change – Contribution of Working Group III to the Second Assessment Report.*
Intergovernmental Panel on the Climate Change (IPCC) (2000). *Special Report on Emissions Scenarios*. Cambridge: Cambridge University Press.
International Energy Agency (IEA) (various years a). *Energy Statistics and Balances of OECD Countries*. Paris: OECD.
International Energy Agency (IEA) (various years b). *Energy Statistics and Balances of Non-OECD Countries*. Paris: OECD.
International Energy Agency (IEA) (various years c). *Energy Prices and Taxes*. Paris: OECD.
International Energy Agency (IEA) (1995). *Energy Policies of the Russian Federation*. Paris: OECD.
International Energy Agency (IEA) (1997a). *Indicators of Energy Use and Efficiency: Understanding the Link between Energy and Human Activity*. Paris: OECD.
International Energy Agency (IEA) (1997b). *Renewable Energy Policy in IEA Countries*. Vol. I and Vol. II. Paris: OECD/IEA.
International Energy Agency (IEA) (1998a). *Enhancing the Market Deployment of Energy Technology. A Survey of Eight Technologies*. Paris: IEA.
International Energy Agency (IEA) (1998b). *World Energy Outlook*. Paris: OECD.
International Energy Agency (IEA) (1998c). *Energy Efficiency Initiative. Energy Policy Analysis*, Vol. 1. Paris: OECD.
International Energy Agency (IEA) (1999). *Non-OECD Coal Fired Power Generation – Trends in the 1990s*. London: IEA Coal Research.
IIASA (International Institute for Applied Systems Analysis) – WEC (World Energy Council) (1995). *Global Energy Perspectives to 2050 and Beyond*. Laxenburg, Austria: International Institute for Applied Systems Analysis.
International Monetary Fund (IMF) (various years). *International Financial Statistics Yearbook*. Washington, DC.
International Monetary Fund (IMF) (various years). *Balance of Payments Statistics Yearbook*. Washington, DC.
International Monetary Fund (IMF) (various years). *World Economic Outlook*. Washington, DC.
International Monetary Fund (IMF) (1993). *Balance of Payments Manual*. Washington, DC.
Inter-Secretariat Working Group on National Accounts (1993). *System of National Accounts 1993*. Brussels, New York, Paris, Washington, DC.
Ishiguro, M. and Akiyama, T. (1995a). 'Energy Demand in Five Major Asian Developing Countries, Structure and Prospects.' World Bank Discussion Paper. Washington, DC: World Bank.
Ishiguro, M. and Akiyama, T. (1995b). 'Electricity Demand in Asia and the Effects on Energy Supply and the Investment Environment.' Policy Research Working Paper 1557. Washington, DC: World Bank.
Jensen, J. *et al.* (2000). 'The Economic Effects of the European Ceilings Proposal.' Paper presented at the Joint EMF/IEA/IIASA Meeting, Stanford University. 20-22 June.
Johansson, O. and Schipper, L.J. (1997). 'Measuring the Long-Run Fuel Demand of Cars: Separate Estimations of Vehicle Stock, Mean Fuel Intensity, and Measured Annual Driving Distances.' *Journal of Transport Economics and Policy*, 31(3), pp.277-292.
Johansson, T.B., Kelly, H., Reddy, A.K.N., and Williams, R.H. (1993). *Renewable Energy: Sources for Fuels and Electricity*. Washington, DC: Island Press.
Jorgenson, D.W. and Wilcoxen, P. (1993). 'Energy, the Environment, and Economic Growth.' In Kneese, A.V. and Sweeney, J.L., eds. *Handbook of Natural Resource and Energy Economics*. Amsterdam: Elsevier.
Johnston, J. (1963). *Econometric Methods*. New York: McGraw-Hill.
Jung, T.Y., La Rovere, E.L., Gaj, H., Shukla, P.R., and Zhou, D. (2000). 'Structural Changes in Developing Countries and Their Implication to Energy-Related CO_2 Emissions.' *Technological Forecasting & Social Change*, 63(2-3).
Kanoh, T. (1992). 'Toward Dematerialization and Decarbonization. Science and Sustainability.' *Selected Papers on IIASA's 20th Anniversary*. Laxenburg, Austria: International Institute for Applied Systems Analysis.
Katseli-Papaefstratiou, L.T. (1979). *The Reemergence of the Purchasing Power Parity Doctrine in the 1970s*. Special Papers in International Economics No. 13. Princeton: Princeton University Press.

Kawashima, Y. (1995). 'The Drafting Process of the Framework Convention on Climate Change and Factor Analysis of the Structure of Confrontation among Nations.' *Kankyo Joho Kagaku Ronbunshu* (J), pp.139-144.

Kelley, A. (1988). 'Economic Consequences of Population Change in the Third World.' *Journal of Economic Literature*, 26(4), pp.685-728.

Kemp, R. (1997). *Environmental Policy and Climate Change*. Cheltenham: Edward Elgar.

Klein, L.R. and Krelle, W., eds. (1983). 'Capital Flows and Exchange Rate Determination.' *Journal of Economics*, Supplementum 3.

Klein, L.R. and Van Peterssen, A. (1973). 'Forecasting World Trade within Project LINK.' In Ball, R.J., ed. *The International Linkage of National Economic Models*, Amsterdam: North-Holland.

Kneese, A.V. and Sweeney, J.L. (Vol. I and II 1985, Vol. III 1993). *Handbook of Natural Resource and Energy Economics*. Amsterdam: Elsevier.

Kolsrud, G. and Torrey, B.B. (1992). 'The Importance of Population Growth in Future Commercial Energy Consumption.' In White, J.C., ed. *Global Climate Change: Linking Energy, Environment, Economy and Equity*. New York: Plenum.

Kosaka, H. and Yano, T. (1999). 'An Empirical Analysis of Carbon Dioxide Emissions: Developing and Energy Model by Panel Data.' Mimeo. Keio University at Shonan Fujisawa Campus.

Krackeler, T., Shipper, L., and Sezgen, O. (1999). 'Carbon Dioxide Emissions in OECD Service Sectors: The Critical Role of Electricity Use.' *Energy Policy*, 26(15), pp.1137-1152.

Kram, T. (2000). *The MATTER Project. Integrated Energy and Materials Systems Engineering for GHG Emission Mitigation*. Petten, the Netherlands: ECN.

Kram, T., Morita, T., Riahi, K., Roehrl, R.A., van Rooijen, S., Sankovski, A., and de Vries, B. (2000). 'Global and Regional Greenhouse Gas Emissions Scenarios'. *Technological Forecasting & Social Change*, 63(2-3).

Kuboniwa, M. (1997). 'Input-Output Analysis of the Russian Economy (1).' (J) *Sangyo Renkan*. Business Journal of PAPAIOS, Vol. 7, No. 4.

Kuboniwa, M. and Gavrilenkov, E. (1997). *Development of Capitalism in Russia: The Second Challenge*. Tokyo: Maruzen Co.

Lashof, D. and Tirpak, D.A. (2000). *Policy Options for Stabilizing Global Climate*. Washington, DC: US Environmental Agency.

Latin American Energy Organization (1999). *Energy-Economic Statistics and Indicators of Latin America and the Caribbean*. Quito, Ecuador.

Lazarus, M.L., Greber, L., Hall, J., Bartels, C., Bernow, S., Hansen, H., Raskin, P., and von Hippel, D. (1993). *Towards a Fossil Free Energy Future: The Next Energy Transition*. A Technical Analysis for Greenpeace International. Boston: Stockholm Environmental Institute Boston Center.

Lind, R.C. (1995). 'Intergenerational Equity, Discounting, and the Role of Cost-Benefit Analysis in Evaluating Global Climate Policy.' *Energy Policy*, 23, pp.379-389.

London Group (1994). *National Accounts and the Environment*. Papers and Proceedings from a Conference, London, England. March. Ottawa: Statistics Canada.

London Group (1995). *Second Meeting of the London Group on Natural Resource and Environmental Accounting*. Conference Papers. Washington, DC: US Department of Commerce, Bureau of Economic Analysis.

London Group (1996). *Third Meeting of the London Group on Natural Resource and Environmental Accounting*. Proceedings volume. Stockholm: Statistics Sweden.

London Group (1997). *National Accounts and the Environment*. Papers and Proceedings from a Conference, Ottawa, Canada. June. Ottawa: Statistics Canada.

Loulou, R. and Kanudia, A. (2000). 'Economic Indicators from a Multi-sector, Multi-region Bottom-up MARKAL Model'. Paper presented at the Joint EMF/IEA/IIASA Meeting, Stanford University, June.

Lutz, C. (1998). *Umweltpolitik und die Emission von Luftschadstoffen: eine empirische Analyze für Westdeutschland*. Berlin: Dunker & Humbolt.

Lutz, W., ed. (1996). *The Future Population of the World: What Can We Assume Today?* 2nd edition. London: Earthscan.

Lutz, W., Sanderson, W., and Scherbov, S. (1997). 'Doubling of World Population Unlikely.' *Nature*, 387(6635), pp.803-805.

Ma, Q. (1996). 'A Bilateral Trade Model for the INFORUM International System.' Unpublished Ph.D. dissertation. Paper presented at the 3rd World INFORUM Conference, Lodz.

MacCracken, C.N., Edmonds, J.A., Kim, S.H., and Sands, R.D. (1999). 'The Economics of the Kyoto Protocol.' *Energy Journal*, special issue.
MacKellar, F.L., Lutz, W., Prinz, C., and Goujon, A. (1995). 'Population, Households and CO_2 Emissions.' *Population and Development Review*, 21(4), pp.849-865.
Maddison, A. (1989). *The World Economy in the 20th Century*. Paris: Development Centre of the Organisation for Economic Co-operation and Development.
Maddison, A. (1991). *Dynamic Forces in Capitalist Development, A Long-Run Comparative View*. Oxford and New York: Oxford University Press.
Maddison, A. (1995). *Monitoring the World Economy 1820-1992*. Development Centre of the Organisation for Economic Co-operation and Development. Paris: OECD.
Maddison, A. (1998a). *Chinese Economic Performance in the Long Run*. Development Centre Studies. Paris: OECD.
Maddison, A. (1998b). 'Measuring the Performance of a Communist Command Economy: An Assessment of the CIA Estimates for the U.S.S.R.' *Review of Income and Wealth*, 44(3), pp.307-323.
Maddison, A. (2001). *The World Economy: A Millennial Perspective*. Development Centre Studies. Paris: OECD.
Mahlman, J.D. (1997). 'Uncertainties in Projections of Human-Caused Climate Warming.' *Science*, 278, November.
Manne, A. (1992). 'Global 2100: Alternative Scenarios for Reducing Carbon Emission'. OECD Economics Department Working Paper No. 111. Paris: OECD.
Manne, A.S. and Richels, R.G. (1992). *Buying Greenhouse Insurance, The Economic Costs of CO_2 Emissions Limits*. Cambridge, MA: MIT Press.
Manne, A. and Richels, R. (1994). 'The Costs of Stabilizing Global CO_2 Emissions: A Probabilistic Analysis Based on Expert Judgements.' *Energy Journal*, 15(1), pp.31-56.
Manne, A. and Richels, R. (1995). 'The Greenhouse Debate: Economic Efficiency, Burden Sharing and Hedging Strategies.' *Energy Journal*, 15(4), pp.1-37.
Manne, A. and Richels, R. (1997). 'On Stabilizing CO_2 Concentrations – Cost-effective Emission Reduction Strategies.' *Environmental Modeling and Assessment*, 2, pp.251-265.
Manne, A. and Rutherford, T.F. (1994). 'International Trade, Capital Flows and Sectoral Analysis: Formulation and Solution of Intertemporal Equilibrium Models.' In Cooper, W.W. and Whinston, A.B., eds. *New Directions in Computational Economics*. Dordrecht: Kluwer Academic Publishers.
Manne, A., Mandelsohn, R., and Richels, R. (1995). 'MERGE: A Model for Evaluating Regional and Global Effects of GHG Reduction Policies.' *Energy Policy*, 23(1), pp.17-34.
McDevitt, T.M. (1996). *World Population Profile*. Report WP/96. US Bureau of Census. Washington, DC: Government Printing Office.
McKibbin, W.J. and Wilcoxen, P.J. (1999). 'Permit Trading under the Kyoto Protocol and Beyond.' Paper presented at the Joint Meeting organized by the International Energy Agency, Energy Modeling Forum, and the International Energy Workshop (IIASA). June, Paris.
Messner, S. and Strubegger, M. (1995). *User's Guide for MESSAGE III*, WP-95-69. International Institute for Applied Systems Analysis. Laxenburg, Austria.
Meyer, B. and Ewerhart, G. (1998). 'Multisectoral Policy Modelling for Environmental Analysis.' In Uno, K. and Bartelmus, P., eds. *Environmental Accounting in Theory and Practice*. Dordrecht: Kluwer Academic Publishers.
Meyer, B. and Uno, K. (1999a). 'Global Econometric 3E-Modelling: The System COMPASS.' In Sydow, A. and Jin-Yi Yu, eds. *1999 International Conference on Mission Earth*. San Diego: Society for Computer Simulation International.
Meyer, B. and Uno, K. (1999b). 'COMPASS – Ein globales Energie-Wirtschaftsmodell.' *ifo-Studien*. 45(4), pp.703-718.
Meyer, B., Bockermann, A., Ewerhart, G., and Lutz, C. (1999). *Marktkonforme Umweltpolitik. Wirkungen auf Luftschadstoffemissionen, Wachstum und Struktur der Wirtschaft*. Heidelberg: Physica-Verlag.
Mitchell, J.V. (1996). *The New Geopolitics of Energy*. Washington, DC: Brookings Institution.
Mori, S. (2000). 'The Development of Greenhouse Gas Emissions Scenarios Using an Extension of the MARIA Model for the Assessment of Resource and Energy Technologies.' *Technological Forecasting & Social Change*, 63(2-3).

Moriguchi, C. (1973). 'Forecasting and Simulation Analysis of the World Economy.' *American Economic Review*, 63(2), pp.402-409.

Morita, T. *et al.* (1994). *Global Carbon Dioxide Scenarios and their Basic Assumptions*. Tsukuba: Center for Global Environmental Research, National Institute for Environmental Studies.

Nakicenovic, N., Nordhaus, W.D., Richels, R., and Toth, E.L. (1996). *Climate Change: Integrating Science, Economics, and Policy*. CP-96-1. Laxenburg, Austria: International Institute for Applied Systems Analysis.

Nakicenovic, N., Grueber, A., and McDonald, A. (1998). *Global Energy Perspectives*. Cambridge: Cambridge University Press.

National Institute for Environmental Studies, Japan. 'Asian-Pacific Integrated Model AIM.' (mimeo). (No publication date given.)

National Research Council (1986). *Population Growth and Economic Development: Policy Questions*. Washington, DC: National Academy Press.

Nishio, K., Fujii, Y., and Yamaji, K. (2000). 'Analysis of the Kyoto Mechanisms using a Global System Model DNE21,' Paper presented at the Joint EMF/IEA/IIASA Meeting, Stanford University, June.

Nordhaus, W. (1994). *Managing the Global Commons, the Economics of Climate Change*. Cambridge, MA: MIT Press.

Nordhaus, W.D. and Kokkenberg, E.C., eds. (1999). *Nature's Numbers, Expanding the National Economic Accounts to Include the Environment*. Washington, DC: National Academy Press.

Nordhaus, W.D. and Popp, D. (1997). 'What is the Value of Scientific Knowledge? An Application to Global Warming Using the PRICE Model.' *Energy Journal*, 18(1), pp.1-45.

Nordhaus, W.D. and Yang, Zili (1996). 'A Regional Dynamic General Equilibrium Model of Alternative Climate-Change Strategies.' *American Economic Review*, September.

Nyhus, D. (1991). 'INFORUM International System'. *Economic Systems Research*, 3(1), pp.55-64.

Nyhus, D., Ma, Q., and Wang, Q. (1996). 'A First Attempt at Linking INFORUM National Models by a Multisectoral Bilateral World Trade Model.' Paper presented at the Fourth INFORUM World Conference, Tokyo.

Oates, W.E. (1995). 'Green Taxes: Can We Protect the Environment and Improve the Tax System at the Same Time?' *Southern Economic Journal*, 61(4), pp.915-922.

Odell, P.R. (1999). 'Dynamics of Energy Technologies and Global Change.' *Energy Policy*, 27, pp.737-742.

Onishi, A. (1999). *FUGI Global Model 9.0 M 200/80: Integrated Global Model for Sustainable Development*. Tokyo: Soka University Institute for Systems Science.

Organisation for Economic Co-operation and Development (OECD) (various years a). *OECD Environmental Data, Compendium*. Paris: OECD.

Organisation for Economic Co-operation and Development (OECD) (various years b). *International Direct Investment*. Paris: OECD.

Organisation for Economic Co-operation and Development (OECD) (country reports, various years c). *OECD Economic Surveys*. Paris: OECD.

Organisation for Economic Co-operation and Development (OECD) (1992a). *Long-term Prospects for the World Economy*. Paris: OECD.

Organisation for Economic Co-operation and Development (OECD) (1992b). *Climate Change: Designing a Tradeable Permit System*. Paris: OECD.

Organisation for Economic Cooperation and Development (OECD) (1994a). *Assessing the Environmental Effects on Trade*. Paris: OECD.

Organisation for Economic Co-operation and Development (OECD) (1994b). *GREEN: The Reference Manual*. Paris: OECD.

Organisation for Economic Co-operation and Development (OECD) (1995a). *The OECD Input-Output Database*. Paris: OECD.

Organisation for Economic Co-operation and Development (OECD) (1995b). *Global Warming, Economic Dimensions and Policy Responses*. Paris: OECD.

Organisation for Economic Co-operation and Development (OECD) (1996). *China in the 21st Century*. Paris: OECD.

Organisation for Economic Co-operation and Development (OECD) (1997a). *The World in 2020: Towards a New Global Age*. Paris: OECD.

Organisation for Economic Co-operation and Development (OECD) (1997b). 'The WorldScan Model: Background and Simulations for Linkages II' (mimeo). Development Centre.
Organisation for Economic Co-operation and Development (OECD) (1997c). *Indicators of Energy Use and Efficiency. Understanding the Link between Energy and Human Activity*. Paris: OECD.
Organisation for Economic Co-operation and Development (OECD) (1997d). *Enhancing the Market Deployment of Energy Technology, A Survey of Eight Technologies*. Paris: OECD.
Organisation for Economic Co-operation and Development (OECD) (1997e). *Energy Technologies for the 21st Century*. Paris: OECD.
Organisation for Economic Co-operation and Development (OECD) (1998a). *Twentyfirst Century Technologies: Promises and Perils of a Dynamic Future*. Paris: OECD.
Organisation for Economic Co-operation and Development (OECD) (1998b). *Eco-Efficiency*. Paris: OECD.
Organisation for Economic Co-operation and Development (OECD) (2000a). *Instruments and Technologies for Climate Change Policy: an Integrated Energy and Materials Systems Modelling Approach*. ENV/EPOC/GEEI(99)15. Paris: OECD.
Organisation for Economic Co-operation and Development (OECD) (2000b). *Assessing the Environmental Effects of Trade Liberalisation Agreement, Methodologies*. Paris: OECD.
Parikh, J., ed. (1996). *Energy Models for 2000 and Beyond*. New Delhi: Tata McGraw Hills.
Parikh, J.K., Parikh, K.S., Gokarn, S., Painuly, J.P., Saha, B., and Shukla, V. (1991). *Consumption Patterns: The Driving Forces of Environmental Stress*. Report prepared for the United Nations Conference on Environment and Development (UNCED), IGIDR-PP-014. Mumbai, India: Indira Ghandi Institute for Development Research.
Parson, E.A. and Fishervanden, K. (1997). 'Integrated Assessment Models of Climate-Change.' *Annual Review of Energy and the Environment*, 22, pp.589-628.
Pearce, D.W. (1991). 'The Role of Carbon Taxes in Adjusting to Global Warming.' *Economic Journal*, 101, pp.938-948.
Peck, S.C. and Teisberg, T.J. (1992). 'CETA: A Model for Carbon Emission Trajectory Assessment.' *Energy Journal*, 13(1), pp.55-77.
Peck, S.C. and Teisberg, T.J. (1995). 'International CO_2 Emissions Control – An Analysis Using CETA.' *Energy Policy*, 23(4), pp.297-308.
Pepper, W.J., Barbour, W., Sankovski, A., and Braaz, B. (1998). 'No-Policy Greenhouse Gas Emission Scenarios: Revisiting IPCC 1992.' *Environmental Science and Policy*, 1, pp.289-312.
Perroni, C. and Wigle, R. (1997). 'Environmental Policy Modeling.' In Hertel, T.W., ed. *Global Trade Analysis, Modeling and Applications*. Cambridge: Cambridge University Press.
Pesaran, M.H., Smith, R.P., and Akiyama, T. (1998). *Energy Demand in Asian Developing Economies*. A World Bank Study. Oxford: Oxford University Press.
Portney, P.R. and Wayant, J.P., eds. (1999). *Discounting and Intergenerational Equity*. Washington, DC: Resources for the Future.
Price, L., Michaelis, L., Worell, E., and Khrushch, M. (1998). 'Sectoral Trends and Driving Forces of Global Energy Use and Greenhouse Gas Emissions.' *Mitigation and Adaptation Strategies for Global Change*, 3(2-4), pp.263-319.
Rayner, S. and Malone, E.L. (1998). *Human Choice and Climate Change*, Vol. 1: *The Social Framework*. Columbus, OH: Battelle Press.
Redclift, M. and Benton, T. (1994). *Social Theory and the Global Environment*. London and New York: Routledge.
Repetto, R. and Austin, D. (1997). *The Costs of Climate Protection: A Guide for the Perplexed*. Washington, DC: World Resources Institute.
Resource Management Association (1996). *RESGEN: Regional Energy Scenario Generator for Asia*. Madison.
Reynolds, T.S. and Cutliffe, S.H., eds. (1997). *Technology and the West*. Chicago: Chicago University Press.
Rogner, H.H. (1996). *An Assessment of World Hydrocarbon Resources*. WP-96-56. Laxenburg, Austria: International System for Applied Systems Analysis.
Rogner, H.H. (1997). 'An Assessment of World Hydrocarbon Resources.' *Annual Review of Energy Environment*, 22, pp.217-262.
Roodman, D. (1996). *Paying the Piper: Subsidies, Politics, and the Environment*. Worldwatch Paper 133. Washington, DC: Worldwatch Institute.

Rose, A. and Stevens, B. (1998). 'A Dynamic Analysis of Fairness in Global Warming Policy: Kyoto, Buenos Aires, and Beyond.' *Journal of Applied Economics*, 1.

Rose, A. and Stevens, B. (1999). 'A Dynamic Analysis of the Efficiency and Equity of Tradable Greenhouse Gas Emission Permits.' Paper presented at the Joint Meeting organized by the International Energy Agency, Energy Modeling Forum, and the International Energy Workshop (IIASA), 16-18 June. Paris.

Rose, A., Stevens, B., Edmonds, J., and Wise, M. (1998). 'International Equity and Differentiation in Global Warming Policy: An Application to Tradable Emission Permits.' *Environmental and Resource Economics*, 12(1).

Rosenberg, N. (1994). *Exploring the Black Box: Technology, Economics, and History*. Cambridge: Cambridge University Press.

Rosenberg, N. (1997). 'Economic Development and the Transfer of Technology: Some Historical Perspectives.' In Reynolds, T.S. and Cutcliffe, S.H., eds. *Technology and the West*. Chicago: Chicago University Press.

Rosenberg, N. and Birdzell, L.E. (1986). *How the West Grew Rich: The Economic Transformation of the Industrial World*. London: I.B. Tauris & Co.

Rostow, W.W. (1990). *The Stages of Economic Growth*. 3rd edition. Cambridge: Cambridge University Press.

Rotmans, J. and de Vries, H.J.M., eds. (1997). *Perspectives on Global Futures: The TARGETS Approach*. Cambridge: Cambridge University Press.

Rotmans, J., van Asselt, M.B.A. *et al.* (1994). *Global Change and Sustainable Development. A Modelling Perspective for the Next Decade*. GLOBO Report Series No. 4. The Netherlands: National Institute of Public Health and Environmental Protection.

Rutherford, T. (1992). 'The Welfare Effects of Fossil Carbon Restrictions: Results from a Recursively Dynamic Trade Model'. OECD Economics Department Working Papers No. 112. Paris: OECD.

Sakurai, N., Ioannidis, E., and Papacontantinou, G. (1996). 'The Impact of R&D and Technology Diffusion on Productivity Growth: Evidence for 10 OECD Countries in the 1970s and 1980s.' Directorate for Science, Technology and Industry. OECD/GD(96)27 Paris.

Sands, P.H., ed. (1992). *The Effectiveness of International Environmental Agreements: A Survey of Existing Legal Instruments*. United Nations Conference on Environment and Development (UNCED). Cambridge: Grotius.

Sands, P. (1995). *Principles of International Environmental Law I. Framework, Standards and Implementation*. Manchester and New York: Manchester University Press.

Sathaye, J. and Ketoff, A. (1991). 'CO_2 Emissions from Major Developing Countries: Better Understanding the Role of Energy in the Long Term.' *Energy Journal*, 12(1), pp.161-196.

Schipper, L., Hass, R., and Sheinbaum, C. (1996). 'Recent Trends in Residential Energy Use in OECD Countries and Their Impact on Carbon Dioxide Emissions: A Comparative Analysis of the Period 1973-1992.' *Journal of Mitigation and Adaptation Strategies for Global Change*, 1, pp.167-196.

Schipper, L., Ting, M., Khrushch, M., and Golove, W. (1997). 'The Evolution of Carbon Dioxide Emissions from Energy Use in Industrial Countries: An End-Use Analysis.' *Energy Policy*, 25(7-9), pp.651-672.

Schmalensee, R., Stoker, T., and Judson, R. (1998). 'World Carbon Dioxide Emissions: 1950-2050.' *Review of Economics and Statistics*, LXXX(1), pp.15-27.

Schneider, S.H. (1990). *Global Warming: Are We Entering the Greenhouse Century*? Cambridge: Lutterworth Press.

Schwartz, P. (1991). *The Art of the Longview: Three Global Scenarios to 2005*. New York: Doubleday.

Sijm, J.P.M. *et al.* (2000). 'Kyoto Mechanisms: The Role of Joint Implementation, the Clean Development Mechanism and Emissions Trading in Reducing Greenhouse Gas Emissions.' Petten, the Netherlands: Netherlands Energy Research Foundation.

Sivasubramonian, S. (1998). 'Twentieth Century Economic Performance of India.' Paper prepared for the 25th General Conference for Research in Income and Wealth, Cambridge, England, 23-29 August.

Skea, J. (1999). 'Flexibility, Emissions Trading and the Kyoto Protocol.' In Sorrell, S. and Skea, J., eds. *Pollution for Sale: Emissions Trading and Joint Implementation*. Cheltenham: Edward Elgar.

Slade, M.E., Kolstad, C.D., and Weiner, R.J. (1993). 'Buying Energy and Nonfuel Minerals: Final, Derived, and Speculative Demand.' In Kneeze, A.V. and Sweeney, J.L., eds. *Handbook of Natural Resource and Energy Economics*. Amsterdam: Elsevier.

Soltes, D. (1999). 'Metadata and Metainformation – Old Concepts and New Challenges.' *IASSIST Quarterly*, 23(1), pp.12-14.

Sorrell, S. and Skea, J., eds. (1999). *Pollution for Sale: Emissions Trading and Joint Implementation*. Cheltenham: Edward Elgar.

Spector, B.I., Sjøstedt, G., and Zartman, I.W. (1994). *Negotiating International Regimes: Lessons Learned from the United Nations Conference on Environment and Development (UNCED)*. London: Graham & Trotman.

State Statistical Bureau of the People's Republic of China, and Institute of Economic Research, Hitotsubashi University (1997). *The Historical National Accounts of the People's Republic of China 1952-1995*. Tokyo: Hitotsubashi University.

Timmer, H. (1998). 'WorldScan – A World Model.' *Quarterly Review Netherlands Bureau of Economic Policy Analysis*, 3, pp.37-40.

Unander, F. and Schipper, L. (1998). 'Past and Future Trends in CO_2 Emissions from Energy Use: The Indicator Approach.' *ENER Bulletin* No. 22. European Network for Energy Economic Research.

Unander, F. and Schipper, L. (1999). 'Which Road from Kyoto? Decomposition of Emission Scenarios.' Paper presented at the Joint Meeting organized by the International Energy Agency, Energy Modeling Forum, and the International Energy Workshop (IIASA), 16-18 June, Paris.

United Nations (various years a). *National Accounts Statistics: Main Aggregates and Detailed Tables*. New York.

United Nations (various years b). *International Trade Statistics Yearbook*. New York.

United Nations (1990). *International Standard Industrial Classification of All Economic Activities, Third Revision*. Series M No. 4, Rev. 3. New York.

United Nations (1991). *Consequences of Rapid Population Growth in Developing Countries*. New York.

United Nations (1993a). *Agenda 21: The United Nations Programme of Action from Rio*. New York: United Nations Department of Public Information.

United Nations (1993b). *Integrated Environmental and Economic Accounting*. New York: United Nations Statistical Division.

United Nations (1997). *Critical Trends: Global Change and Sustainable Development*. No. ST/ESA/255. New York.

United Nations (1998). *World Population Projections to 2150*. UN Department of Economic and Social Affairs. New York: United Nations.

United Nations Development Programme (UNDP) (1997). *Human Development Report 1997*. New York: Oxford University Press.

UNFCCC (1992). *United Nations Framework Convention on Climate Change. Convention Text*. UNEP/WMO Information Unit of Climate Change (IUCC) on behalf of the Interim Secretariat of the Convention. Geneva: IUCC.

UNFCCC (1997). *Kyoto Protocol to the United Nations Framework Convention on Climate Change*. FCCCC/CP/L7/Add.1. December. New York: United Nations.

United States Bureau of Census (1996). *World Population Profile 1996*. WP/96. Washington, DC: US Government Printing Office.

United States Environmental Protection Agency (1990). *Policy Options for Stabilizing Global Climate*. Report to Congress. Washington, DC.

Uno, K. (1995). *Environmental Options: Accounting for Sustainability*. Dordrecht: Kluwer Academic Publishers.

Uno, K. (2000). 'Environmental Effects of Trade Liberalization Agreements: the COMPASS Approach.' OECD Proceedings. *Assessing the Environmental Effects of Trade Liberalisation Agreements, Methodologies*. Paris: OECD.

Uno, K. and Bartelmus, P., eds. (1998). *Environmental Accounting in Theory and Practice*. Dordrecht: Kluwer Academic Publishers.

Van Asselt, M., Rotmans, J. *et al.* (1995). *Uncertainty in Integrated Assessment Modelling: a Cultural Perspective Based Approach*. GLOBO Report Series No. 9. The Netherlands: National Institute of Public Health and the Environment.

Van der Zwaan, B.C.C., Gerlagh, R., Klaassen, G., and Schattenholzer, L. (2000). 'Endogenous Technological Change in Climate Change Modelling.' Paper presented at the Joint EMF/IEA/IIASA Meeting, Stanford University, June.

Viguier, L. (1999). 'Emissions of SO_2, NOx, and VO_2 in Transition Economies: Emission Inventories and Divisa Index Analysis.' *Energy Economics*, 20(2), pp.59-75.

Vouyoukas, E.L. (1992). 'Carbon Taxes and CO_2 Emissions Targets: Results from the IEA Model'. OECD Economics Department Working Papers No. 114. Paris: OECD.

Vouyoukas, E.L. (1993). 'IEA Medium Term Energy Model.' OECD, *The Costs of Cutting Carbon Emissions: Results from Global Models*. Paris.

Vouyoukas, E.L. (1996). 'World Energy Outlooks.' In Lasourd, J.-B., Percebois, J., and Valette, F., eds. *Models for Energy Policy*. London: Routledge.

Wallace, D. (1995). *Environmental Policy and Industrial Innovation: Strategies in Europe, the USA, and Japan*. London: Earthscan.

West, G.R. (1995). 'Comparison of Input-Output, Input-Output + Econometric and Computable General Equilibrium Impact Models at the Regional Level.' *Economic Systems Research*, 7(2).

Wexler, L. (1996). *Improving Population Assumptions in Greenhouse Emissions Models*. WP-96-099. Laxenburg, Austria: International Institute for Applied Systems Analysis.

Weyant, J.P. (1997). 'Technological Change and Climate Policy Modelling.' Paper prepared for the Workshop on Induced Technological Change and the Environment. Laxenburg, Austria: International Institute for Applied Systems Analysis.

Weyant, J.P., ed. (1999). 'The Costs of the Kyoto Protocol: a Multi-Model Evaluation.' *Energy Journal*, special issue.

Weyant, J.P. and Hill, J.N. (1999). 'Introduction and Overview.' *Energy Journal*, special issue, pp.vii–xliii.

Whalley, J. and Wigle, R. (1992). 'Results for the OECD Comparative Modelling Exercise from the Whalley-Wigle Model'. OECD Economics Department Working Papers, No. 121. Paris: OECD.

White, J.C., ed. (1992). *Global Climate Change: Linking Energy, Environment, Economy, and Equity*. New York and London: Plenum Press.

Wier, M., Lenzen, M., Munksgaard, J., and Smed, S. (2001). 'Effects of Household Consumption Patterns on CO_2 Requirements.' *Economic Systems Research*, 13(3), pp.259-274.

Wigley, T.M.L. and Paper, S.C.B. (1993). 'Future Changes in Global Mean Temperature and Sea Level'. In Warwick, R.A., Barrow, E.M., and Wigley, T.M.L., eds. *Climate and Sea Level Change Observations, Projections and Implementation*. Cambridge: Cambridge University Press.

Wigley, T.M.L. *et al.* (1994). *Model for the Assessment of Greenhouse-gas Induced Climate Change*, Version 1.2. Climate Research Unit, University of East Anglia, UK.

Wigley, T.M.L., Richels, R., and Edmonds, J.A. (1996). 'Economic and Environmental Choices in the Stabilisation of Atmospheric CO_2 Concentrations.' *Nature*, 379, pp.240-243.

World Bank (various years a). *Global Economic Prospects and the Developing Countries*. Washington, DC.

World Bank (various years b). *World Development Report*. Oxford: Oxford University Press.

World Bank (various years c). *Global Development Finance, Analysis and Summary Tables*. Washington, DC: World Bank.

World Bank (1993). *The Asian Miracle, A World Bank Policy Research Report*. Washington, DC: World Bank.

World Bank (1996). *Long-term Trends of World Population '94/95 (1990-2150)*. Washington, DC: World Bank.

World Bank (1998). *Global Development Finance* (Formerly published as the *World Debt Tables*). *Analysis and Summary Tables. Country Tables.* Washington, DC.

World Business Council for Sustainable Development (WBCSD) (1997). *Exploring Sustainable Development: WBCSD Global Scenarios 2000-2050*. London: WBCSD.

World Commission on Environment and Development (1987). *Our Common Future*. Oxford and New York: Oxford University Press.

World Energy Council (WEC) (1998). *Survey of Energy Resources*. London: WEC.

World Energy Council and the International Institute for Applied Systems Analysis (1995). *Global Energy Perspectives to 2050 and Beyond*. Laxunburg, Austria: IIASA.

World Resources Institute (WRI) (1997). *Resource Flows: The Material Basis of Industrial Economies*. Washington, DC: WRI.

Wu, H.X. (1998). 'An Alternative Estimation of Chinese Industrial Performance, 1949-95.' Paper prepared for the 25th General Conference of the International Association for Research in Income and Wealth. Cambridge, England.

Yoshitomi, M. and Shirai, S. (2000). *Technical Background Paper for Policy Recommendation for Preventing Another Capital Account Crisis*. Tokyo: Asian Development Bank Institute.

Zaba, B. and Clarke, J., eds. (1994). *Environment and Population Change*. Liege, Belgium: Derouaux Ordina Editions.

COMPASS working papers

Koltsov, A. and Volkov, V. (1998). 'Comparison: SNA and IO Table for Russia, 1995.' December 1998 Workshop, Keio University.

Lutz, C. (1998). 'The Linkage of Trade and 18 IO Country Models: A Historical Price Simulation.' December 1998 Workshop, Keio University.

Lutz, C. (1998). 'User Manual for the Macroeconomic System.' December 1998 Workshop, Keio University.

Meyer, B. and Ewerhard, G. (1998). 'Multisectoral Policy Modelling for Environmental Analysis.' In Uno, K. and Bartelmus, P., eds., *Environmental Accounting in Theory and Practice.* Dordrecht: Kluwer Academic Publishers.

Meyer, B. and Uno, K. (1998). 'The Internationally Linked Dynamic Input-Output Model COMPASS: The Logical Structure.' Paper presented at the Twelfth International Input-Output Conference, held 18-22 May, 1998, New York.

Meyer, B. and Uno, K. (1999). 'Global Econometric 3E-Modelling: The System of COMPASS.' Paper presented at 1999 Western MultiConference, San Francisco, January 17-20, 1999. Sydow, A. and Jin-Yi Yu, eds., *1999 In International Conference on Mission Earth.* San Diego: The Society for Computer Simulation International.

Meyer, B., Bockermann, A., and Lutz, C. (1999). 'The Endogenization of Trade Shares in the Global Model COMPASS.' APEC Energy Project, Keio University.

Miketa, A. (1998). 'Price Dependency of Energy Input: Experiment Based on OECD IO Tables and Energy Balances.' Paper presented at the Twelfth International Input-Output Conference, 18-22 May, New York.

Miketa, A. (2000). 'Maddison Data Update, 1991-1997/1998.' Uno Laboratory. Keio University.

Umehara, Y. (1998). 'The Energy/Environmental Model for 3E-COMPASS.' December 1998 Workshop, Keio University.

Umehara, Y. (1999). 'Modeling Energy Balances for Policy Simulation.' Paper presented at a joint meeting of the International Energy Agency, The Energy Modeling Forum, and the International Energy Workshop, held 16-18 June, Paris. Revised September 1999.

Uno, Kimio (1998a). 'Identifying Research Priority.' In Uno, K. and Bartelmus, P., eds., *Environmental Accounting in Theory and Practice.* Dordrecht: Kluwer Academic Publishers.

Uno, K. (1998b). 'Policy Application of Global Economy-Energy-Environment (3E) Accounting and Modelling.' Paper presented at the Twelfth International Input-Output Conference, 18-22 May, New York.

Uno, K. (1998c). 'From Single-Country Accounting to Global Modelling.' Paper presented at the Fifth Meeting of the London Group on Natural Resources and Environmental Accounting, Fontevraud, 25-29 May, 1998.

Uno, K. (1999a). 'Energy Projection: Comparison of Methodologies (preliminary).' Paper presented at a joint meeting of the International Energy Agency, The Energy Modeling Forum, and the International Energy Workshop, 16-18 June, Paris. Revised October 1999.

Uno, K. (1999b). 'Environmental Effects of Trade Liberalization Agreements: The COMPASS Approach.' Paper presented at the OECD Joint Workshop on Trade and Environment, 26-27 October, Paris.

Uno, K. (1999c). 'Environmental Accounting, Policy Analysis, and Global Governance.' Paper prepared for the Sixth Meeting of the London Group on Natural Resources and Environmental Accounting, 15-19 November, Canberra.

Uno, K. (2000). 'An Analysis of Global Ecological Linkages using an Input-Output Database.' Paper presented at an International Workshop Globalization and the World Economy: Changes and Challenges, 14 December, Hitotsubashi University.

Uno, K. (2001). 'Very Long Term Demographic and PPP-Adjusted Economic Data for the World Community.' Project Report. Keio University SFC Research Institute.

Uno, K. and Meyer, B. (1999). 'Data Structure and Logical Flow of the 3E (Economy-Energy-Environment) Model COMPASS (Comprehensive Model for Policy Assessment).' Paper presented at a joint meeting of the International Energy Agency, The Energy Modeling Forum, and the International Energy Workshop, 16-18 June, Paris.

Uno Laboratory (1999). '3E COMPASS, Economy-Energy-Environment Comprehensive Model for Policy ASSessment.' COMPASS-single-manual. Keio University.

Vanwynsberghe, D. and Wang, Y. (1996). 'Quickdyme, Installation Files.' Center of Excellence, Keio University.

Vanwynsberghe, D. and Wang, Y. (1996). 'Quickdyme Modelling, the Naming Conventions.' Center of Excellence, Keio University.

Vanwynsberghe, D. (1997). 'Naming Convention, the Quickdyme Approach.' Center of Excellence, Keio University.

Vanwynsberghe, D. (1997). 'The Integral Project User Manual Time Series.' Keio University.

Vanwynsberghe, D. (1998). 'The Sector, the Branch, the SNA and the Financial Definition.' COMPASS Modelling Project, Keio University.

Vanwymsberghe, D. (1998). 'Object-oriented Approach in International Input-Output Models.' Paper presented at the Twelfth International Input-Output Conference, 18-22 May, New York.

Vanwynsberghe, D. (1998). 'COMPASS Regions and Branches.' December 1998 Workshop, Keio University.

Vanwynsberghe, D. (1998). 'World Trade Branches and SITC Codes.' December 1998 Workshop, Keio University.

Vanwynsberghe, D. (1998). 'COMPASS Financial Account Framework and Variable Names.' December 1998 Workshop, Keio University.

Vanwynsberghe, D. (1998). 'From Input-Output over Macro Economic Aggregates to Financial Accounts FIOM.' December 1998 Workshop, Keio University.
Vanwynsberghe, D. (2001). 'Long-Term Data on Economy, Demography, and Energy.' Uno Laboratory, Keio University.
Vanwynsberghe, D. and Wang, Y. (1997). 'Quickdyme, Make Country SNAIO Database System Manual.' Center of Excellence, Keio University.
Vanwynsberghe, D. and Wang, Y. (1997). 'Macro Models, Make Country Macro Models.' Center of Excellence, Keio University.
Wang, Y. (1996). 'GETAIJ & PUTAIJ, a Program for Analysing the Elements of a Matrix.' Center of Excellence, Keio University.
Wang, Y. (1996). 'MUDAN III, A Multi-sector Model for China.' Center of Excellence, Keio Univertity.
Wang, Y. (1997). 'Quickdyme, Country Model Software.' Center of Excellence, Keio University.
Wang, Y. (1998). 'From Data Preparation to Construction of Simulation Models.' Paper presented at the Twelfth International Input-Output Conference, held 18-22 May, 1998, New York.
Wang, Y. (1998). 'What's New in Quickdyme Version 2 and the Present Situation of IO Models.' December 1998 Workshop, Keio University.
Wang, Y. (1999). 'An Approach to Determine Gross Output Price Index from Value Added Price Index.' APEC Energy Project, Keio University.
Wang, Y. (1999). 'Price Equation for the National Model Linked with Trade Model.' APEC Energy Project, Keio University.

Notes on Contributors

Frank Hohmann has studied computer science and business administration at the University of Osnabruuck (Diplom Betriebswirt 1994). Since 1996 he supported GWS mbH as a freelance consultant and since 1999 he has been working as a databank and software expert for GWS.

Alexey Koltsov is doctor of economics and serves as the Head of Division, Summarizing Department for Industry and Science and Technologies, Russian Federation. He is responsible, among other things, for structural and medium-term analysis and forecasting of the Russian economy.

Zuo Li is the Director, the Development Forecasting Department of Jiangsu Information Center, China. She graduated from Northeast Univerisity in 1982 and started economic modelling and analysis in 1985, and has been working in the field of multi-sector modelling since 1995.

Christian Lutz is vice-director of GWS mbH. He studied economics at the University of Tuubingen (Diplom Volkswirt 1993) and at the University of Osnabruuck (PhD in economics 1997). Since 1997 he has been research fellow at the GWS. The focus of his research has been economics of energy, air emissions, land use, and international trade.

Bernd Meyer is professor of macroeconomic theory at the University of Osnabruuck and director of the GWS mbH. In 1997 he was visiting professor at the Department of Policy Management at Keio University, Japan. Since 1996 he has been a member and since 1999 chairman of the Council of Advisors for Economic and Environmental Accounting for the German Ministry of Environment. His research in the last years concentrates on the construction and application of macroeconomic and interindustry simulation and forecasting models with a focus on environmental questions.

Yumiko Umehara is the National Program Director of LEAD (Leadership for Environment and Development) Japan Program. She is a fellow of the LEAD Program that is headquartered in London. She obtained MA from the Graduate School of Media and Governance in 2000 and has been a researcher at the Keio University Research Institute at Shonan Fujisawa Campus (SFC). Her academic papers on multi-sector energy balance model have been presented at Energy Modelling Forums at IEA (1999) and Stanford University (2000), and OECD Joint Workshop on Trade and Environment (1999).

Kimio Uno is Professor of Media and Governance at Keio University, Japan. Previously he taught at the Graduate School of Environmental Science at University of Tsukuba. He obtained PhD in economics from University of Illinois in 1973 and was a visiting fellow at Yale University. He served as a consultant to the United Nations Conference on Environment and Development and the United Nations University on various occasions. He worked as research associate at the International Institute for Applied Systems Analysis (IIASA). He chairs a government committee on environmental accounting.

Dirk Vanwynsberghe is a consultant in system integration at Economic Research International in Belgium. He holds MA in applied economics and commercial engineering from University of Leuven. With his expertise in energy areas and database construction, his international experience includes assignment at UNIDO, Chase Econometrics, and consulting and teaching in various parts of the world.

Vladimir Volkov is doctor of economics and currently the Deputy Head of Macroeconomic Analysis and Forecasting Department, Ministry of Economic Development and Trade, Russian Federation. His responsibility focuses on short-term macroeconomic analysis.

Wang Yinchu is Vice-Director of the Economic Information Center, Jiangsu Province, China. He graduated from Nanjing University in 1977 and has been working in the field of economic modelling and analysis since 1977. He has worked on multi-sector modelling and analysis for China since 1992.

Disclaimer

Project Support: The project has been supported by the following sources.

Institute of Energy Economics: 'Construction of APEC Energy Network', Keio University Shonan Fujisawa Campus Research Institute, 1995-1999.

Japan Society for the Promotion of Science, Grant-in-Aid for International Research Collaboration: 'Global Scenario on Energy and Environment', 1998-2000.

Ministry of Education, Database Construction Subsidy Scheme: 'Comprehensive Model for Policy Assessment', Keio University, 1997-1999.

Ministry of Education, Center of Excellence: 'Creative Digital Media, Its Impact on the New Century', Keio University Shonan Fujisawa Campus Research Institute, 1996-2000.

Economic Planning Agency, Japan and Ministry of Economy, Russian Federation: 'Joint Research Project on the Russian Economic Reform', 1996-2000.

3E Research Institute: 'Energy, Environment, and Economy'. Keio University, Japan and Tsinghua University, China, project supported by the JETRO and the NEDO of the Japanese Government, 2000-2004.

Keio University Shonan Fujisawa Campus Research Institute research grant: 'Very Long Term Demographic and PPP-Adjusted Economic Data for the World Community', 2000.

Ministry of Education and Science, Hightech Research Center Project: 'Next Generation Cyber Space', 2000-2002.

Japan Society for the Promotion of Science, Grant-in-Aid for Publication of Scientific Research Results, Database: 'Global Economy Energy Environment Database (Global 3E)', 2001.

Japan Society for the Promotion of Science, Grant-in-Aid for Publication of Scientific Research Results, Academic Publication: 'Economy-Energy-Environment: A Global Perspective', 2001.

Index

Economy & Environment

1. F. Archibugi and P. Nijkamp (eds.): *Economy and Ecology: Towards Sustainable Development.* 1989 ISBN 0-7923-0477-2
2. J. Bojö, K.-G. Mäler and L. Unemo: *Environment and Development: An Economic Approach.* 1990 ISBN 0-7923-0802-6
3. J. B. Opschoor and D. W. Pearce (eds.): *Persistent Pollutants: Economics and Policy.* 1991 ISBN 0-7923-1168-X
4. D.J. Kraan and R. J. in 't Veld (eds.): *Environmental Protection: Public or Private Choice.* 1991 ISBN 0-7923-1333-X
5. J.J. Krabbe and W.J.M. Heijman (eds.): *National Income and Nature: Externalities, Growth and Steady State.* 1992 ISBN 0-7923-1529-4
6. J. Bojö, K.-G. Mäler and L. Unemo: *Environment and Development: An Economic Approach* (revised edition). 1992 ISBN 0-7923-1878-1
7. T. Sterner (ed.): *Economic Policies for Sustainable Development.* 1994 ISBN 0-7923-2680-6
8. L. Bergman and D.M. Pugh (eds.): *Environmental Toxicology, Economics and Institutions.* The Atrazine Case Study. 1994 ISBN 0-7923-2986-4
9. G. Klaassen and F.R. Førsund (eds.): *Economic Instruments for Air Pollution Control.* 1994 ISBN 0-7923-3151-6
10. K. Uno: *Environmental Options: Accounting for Sustainability.* 1995 ISBN 0-7923-3513-9
11. K. Uno and P. Bartelmus (eds.): *Environmental Accounting in Theory and Practice.* 1997 ISBN 0-7923-4559-2
12. J.C.J.M. van den Bergh, K.J. Button, P. Nijkamp and G.C. Pepping: *Meta-Analysis in Environmental Economics.* 1997 ISBN 0-7923-4592-4
13. S. Faucheux, M. O'Connor and J. v.d. Straaten: *Sustainable Development: Concepts, Rationalities and Strategies.* 1998 ISBN 0-7923-4884-2
14. P. Kågeson: *Growth versus the Environment: Is there a Trade-off?.* 1998 ISBN 0-7923-4926-1
15. J.C.J.M. van den Bergh and M.W. Hofkes (eds.): *Theory and Implementation of Economic Models for Sustainable Development.* 1998 ISBN 0-7923-4998-9
16. J.N. Lekakis (ed.): *Freer Trade, Sustainability, and the Primary Production Sector in the Southern EU: Unraveling the Evidence from Greece.* 1998 ISBN 0-7923-5151-7
17. M. Boman, R. Brännlund and B. Kriström (eds.): *Topics in Environmental Economics.* 1999 ISBN 0-7923-5897-x
18. S.M. de Bruyn: *Economic Growth and the Environment.* An Empirical Analysis. 2000 ISBN 0-7923-6153-9
19. C. Kraus: *Import Tariffs as Environmental Policy Instruments.* 2000 ISBN 0-7923-6318-3
20. K. Uno (ed.): *Economy - Energy-Environment Simulation. Beyond The Kyoto Protocol.* 2002 ISBN 1-4020-0450-8

Economy & Environment

21. J. Beghin, D. Roland-Horst and D. Van der Mensbrugghe (eds.): *Trade and the Environment in General Equilibrium: Evidence for Developing Economies.* 2002
ISBN 1-4020-0479-6

KLUWER ACADEMIC PUBLISHERS – DORDRECHT / BOSTON / LONDON

MIX
Papier aus verantwortungsvollen Quellen
Paper from responsible sources
FSC® C105338

If you have any concerns about our products,
you can contact us on
ProductSafety@springernature.com

In case Publisher is established outside the EU,
the EU authorized representative is:
Springer Nature Customer Service Center GmbH
Europaplatz 3, 69115 Heidelberg, Germany

Printed by Libri Plureos GmbH
in Hamburg, Germany